Corridors to Extinction and the Australian Megafauna

Corridors to Extinction and the Australian Megafauna

Steve Webb
Bond University
Gold Coast, Queensland
Australia

ELSEVIER

AMSTERDAM • BOSTON • HEIDELBERG • LONDON
NEW YORK • OXFORD • PARIS • SAN DIEGO
SAN FRANCISCO • SINGAPORE • SYDNEY • TOKYO

Elsevier
32 Jamestown Road, London NW1 7BY
225 Wyman Street, Waltham, MA 02451, USA

First edition 2013

British Library Cataloguing-in-Publication Data
A catalogue record for this book is available from the British Library

Library of Congress Cataloging-in-Publication Data
A catalog record for this book is available from the Library of Congress

ISBN: 978-0-12-407790-4

For information on all Elsevier publications
visit our website at store.elsevier.com

This book has been manufactured using Print On Demand technology. Each copy is produced
to order and is limited to black ink. The online version of this book will show color figures
where appropriate.

Contents

A Prologue to Extinction

In his book *Wonderful Life*, Stephen Jay Gould cogently, and with a hint of mischief, explores the miniscule chances that humanity would have developed just through evolutionary inevitability. His use of rerunning the tape of life through various possible combinations of chance evolutionary pathways, episodes of possible extinction and alternative corridors of success against different environmental backgrounds shows clearly the gross unlikelihood that I would be here writing this book – or that 7 billion people (one can always hope!) would be reading it. Humans do not top the tree of life as a natural progression or preordained fact but as a grand collection of chance events that not only acted for us but also acted against many other species. So, we are here because others – many others, whether they were extremely successful or poorly adapted – became extinct through a series of multiple extinctions thus giving us the chance to shine. Opportunities were taken by life on the Earth that could easily have ended up as blind alleys, and many did. Turning evolutionary corners was also important in building complexity within already complex life, so much so that life eventually emerged that had an extraordinary awareness of its surroundings. That was paid for by extraordinary loss of life and the massive and widespread sweeping of life forms from the Earth that may also have led down their own corridors to organisms equivalent to those that exist now. These are life history's mechanisms of chance.

When we read the word *extinction* we almost always think first of the loss of an animal or group of animals or organisms in the past. In fact, *extinction* for most people immediately conjures up the dinosaurs, or more correctly, their demise. We also think of the loss of our animal compatriots with whom we share the Earth today, although not always beneficently. I would suggest that few of us ever think of ourselves as being part of their world or the natural processes to which they belong: going *extinct* always happens to some other animal somewhere else. We are, however, as intertwined with the natural world as any other organism but more often than not we just don't see it. It is as though we sit in a bubble observing the world around us, perhaps studying it, moving along with life but somehow being removed or protected from it; we seem to feel as though we are not really there, lightly floating above the realities of planetary biology and often believing that what affects other organisms won't affect us. There is also an element of *It is there, it will always be there, but what has it got to do with us, really?* Many approach climate change that way.

Modern life, everyday living and particularly living among the concrete ramparts of cities or fields of suburbs has removed us mentally and physically from real nature and its reality and, thus, has obscured the reality of our position in the web of life. Generally speaking, we no longer see our life-support system in the way Neanderthals and thousands of generations of hunter-gatherers starkly and probably nervously did before us. They saw nature at its best and worst everyday and they experienced its threats to their very existence very much in their faces and for the most part constantly. Everyday they awoke there was no guarantee they would last till sunset. However, we view the hunter–gatherers of the Ice Age as far removed from us and, of course, they are, in time at least. We often do not recognise, however, that everybody on the planet was a hunter–gatherer 8,000–10,000 years ago, a mere blink in evolutionary and geological time, and we carried on that lifestyle for 99% of our existence as a species. They were inexorably tied to their environment and surroundings in day-to-day battles for survival as part of the environment, linked into its every crinkle, tempest and drought; they were mapped into it inextricably. They had to know their environment and how it worked, the animals, plants and weather, where food could be found and how to get it or else they perished. The constant search for food was central to their day-to-day activities and this pursuit tied them to the world around with fierce bonds – they had to know it intimately and respect it for its power. To get food, they had to know what was available, when it was available and how to collect or hunt it before it disappeared or rotted or something else ate it. Even a few berries could easily disappear as bears, birds and weather attacked them almost before they could ripen in many cases. Getting there first was the priority, but being first depended on knowledge honed by years of experience of environmental change and its contrary nature, animal behaviour, climatic conditions and the hunting and hunger of other people and things.

Knowing what was around to eat, however, was not enough: the next step was to have the skills to go out and get it, including a drive to be successful that more often than not came from hunger and the requirement for energy and strength to finish the job and return with your hard-earned sustenance. That required an intimacy with and knowledge of the environment that almost all of us have totally lost as the 21st-century humans surrounded by electronic devices and the world they conjure before us; because of this, we will come close to becoming extinct in the next ice age – and the next ice age *will* come …. Our technology won't work in such conditions but hunter–gatherer technology and skills probably will, as they did in the past.

The vast majority of humanity, with some recognised exceptions, has now been placed in a new world. Instead of the cold caves of Neanderthals we now occupy an electronic cave. It is warmer and safer but not as interesting, and it has physically separated us from the planet. We have developed a tunnel view through the computer screen 'cave entrance', looking at the world safely and dispassionately across the threshold of the Web and from behind the electronic fire that keeps danger at bay. We are in our bubble: we can explore where we like in complete safety. Unlike previous cave dwellers, however, nothing is beyond our view of the world: the problem is that there is so much knowledge out there that we have neither the time nor the intellect to take it all in. Moreover, most of it is totally irrelevant to most of us. But at the same time, we should ask ourselves how much of it is worth taking in. The present computer creed seems to be 'knowledge for knowledge's sake', however shallow, poorly assembled and dubiously correct that knowledge might be. Nevertheless, the office and lounge room hold the chairs of exploration through computer technology: the world is literally at our fingertips and feet and we can explore and find out about anything in perfect safety. The downside is that direct contact with our world is in itself becoming extinct and with it the last vestiges of feeling and living in the environment for what it is, for what it does for us and learning how it works in a practical sense. How many times can you ask someone: Have you been to or seen such and such? No, comes the answer, but I saw something about it on the Web....

We have lost direct touch with our world and in my view this goes part of the way to explain why so many people know so little about it. It may also explain why we shy away from and become suspicious of science, the scientific method and explanations derived from those. Somehow the computer tells all, but where are the scientists in all that knowledge? They are missing in the eyes of the public; the information on the Web just gets there – but how? Worse, the answer seems to be: Who cares? It's all a bit analogous to the way that food comes in tins or frozen packets and milk comes in plastic bottles or cartons and it's all found in large buildings called supermarkets – but the cows and farmers are lost in the process. Does this mean that cows are becoming extinct on the level of popular culture?

There is now a widespread perception that we as humans exist beyond the realm of the natural world and are not part of the vast planetary network of interdependent organisms and Earth systems: rather, we are just observers living in an electronic cocoon, viewing the pageant of the world turning around us. But we *are* here, and many of us, although I would argue not enough, have begun to look at our world with an inquiring eye. We can study our planet as intensely as we like but unless we see and understand it we cannot bring to bear all the wisdom and ingenuity we have gathered around us over the centuries and put it to good use. Part of the problem seems to be our insulation within our world, underpinned by the way in which most of us live that has helped build that bubble around us. The world that many live in comprises the block of humanity that largely controls the world, namely the developed First World from where the Earth's resources are organised, mined and liberally used and from where a vast and aggressive economic imperative is driven. That is something, I maintain, that cannot last even though many will totally deny this against all the evidence. The profligate use of resources and the disturbance of environments and habitats for over 250 years have now brought us to a watershed with respect to our impact on the Earth's ecosystems and its climate. It translates into a level of extinction among species and organisms that can and probably will begin to break down the vast natural networks and systems that make our uniquely biological planet function. My guess is that we will not draw back from that headlong course; rather, we will continue to develop because of the cultural and financial imperatives that drive us ever onward, and we will only stop when our species begins to die as a consequence.

To exemplify this, a report published in the popular science journal *New Scientist* (27th February, 2010, p. 30) offered some stark insights into the dangers the world faces at present. It is not necessarily a unique report; similar articles have appeared in similar journals for the last four decades. Most are not read by the general public, of course, but they are there for all to view. (Perhaps this information should replace some of the cooking and game shows and diet programs that festoon our media. Ah, but that would replace the nightly anaesthetic – not good.) The article shows that our environmental situation is worse than four decades ago because the problems have accumulated exponentially since then – and so have we. The latest report highlights 10 central areas of concern including ocean acidification, chemical pollution, atmospheric aerosol loading, stratospheric ozone depletion, damage to the phosphorous and nitrogen cycles upon which all the living things rely, climate change, the loss of biodiversity and changes to land and fresh water use. These cover all the ingredients that drive the world's biological functions of which we are a part, and demonstrate that, basically, the world's metabolism is critically dysfunctional or, to put it more pointedly, sick.

Probably the most important problem, and the one that sums up all the others, is the impact of humanity on biodiversity and the resulting losses that this causes throughout the natural world. The rate of extinction has now reached 10 million species annually, 10 times the natural rate, while our population is increasing exponentially by 100,000 people per day. Some might argue that such a loss of species is an overestimate, but even if we halve it, the total is a staggering number. The report points out that the present extinctions are not necessarily focussed on large species but rather on those that service the Earth through recycling wastes, cleaning water, absorbing carbon and maintaining the correct balance of oceanic chemistry. In other words, the world's blood chemistry and metabolic functions are changing, and we all know what that would signal for an individual human or for other complex organisms. Put more bluntly, what if you had such a diagnosis: would you seek advice and make the necessary changes or carry on as usual? The article finishes by suggesting that such a global condition may lead to a catastrophe as large as any of the six largest mass extinctions of the past. As with all those, this latest extinction will bring a radical change in planetary biology and its ecosystems, resulting in a new world order of life forms and their hierarchy (probably all small and very crawly); the Earth will face another massive biological turn, leap or upheaval, whichever way you look at it. As for us, we may not have to worry about the next ice age because, to paraphrase Paul Ehrlich, *whatever your cause, it's a lost cause if the world's population keeps growing....*

In the end, the world will survive, as it did in the past, but we will not. And why are we so special that we should? We are just another species that came and will go as many before us did and will do after us, although we may very well be one of the fastest to have evolved and disappeared – a dubious honour. It would be an extremely sad waste of evolution if we move on this magnificent planet and, with all our intelligence and ability, leave nothing more significant than any of the life forms that preceded us – just our fossils, if we are lucky. We will also be the only life form that created its own extinction.

The phrase 'extinction is forever' has been proven over and over again. In the next few chapters we are going to take a look at some of the evidence for extinction in the past and how it comes about. The reader might notice that while some extinctions had devastating causes, others were part of a process that just crept upon those who were involved. Some climate change here, a dash of environmental change there and pop! another species disappears.

Just a note: in dealing with dates I will often use the phrase 'million or thousands of years ago'. At other times I will use the standard indicator of Ma (millions of years), ka (thousands of years) or Ba (billions of years). For those unfamiliar with the term '*ca.*' I use that for 'around' as approximating a date.

List of Figures & Tables

Figures

Colour versions of selected figures and graphs can be found at http://booksite.elsevier.com/9780124077904

Tables

Acknowledgements

I would like to thank Bond University for the financial and logistic support it has given me over the years to carry out my research. I am also grateful for the help I have received from my longtime colleagues Giff Miller and John Magee as comrades in the field and adversaries in view regarding megafauna. They have been a great source of support added to that of others.

Of course, most authors have a tolerant family behind them and I want to thank my wife Beata for her support in this endeavour as well as her patience with the weeks I spent away in the field. My daughter Ali made an unusual contribution on many occasions, needing to be taken somewhere, and so taking me away from my sometimes fanatical adherence to my chair and computer screen. She also makes me smile even when she is not actually near me, a vital component to staying sane.

1 The Big Five or Six or More ...

Introduction

The new science of Earth Systems emerged to view the world in a holistic manner in terms of how it works. The purpose of this book is similar. It will try and explain how natural forces working holistically affect animals and why that often results in their extinction. That principle is not unique but applied in the way used here it could be because I want to focus on the Australian megafauna. Arguments raised on the issue of megafauna extinctions do not often involve looking at the parsimonious ways these animals could have disappeared or what it might have taken for them to do so. Nor do they invoke the natural phenomena of climatic and environmental change as main causative agents in these past extinctions. The arguments that will later be discussed focus particularly on the Quaternary Ice Ages, the climate extremes brought about by those events and the environmental consequences Australia underwent during the Quaternary. We can now appreciate, perhaps more than ever before, what climatic change could bring to put animals in danger: each day brings new forecasts of such consequences for the natural world as our present climatic change unfolds.

However, before plunging into the world of Quaternary Ice Ages and extinctions in Australia, as the title of this book suggests, I want to use a couple of chapters to recall some of the famous extinctions, wander through the processes that brought them about and look at the forces that are always waiting in the wings to act to eliminate animals. Those forces took millions of past life forms and that included Australia's giant marsupials. More often than not, extinction does not have a single cause; rather, as many will appreciate, it emanates from many causes and depends on myriad factors including the biogeographic circumstances of the animals themselves. While today we are well aware of the part that humans have played and continue to play in the demise of modern species, there was, in times past, a natural armoury of forces aligned against the Animal Kingdom that sent many to their doom no matter how adapted they were to their surroundings or how quickly they could adapt. Those forces were there during the last 2 million years and they are still there, but today, unfortunately, many have speeded up, been masked or been beaten to the draw by the acts of humans of one kind or another. Thus the natural slow processes or long corridors that animals pass through before reaching extinction are often masked by the rapid events of human population growth, land clearing, general development and the movement of humans into ever greater expanses of our planet, usually to the detriment of the animals that live there. Therefore, it is worth looking at these natural forces, and examining those corridors of various lengths and the forces that operate within them.

Corridors to Extinction and the Australian Megafauna. DOI: http://dx.doi.org/10.1016/B978-0-12-407790-4.00001-X

We will review the causes of the main extinction events of the past first as exemplary measures of why extinctions take place before focussing on Australia and its comparatively recent extinction event, the megafauna extinctions. We will next ask: Is there a positive side to extinction? If there is, what is it and how can that be possible?

What Has Extinction Ever Done for Us?

The title of this section is a parody of the line from the Monty Python film *Life of Brian*. The Chairman of the revolutionary committee of the People's Front of Judea (John Cleese) addresses his very small audience and, folding his arms defiantly, asks sarcastically, 'And what have the Romans ever done for us?' in an effort to stir up his very small band of followers against his country's invaders. Of course, he is insinuating that the Romans have never done anything for them – how could they as invaders? However, rather than hearing a unanimous 'nothing' from his audience, he is nonplussed to hear various members of the group begin to list all the positive things the Romans have done for them: better roads, medicine, education, law and order and *the aqueduct*! Similarly, the unexpected dire consequences of extinction, as we know them, have also brought benefits to the world's biological organisms and we humans would definitely not be here without them.

As humans, surrounded by our gadgets and gizmos, overpowering technology and vast civilisations and seemingly an indestructible presence, we commonly perceive that we are the end product of an evolutionary process that has *inevitably* led to us. That could not be further from the truth and I am not the first to point that out. Because we perceive who we are, think about and manipulate the world around us, plan, reason about, gather knowledge of and examine our circumstances in a way that, as far as we know, no other animal does, we see ourselves as special and good enough to be the rightful custodians of the planet and, naturally, the top cats of the evolutionary process. This is not necessarily a deliberate thought, but just comes naturally from our large and complex brains as though it is just built into our biological and neurological structures like a nose. We might even see ourselves as the top predator in the same way *T. rex* was, as an unassailable and powerfully equipped predator–scavenger that could tramp the world unchallenged, able to overcome and take anything it wanted. It is likely, however, that *T. rex* did not see itself like that or see itself as anything at all. Although most of us never even contemplate such a thought, evolution does not care about us and neither does planet Earth. Why should it; how could it? Evolution is a process, not a thinking organism, and we are subject to it as much as any other creature: this is the level playing field that brings us back to Earth. We are here because of a series of circumstances, incidents and accidents and a chain of ancestral beings of one type or another being lucky. But our species has had its own series of extinctions (Neanderthals, Denisovans and *Homo floresiensis* being only some) and, seemingly, we *Homo sapiens sapiens* are the last in the line. Being whatever we perceive ourselves to be, we march forward balanced on the random acts of astronomical mechanics, on the repercussions of terrestrial geophysics, on climate variation and on the survival of one group of genes over another.

The outcome is evolution, with a few other things thrown in, like adaptation and *extinction*. Ah, there it is. Although there is the possibility that *T. rex* was on its way towards extinction, together with other dinosaur species, when it disappeared, its might was absolutely cut short by one of those random acts and, for all its power and terrifying presence, it went extinct almost in the blink of an eye – just blown away like a leaf in a cloud of dinosaurian debris, darkness and climatic catastrophe. If we had been there we would have disappeared also.

Extinction is an emotive word and one that conjures up disaster among many other adjectives. It immediately invokes negative ideas focussed on loss of life, sometimes massive losses, among the planet's life forms and the demise of fascinating species that will never be seen again, at least 95% of which have never been seen alive by humans. In that way extinction can be regarded as a terrible disaster that brings devastation to the biological world on various levels and that is, therefore, not something to be encouraged. It also intrigues us because it means the loss of creatures we know little or nothing about, and never will, and so they become mysterious to us as the megafauna do. However, the truth is otherwise, as in the case of the positives that came from the presence of the Romans in Judea. Yes, extinction does cause loss of unique life forms but our planet would be crowded without it. It would also mean that creatures we know and love would not be with us today. The planetary development of life can be seen as a two-sided coin working in dialectic rhythm with evolution on one side and extinction on the other, with life lurching between the two. The two forces play and compete against each other, often in complex manoeuvres that are still not completely understood. Of course, they are part of the same coin and extinction has been as much a force for change as evolution has. Indeed, they are intertwined in a sort of stop–go swirling dance of production and change through time, directing life forms this way and that, for good and bad. They produce dead ends and brilliant sparkles of living miracles and life forms that live 5 min, metaphorically speaking, while others span hundreds of millions of years.

As for the speed of effect between these two mechanisms of change, this is now more debatable than ever. Evolution is usually seen as taking place through many small, incremental alterations in an organism over great amounts of time as genes change and mutations accumulate, producing beneficial forms that gradually become incorporated into the form and function of planetary life. Stephen Jay Gould and Niles Eldridge argued for a rather spasmodic series of speeding up and static cycles in what they termed 'punctuated equilibrium' where evolution moved in step-like jumps that sped up and then slowed down at spaced intervals. Possibly, but whatever its mode of production, evolution as a term evokes the ponderous: a process taking many generations and vast amounts of time to produce observable change. But now it is blatantly apparent that there is another form of evolution that has been documented for nearly as long as Darwin's theory: *rapid evolution*. There are now many cases where fish, moths, weevils, Cane Toads (*Bufo marinus*) and Galapagos Finches show that the road to speciation can be a very short one, measured in decades, not millennia. Whether fast or slow, or a combination of both, the natural processes of evolution are the mainstay of biological change and they largely created the world we know today. Those processes can be seen operating in both the field and the laboratory. We have learnt much about evolutionary mechanisms and developed the study of genetics that proves the biological processes involved. This has gone hand in hand with technology, particularly vast computer power that has allowed sequencing of many genetic codes from many organisms, including us. It is in this realm that we now have irrefutable proof of the existence of evolution and its ways of changing life forms which cannot be disputed. Indeed, genetic sequencing has allowed us to study animals and ourselves in a way not possible before, placing them and us in both space (relationships between life forms) and time (when these entities split from one another). Evolution is no longer a mystery, nor is it a *theory*, at least not in the pejorative way it is used usually by those of a religious fundamentalist persuasion. Rather, it is a theory composed of facts – although that does not necessarily mean there is not more to learn about the processes that make up a particular theory. The word 'theory' has been used by some to dismiss something that is not believed by them and has, therefore, become a weak link, or the dead giveaway: if it's a theory it only exists as a belief. Wrong! I will let Steve Gould reply in the neatest possible way:

> *Evolution is a theory. It is also a fact. Facts and theories are not rungs in a hierarchy of increasing certainty. Facts are the world's data. Theories are structures of ideas that explain and interpret facts. Facts do not go away while scientists debate theories to explain them. Einstein's theory of gravity replaced Newton's, but apples did not suspend themselves in mid air pending the outcome.*
> (Interview published in *Discover*, Winter 1987)

For many, religious mythology and mysticism explains the emergence and variety of organisms in the Animal Kingdom. The truth is every bit as fascinating as human-contrived mythologies and gods, unless you require a ready-made answer for everything even if that answer is not true. The incredulity expressed by some at the 'sudden' appearance of various life forms requires, for them, invoking the mythology of God's Creation as the only way it could have been achieved. For the most part, however, the process was far more exciting, complex, exhilarating and long-winded than that, and we continue to learn more about it as we go. Having an imaginary friend to tell you the answer to everything is not fun: fun is *not* knowing and working within the realms of science to find out how the world really works. If we knew everything, there would be nothing to do except impose exceptionally conservative and restrictive rules on a naturally inquisitive and gregarious animal: us! How boring. *Finding out* is what drives humans and it makes us special in the animal world: this is the realm of the researcher. It boosts enthusiasm for learning about the natural world – what will it tell us next, what amazing evidence will turn up about this and that and what are the mysteries that confront us? I believe learning gives most researchers a far greater high than any artificial drug induced habit ever could. I would also like to place a bet that the vast majority of those involved in research began their fascination when young, reading about dinosaurs and 'cave men', and collecting fossils, rocks, beetles and all sorts of other natural objects that can be categorised, studied and look good in a box that was usually kept under the bed. I even collected bird's eggs – what a crime! But my boyhood was in a world without the word *conservation*. I knew only that dinosaurs had gone extinct, not other animals and birds, and I had no idea that we played a part in extinctions. Big game hunting was still a widely accepted 'sport' and one of my old books had a picture of a hunter with his foot on the head of a dead lion.

Extinction has played a vital role in the layout of our present biological world. In a peculiar reversal of these two processes, just as we now know evolution can move rapidly, so extinction can move very slowly unless caused by sudden catastrophe. It acts in a similar way to evolution by pushing change forward and clearing the decks for the emergence and sorting of new species. That is particularly so when it comes to mass extinctions: they drive forward selection and introduce new chapters for life across the world usually in a geological instant. In fact, it all took an almost incomprehensible length of time to work through the increasing complexity of life on the Earth and it required all the geological and climatic ingredients the world could muster to enable that story to develop the way it has. The processes of making the building blocks of the Earth we know today took enormous amounts of time. They did not appear overnight: the era of bacteria alone was almost half the age of the Earth, the atmosphere took at least 1.8 billion years to form and the first fish appeared only after 89% of the Earth's age had passed. These facts seem to support the old adage that, given enough time, anything can happen.

However, the Hindu myth of the world being supported on the backs of four elephants standing on a turtle is a good metaphor for the truth. What we do know is that there are four major interconnected terrestrial forces or systems (lithosphere, atmosphere, biosphere and hydrosphere) and they stand as cornerstones working together to make the world tick. There is a fifth, the cryosphere, but ice has been with us for only a small percentage of Earth's past so we will forget it for the moment, although it becomes extremely important in later chapters. These four forces could represent the four elephants. They have been the interconnected driving forces behind the evolving Earth for the last 4.5 billion years and they have often joined forces to eliminate life on many occasions. An astronomical force (the turtle) includes the position of the Earth in the Solar System, the type of system we are in, the type of star we revolve around and the *Goldilocks* position of the Earth's orbit in the Solar System, all of which enable the terrestrial forces to function. Besides the fact we are made of star dust, the universe around us naturally affects our planet in subtle and not so subtle ways. The Sun is the major driving force of our climate and the resulting environmental conditions here on Earth. Changes in our the orbit around the Sun, the Earth's axial angle in relation to the Sun and the spin of that axis all play a part in how hot or cold we become. Electromagnetic radiation and meteorites also have an effect on our disposition as they bombard the planet's surface, but thankfully our earliest life forms built us an atmosphere to help guard against impacts and deadly ultraviolet radiation. Fortunately, the meteoritic debris of the Solar System seems to be slowly running out of much of the debris that was left over from its early beginnings 5 billion years ago.

The four cornerstones constantly interact with one another through many feedback mechanisms (including astronomical ones). They sculpt the planetary ecosystems that drive life onwards in its many forms. At the same time, changes to any of the five can cause extinctions big and small. The complex and convoluted interactions among these major planetary drivers form such an intricate web that teasing them apart is difficult; we are still in the process of untangling them and probably will be for some considerable time. It is doubly hard to do this when looking into the past. Indeed, the solid action and reaction of climatic and geological forces cannot be separated in assessing their respective contributions for the continuation of life on the Earth. Geological and climate changes have done the lion's share of moulding life here and encouraging it to spread and adapt in the myriad ways it has. But occasionally those same two processes clear out life in various ways and to different degrees and these have taken the biological Earth on a broad and very convoluted pathway that eventually led to the emergence of humans. Extinction was an integral part of this and it also allowed a certain set of circumstances to emerge that particularly favoured our early mammalian and then primate ancestors. But our biggest break was the extinction of the dinosaurs as it was for many, many other mammalian species.

The conservationist hymn says that *extinction is forever* and that is quite right. The loss of an animal, plant, bug, worm or forget-me-not is the complete loss of a unique set of chromosomes bearing an extremely large and unique genetic sequence with a unique array of assembled base pairs in that now-familiar A–T (adenine–thymine) and C–G (cytosine–guanine) combination constituting the unique DNA sequence profile of an individual. It is also the loss of an organism that was associated with, dependent on or beneficial to other organisms. The loss, therefore, shuffles its immediate world till something else takes its place. The only thing not unique is the famous double-helix DNA structure common to all living things. All genetic assemblages in the world today took millions of years to sort and resort and gather mutations to produce a particular individual. Each species on the planet has a unique set of DNA sequences and they represent the end product of a wide range of life histories that went before and which had their bad and good times. But the evolutionary road was peppered with climatic and geological changes that altered environments, brought predators and provided just the right food resources and other forms of sustenance and protection at the right time. The end product represents the assortment history of DNA, which also depended on a variety of climatic and environmental factors as well as interactions between an organism and its competitors. That is evolution: it has no goal, it does not work for a preordained purpose or to produce a particular life form; rather, it just makes them from fortuitous natural and climatic ingredients, the meeting and assortment of DNA, random mutation and a soupçon of the right environment for success. That is probably an oversimplification but the essence is certainly true. The evolutionary sequence of events that take place, whether they are good or bad, is just another mixing of DNA material that will produce another or many other organisms in the future no matter how long that takes to accomplish. Another creature similar to the one that goes extinct may emerge in the future. It will certainly consist of the same four base pairs but any lost creatures will never come back or evolve in quite the same way with their unique genetic sequence among all those that have gone before as well as those that will come after it.

It is said that 99.99% of all life forms that ever lived on the Earth have gone extinct. The fossil record shows us how many family trees have been biologically assembled and lost. Turnover of life is not only essential: it will continue, all things being equal. It is worth viewing our own survival from this platform and wondering how long humans have left. One pattern that often emerges when viewing the palaeontological record is that many animals have gone through an evolutionary bubble rather than a bottleneck. The fossil record shows many cases where a particular animal type emerged then evolved into a number of genera and species within those genera that then slowly went extinct till only one representative species was left; humans happen to be an example of that. The extinction of the last species marks the end of the group; it then becomes just another 'type' that once lived here. Humans have gone through this process, probably starting with the gorilla–chimp split between 8 and 10 million years ago, then continuing with the chimp–hominid split around 7 million years ago, and then developing into multiple genera, each with several species, during the period between 3 and 6 million years ago as the Earth cooled and became drier. The question of how many species has confused palaeoanthropologists and caused numerous arguments among them for decades, but the big question is which branch led to us. The variety of hominin types then began to be reduced *ca.* 1.5–2.0 Ma, so that by 1 million years ago only two or three species existed, or so they say. I myself think they were all 'us'. By 100,000, years ago all had gone extinct except modern humans, Neanderthals, and perhaps three or four sub-species including Denisovans and *Homo floresiensis*. By 25,000 years ago, Neanderthals and all archaic groups had become extinct and by 10,000 years ago, so had *florisiensis*. We were then on our own, the only species of the only genus left. That whittling-down process is a familiar one in the extinction process. How long have we got now?

Background Extinction

Extinction comes in various forms. It can mean the loss of single or multiple species or even Orders of animals during mass extinctions. We take a look at mass extinctions below but the subtle form of *background* extinction has always taken place and continues to do so even though the diversity of species continues to grow. The trouble is that today it is difficult to separate background extinction from that derived from human activity of various kinds. Background extinction usually takes the form of species loss over a long time; indeed, one of the ways we determine whether there is an abnormal extinction event is to measure the loss of species at any particular time against what might be an expected background extinction rate. That can vary between 10% and 20% of species per million years. Another way of looking at it is that a species usually lasts from between 0.1 and 15 million years with an average life expectancy of 5 million years, depending on the size and complexity of the organism. Background extinction can also be called *natural extinction* because it is part of the natural turnover of species and as such is that part of the evolutionary process that makes up broad trends in animal types over hundreds of thousands or millions of years. Background extinction occurs when climatic or environmental circumstances change or when new mixes of animals take place. The new mix may bring conflict or competition that was not there before or introduce animals that may be more efficient in the same ecosystem as the endemic group or that need to share the same food resources. It can also mean the introduction of a new predator. The same process can also benefit others, letting an animal in where it might not have been given the chance previously or easing its everyday living requirements in one way or another, providing it with an advantage. Background extinction can also occur when a particular population has become too small or isolated or its habitat or preferred food resources change. Even slight changes in climatic variables may also predispose hardship on one creature while favouring another. Competition may increase or geographic changes may bring two groups of animals together that have similar habits or behaviours. Such changes occur all the time in the face of both long- and short-term climate and sea level change, and even when plate tectonics bring whole continents together or pull them apart. These principles are discussed in later chapters in more detail, but they add up to the *biogeographic* extinction processes. These processes occurred many times during the Tertiary when various mammal populations changed and mixed together with some going extinct, as we shall see in Chapters 2 and 3. Whatever the cause, background extinction has always been with us and is the main driving force for change and turnover in the animal world in the absence of mass extinctions.

Mass Extinction

The year 2009 marked the 200th anniversary of the birth of Charles Darwin and the 150th anniversary of the publication of his *Origin of Species*. It is sad that Darwin died (only just) before the advent of the first genetic studies. I am sure he would have had a very demure but broad grin through that large white beard had he been told what data lay in the nucleus of white blood cells and that it would confirm his assertions of a hereditary component which conferred adaptive features and qualities on succeeding generations. However, in a 'pre-genetics' world he and the co-discoverer of evolution, Alfred Russel Wallace, went as far as they could in realising there was a hereditary element to natural selection, the driving force behind evolution. The intricacies of that hereditary element were yet be sorted out, but for Darwin and Wallace extinction was also an important part of evolution that marked the natural turnover of animals losing out on their journey, unfit to continue for one reason or another and, therefore, moving aside for others. Indeed, the discovery of extinct life forms was one of the prompting mechanisms for their thoughts on evolution. It was those animals that were just that bit more suited or able to adapt to changing conditions, or that were slightly quicker in their reactions to forces posed against them that survived till they, in turn, became outmoded or outcompeted by others. Those that could not change or adapt fast enough or further because they had reached the end of their abilities to change disappeared. It has now been suggested that while we have long had the phrase 'survival of the fittest' as the main evolutionary selective force, variety within species may play an equally important part in survival – hence the term 'survival of the varied'. Those species possess variety or behavioural breadth and are more able to withstand the fickle vagaries of selection because of their wider genetic qualities compared to species that possess only a conservative, narrow genetic variety. For example, those having a wide variety of height, with shorter and taller members, will be successful if there is a selection that requires movement towards either end of the scale. All this is fine, and biological or behavioural change can help avoid extinction, but from time to time there have been forces that the biggest, meanest and most broadly adapted beasts could not avoid: mass extinction.

In Darwin's day, the details of mass extinctions and their temporal sequences were still largely unknown, although the broad ideas were rapidly emerging. Earth scientists working before the 19th century were labelled 'catastrophists' because of their belief that before the concept of evolution they could see the mass turnover of animals at different times in the past due to catastrophic events. To them these events represented the complete elimination of all animals each time, which left the whole process to begin all over again. In the middle of the 19th century, the English naturalist John Phillips recognised that the fossil record suggested three major eras: the Palaeozoic, the Mesozoic and the Cainozoic ('ancient', 'middle' and 'early' life) each divided by the extinction of major groups of animals. He knew this from the confinement of certain species to certain stratigraphic layers of rock which clearly showed through the layer cake of superposition that they contained different groups that must have lived at different times. These were the first indications that there had been several great turnovers of animals during the past and that they had been caused by definable catastrophes.

While catastrophic in their outcome, mass extinctions have been extremely important in the evolutionary equation. They brought major alterations to the direction the Earth's biosphere took and they had nothing to do with natural selection, competition or gradual evolutionary change. They took place as catastrophic forces that were totally external to normal evolutionary processes and mechanisms when astronomical or geological factors completely and suddenly overwhelmed the running of things. In some cases such events overtook the whole of the Animal Kingdom and it was only a mixture of sheer chance and being in the right place at the right time that saved those few who survived them. They remain signposts for hard turns to the left and right in Earth's biological history that would never have occurred through the slow processes of natural selection. To paraphrase Steve Gould, it was not just the fittest that would survive: it was the luckiest!

The Big Five, Six, etc.

The term 'big five' has been applied to the largest mass extinctions in history. The sixth is said to be those taking place at present. Frankly, I am confused by the numbering because when one looks at the various major extinctions, there are certainly more than five. The number is defined by how large an extinction must be to qualify,

Table 1.1 Ten Major Extinctions of the Past

Date (Ma)	Major Victims
65	**Dinosaur extinctions at K–T boundary** *(Pterosauria, Theropoda, Saurapodomorpha, Ornithopoda, Ceratopsia, Stegosauria and Ankylosauria, ammonites).*
125	**Early Cretaceous extinctions.** *Primitive amphibians (Temnospondyles).*
144	**Late Jurassic extinctions.** *Synapsida –mammal-like reptiles, sauropods begin to decline.*
180	
200–220	**Mid-Triassic and Triassic–Jurassic (T–J) extinctions.** *Procolophonidae, Younginiformes and Kazanian proto-mammals.*
245	**Late Permian extinctions.** *Ninety per cent of all complex life went extinct. Some families included Nectridea, Microsauria (?265Ma), Anthracosauroidea, Seymouriamorpha, Captorhinidae, Pareiasauridae, Trilobites.*
365–400	**Late Devonian extinctions.** *Comprised a series of extinction events in which Icthyostegadae, brachiopods, stomatoporoids, placoderms and jawless fish disappeared.*
440–450	**Late Ordovician extinctions.** *One hundred families became extinct.*
530	**Late Cambrian (Vendian or Ediacaran) extinctions.** *Seventy per cent of species go extinct, sea pens, jelly fish, sea mice, segmented worms disappeared.*
650	**Pre-Cambrian extinctions ('Snowball Earth').** *Extinction of microorganisms i.e. cyanobacteria and other extremophilic life forms.*

as well as by its definite separation from other extinctions that happened very close together. Some do seem to blend into one another but that is probably because when we look back, 10 million years between two events looks very close but in our present time scale it is a very long period of time. There have been smaller events that, while being smaller than the largest ones, were in their own right very large and widespread (Table 1.1). They eliminated dozens of species, which cannot really be considered uneventful, and they were certainly not the gradual moving over of one species for another. Mass extinctions could, however, be regarded as an extremely radical type of selective process occupying one end of the natural selection scale, with survival based more on pure chance than anything else. These natural events were usually caused by exaggerated combinations of standard planetary forces and Earth's geological forces upon which the continuation of most life forms depend. They moved many families and genera out of the way in a geological instant, allowing others to gradually take over. But they also marked a very significant milestone in the direction Earth's biosphere took afterwards.

Categorising the worst as *seriously catastrophic* depends on the number of species lost, but those considered major usually contain the loss of at least 40 genera. This is only a guideline, however, because, as usual with biology, there is a range of how many genera and species go extinct, and in most cases we are still unsure about numbers. Moreover, what we might consider a major loss today might only constitute a small spike on the long road of extinction history. On the practical side, an accurate tally of every genera and species that disappeared has still to be completed. How often do we hear that a new species of dinosaur has been discovered? Well, that is another species that went extinct. Perhaps we will never know how many species have lived on the Earth and disappeared in the mists of extinction. There is no doubt that many, many have come and gone and we know nothing at all about them; in fact, they are yet to be discovered. So extinctions can be graduated from small to large (the latter usually involves the loss of over 50% of species); however, I don't think it is helpful to get into these sorts of ranking categorisations based on a supposed number of lost species which we do not understand fully.

The major difference between catastrophic extinction and natural selection is that the former includes major biological loss involving whole genera; it usually occurs in a geological flash of time and can be attributed to a natural phenomenon that overtakes the planet, and there are different kinds that strike in varying magnitude. They include, for example, widespread volcanic activity, asteroid impacts, tectonics, sudden changes in oceanic and atmospheric currents, global cooling and warming (rapid or otherwise), and changes in atmospheric and marine chemistry. So I now want to take a brief look at the major events that have taken place during the last half billion years, avoiding a detailed analysis of the events. Nothing I say is particularly new but they are worth briefly reviewing so as to get some perspective on their history and to lay the foundations of what comes along in later chapters.

Pre-Cambrian, Cambrian and Ordovician Extinctions

The Cambrian is noted for the 'Cambrian Explosion', a term describing the emergence of a great variety of life forms that blossomed between 570 and 530 Ma. It was also a major step in the development and evolution of life but, ironically, it followed the time of the first mass extinctions ca. 650 Ma. I hinted before that dealing with these events placed so far in the past has always been difficult because time scales are 'squashed' that far back. In that case, the details of evolutionary processes become a little mysterious so that apparent 'jumps' are seen in the fossil record and stages between organism development are difficult to trace and are lost. Moreover, places with geology of the right age are not plentiful and those that have it are usually placed far apart in the modern world, adding to the confusion of tying up relationships between organisms and places of emergence. Gathering evidence takes time and the harder it is to gather, the longer the process of putting together pictures in such ancient time depths. Causation at this distance has been only general and broadly developed but evidence is slowly being gathered; however, while eagerly awaited, different forms of evidence can often complicate matters into a confused amalgam.

The main victims of the extinction were microorganisms, particularly cyanobacteria. This was the time of 'Snowball Earth', a planetary freeze-over when massive ice build-up occurred on all continental land masses, small as they were at that time. It has been suggested that ice formed as far as the Equator but many now feel that at least some areas probably remained slushy rather than forming solid ice. Sea levels fell before the oceans themselves froze down to a depth of 1 km. This resulted in a steep temperature drop, cooling the shallow warm water that cyanobacteria favoured as well as withdrawing the shallow seas they lived in as oceanic waters shrank.

All life at this time lived in the oceans, probably in the more temperate, shallow seas near coasts. What happened to the oceans, therefore, happened to all planetary life within it and it was here that the first delicately built complex creatures evolved. Ironically, it seems to have begun just after the cyanobacterial extinctions in the time of Snowball Earth which took place between 730 and 705 Ma and 660 and 635 Ma. As those extinctions were taking place, the Earth's atmospheric oxygen was increasing, between 800 and 542 Ma (Fike et al., 2006). This led to increased oxygenation of the oceans in three stages; eukaryotic organisms appeared and diversified during the last two stages. The subsequent oxidation of large reserves of organic carbon in the oceans may have played a key role in the development of some of the first complex life forms like the Ediacaran Fauna. They emerged in the Vendian or Ediacaran Period (~635 to ~548 Ma) as a group of soft-bodied creatures such as sea pens, jellyfish, bottom-dwelling sea mice and segmented worms. The Ediacaran Fauna was first discovered in 1948 by geologist Reginald Sprigg, a student of the Antarctic explorer Sir Douglas Mawson, in the Ediacaran Hills of the Flinders Ranges in South Australia. Over the last 60 years, other sites of similar early life forms have been discovered in ranges close by. Many of these were caught up in another extinction ca. 530 Ma which eliminated 70% of species. Details of this extinction are, however, still sketchy. The extinction may not have been entirely caused by Snowball Earth; one complication that has arisen is one that becomes familiar much later in Earth's history: meteorite impacts. Evidence is being gathered for the presence of a series of impacts across Sweden, Russia, Estonia and Norway, representing the fragmentation of a large asteroid, each fragment big enough destroy life. A similar event happened during Snowball Earth when a 4.8-km-wide asteroid ploughed into the Gawler ranges of South Australia just west of the Flinders Ranges. This was half the size of the Chicxulub impact that finished the dinosaurs and is known as the Acraman impact. The remnant crater has totally filled in and is now a dry playa lake in the South Australian outback.

In reality, the 'Cambrian Explosion' continued to the next epoch, the Ordovician, as even in the face of the two previous extinctions life continued to thrive and blossom as marine clades diversified, particularly crinoids, graptolites, bryozoans, cephalopods and bivalves. With more genera around, a complex ecological community began to develop. Ironically, the Ordovician also saw another great extinction which affected only marine genera: with a total lack of terrestrial fauna, that was all there was. It took place between 450 and 440 Ma and was the second most devastating event among marine life in Earth's history. Although over 100 families of marine invertebrates disappeared, it does not rank as a sudden event even if species loss was widespread. It is worth remembering that while this looks like a rather narrow window that far back in time, 10 million years is a long time and the extinction process may have been relatively gradual before its culmination, a theme that will become familiar as we move along in the story of extinctions. Stratigraphic evidence from the Ordovician is not accurate enough, however, to enable us to tease out the details of extinction speed. Consequently, species extinction may have been more rapid than is apparent at present. Also, both seem to have been precipitated by global cooling and the lowering of sea levels, exposing continental shelves where many marine species resided, a situation similar to that of the Cambrian Extinction. However, once again the details of these early events are few and the subject of considerable debate.

Devonian Extinctions

The next great extinction took place in the Devonian and also involved mostly warm water marine genera. Diversification occurred once again just as it did in the Cambrian as surviving groups once more took hold and spread among vacated niches, a process that had begun after the Ordovician extinctions, continued throughout the Silurian and into the early-middle Devonian. The occupation of vacated niches is another familiar strategy that organisms undergo following extinctions. The diversification that followed saw the first true land plants that would give rise to the first forests in the form of giant club mosses. Moreover, life was spreading onto the land as sharks, bony fish and ammonites began to dominate the marine environment. The process of the recapitulation of species that occurred at this time can be seen repeatedly following the mass extinctions that were to follow – a true example of survival of the fittest if ever proof was needed. The Devonian extinction consisted of a series of separate events spread between 365 and 400 Ma, although the largest one took place in the middle of the late Devonian, *ca*. 375 Ma. There were eight to ten separate events spread over about 35 million years. During that time 19 families and 50 genera disappeared.

The main victims of the Devonian event were the stromatoporoids, which had been responsible for massive coral building around the world. Other species included the fearsome armoured fish, the placoderms, as well as the much more harmless jawless fish, brachiopods and trilobites. Again, this extinction has been attributed to global cooling when ice formed on the mega-continent Gondwana. Falls in global temperatures with widespread glaciation followed, with the same effects as in the Ordovician event. But asteroid impact has again been posited as a contributing force, although, again, this is disputed, and general uncertainty persists in making a final determination of where to put the blame. One thing is very possible, however: these extinctions, as with many others, probably had a multiplicity of contributors.

Permian Extinctions

The biggest extinction in the history of the planet, and one that really deserves the label 'mass extinction', took place in the Permian. It eliminated 80–95% of marine and 70% of terrestrial species, which almost took the Earth back to pre-complex life times; this would have been a shame after the 4-billion-year struggle it took for it to get this far. It would have been a sad end, but perhaps this is what happened on Mars between 3.5 and 4.0 Ba, assuming no life (microscopic or otherwise) exists there today. By the time of the Permian Extinctions, Trilobites had existed for 300 million years and giant sea scorpions (eurypterids) had been around for 250 million years but both disappeared at this time, together with many other animals and plants.

The extinction is placed fairly accurately at 241 Ma, but there is evidence from the geological records from Europe Greenland Asia that two smaller, previous extinctions may have taken place at 270 and 260 Ma. The main extinction saw the marine environment fare worse than the terrestrial one, with brachiopods, ammonites, gastropods, crinoids, various sharks and bony fish all experiencing various degrees of loss. Ten genera of Permian sponges as well as the tabulate and ruggose corals disappeared. That was accompanied by a great reduction in other reef organisms, with the bryozoans losing 100 genera. On land, the swamp-based club mosses and horsetails were reduced in the face of a warming and drying planet and an analysis of over 8500 specimens from the Urals and the South African Karoo region dated 260–242 Ma shows that 78% of terrestrial animals perished. Members of the mammal-like reptiles (therapsids), reptiles, amphibians, pelycosaurs and insects went the same way but Lystrosaurus, an ancient mammal relative, was one of the comparatively few animals that came through this time. The 'Permian Extinctions' not only devastated the planet's biosphere, but they also may have taken place extremely quickly, perhaps only in the space of 165,000 years!

The cause of such sweeping devastation is, again, not entirely clear but it is believed that a sequence of events caused it rather than a single factor – and that is where the confusion lies. Nobody knows exactly the contribution of each. Possible causes include volcanic activity, plate tectonics, global warming and cooling and changes to the 'ocean conveyor' current that distributes heat, oxygen and nutrients around the world's oceans. The extinctions probably began with massive volcanic activity known as the Siberian Traps that spewed out 3 million cubic kilometres of magma, covering 7 million square kilometres of flood basalts, enough to cover the Earth to a depth of 20 m (Reichow et al., 2009; Saunders & Reichow, 2009). Enormous amounts of sulphur dioxide, fluorine and carbon dioxide were released into the atmosphere, blocking sunlight and cooling the Earth. The sulphur dioxide underpinned that cooling in the short term and killed plants and fauna. Trapped hot air then began to build up under the continuing cloud cover and the thickening volcanic gases of fluorine and carbon dioxide then produced a Greenhouse Effect, heating up the world. Temperatures are thought to have risen anywhere between 10°C and 30°C during this time, initiating a breakdown of food webs and bringing both abrupt and gradual rates of extinction among terrestrial and marine plants and animals. The world's tectonic plates then performed an amazing feat, coming together to form the mega-continent Pangaea. The environmental consequences of this event produced a more extreme climate that brought aridity to the centre of the new continent. This disrupted environments and ecosystems, disadvantaging animals and plants that were not adapted or could not adapt to these changes. The hot planet then began to cool, promoting aridity, while formation of ice across much of the continent caused falling sea levels and exposing shallow coastlines where many marine species had lived and sheltered. During this turmoil of vast climatic shifts, massive extinctions took place among all living groups around the world.

Lowering of sea levels as ice built up brought reduced habitat, increased climatic seasonality and ecological instability. In a rather subtle consequence, it also increased the oxidation of organic carbon that was previously submerged beneath the oceans but was now exposed. Changes in ocean currents accompanying the Pangean formation added to this by releasing large amounts of CO_2 that had been taken up during the earlier volcanic activity. Their release saw the rise in atmospheric CO_2 that drove global temperatures higher. The oceans began to warm up and this is where the bulk of extinctions took place when they reached a level that stopped the ocean conveyor. The conveyor circulates warm and cold water around the planet, taking about 1000 years to complete one circuit. It disperses nutrients as well as warm water, both vital to life; thus, it acts as the Earth's arterial system. Around 251 Ma it completely stopped: in effect the world had a heart attack! The conveyor operates by the cooling of currents as they reach high latitudes. They become saltier and heavier so they sink and return to their origin, travelling at lower depths through the ocean. Oceanic warming prevented that mechanism from operating because the oceans were now too warm for the cold water to sink and the nutrients, minerals and oxygen that the conveyor also carries were no longer being distributed throughout the oceans. In effect, the ocean died: it became anoxic, devoid of oxygen, and almost everything that lived in it perished. But the extinctions were not finished. Deadly hydrogen sulphide that had built up in the anoxic oceans was released into the atmosphere, killing most of the creatures that inhabited the land. Evidence for the mass extinction can now be found in the compressed, black dolomite layers found in places such as the naturally named Dolomites between Italy and Austria. It probably took about 20 million years for the Earth to recover from this catastrophic period which, in light of the temporal sequences we are looking at, is recovery in a flash. But this was a truly remarkable recovery given that the planet had almost perished. What was that catch-cry I hear – *save the planet*? From its record, the planet is quite capable of saving itself.

Late-Triassic and Triassic–Jurassic Extinctions

Another mass extinction took place at the Triassic–Jurassic (T–J) boundary spanning 200–220 Ma. This caused an eclectic spread of disappearances but mainly in the ocean. Some of those that disappeared at this time included reef-building organisms, pelagic conodonts, nautiloids, various types of bivalves, all the ceratites, placodonts and various amphibians and reptiles. The cause was not well understood until recently but volcanic activity on a grand scale was again believed to have been involved together with the Earth's old enemy: meteorite collision. As an example of the way we constantly learn about the world and try to catch up with developments, one has happened as I write. Jessica Whiteside of Brown University and her colleagues (2010) have published their findings that show firm evidence of a massive volcanic event that occurred in the Rhaetian Age (204–200 Ma). It was associated with the break-up of the supercontinent Pangaea and spreading of the mid-Atlantic ridge. It involved the spewing out of 2 million cubic kilometres of lava by the Central Atlantic magmatic province near Iceland: it was two-thirds the size of the Siberian Traps and it may have happened in less than 1 million years. The resultant emission of sulphur and carbon dioxide sent the world on a climatic roller coaster as it did in the Permian. Volcanic emissions into the atmosphere once again caused abrupt planetary cooling, particularly by sulphur dioxide build-up. They also blocked sunlight, exacerbating warming as atmospheric CO_2 built up. As a result of this atmospheric chemical see-saw, extinctions occurred in all the ecosystems, causing the synchronous demise of between 50% and 70% of both marine and terrestrial species alive at the time (depending on which author you consult); this was also the time at which the ancient ammonites became extinct. But what was not known till now was that volcanic activity alone was probably the sole cause of this large extinction; previously, the combination of meteorite impact and volcanics had been thought to be the cause.

There is a footnote to the T–J extinctions, however. Once again, there seems to have been a gradual turnover of species during this event. The extinction itself does not seem to have been a sudden mass event but rather one happening over a long period of time, 4 million years or more. It must also be noted that smaller extinctions occurred at 220 and 210 Ma, possibly smaller preliminaries to the major event not linked to the later volcanic event and similar in pattern to the Permian Extinctions. But the importance of the T–J extinction was its bias towards certain families of animals which gave the dinosaurs their big chance to become the dominant animal group.

Small to medium extinctions punctuated the very large major events briefly described above. Remember that there is a constant background buzz of extinctions taking place beyond the big ones, the ones that Darwin knew all about. Different levels of species turnover have been occurring for hundreds of millions of years. It is what nature does: the slow and the subtle and the fast and the furious, many with causes that are much too faint to be picked up in the fossil and geological record and becoming even more difficult to detect the farther back in time we travel with the effect of time squashing mentioned earlier. There must also be times at which the subtle background extinctions change frequency: speeding up and slowing down again, perhaps in response to the vagaries of brief and/or rapid climate instability, minor or localised volcanic events and tectonic activity which occur all the time. It might be these that predispose a particular species or genus to larger or more visible events which then tip the balance, sending them extinct. Once again, in order to see such processes in action we require a fossil record that is complete or nearly complete and an accurate chronological record of that assemblage. A complete fossil assemblage, however, is a Holy Grail and one not expected to be found soon: both of these are almost impossible to acquire from the past and the farther the record goes back the harder the task becomes. But the same could be said for those of recent times, such as the megafauna extinctions, a mere yesterday even compared to those above but still lacking vital clues, as we shall find out in later chapters.

End-Cretaceous (K–T) Extinction

The final big extinction that I want to mention is, of course, the mass extinction that everyone knows about. That occurred at the end of the Cretaceous, and is more often referred to as the K–T extinction of the dinosaurs that took place *ca.* 65 Ma. It is so well known I will not describe it in any kind of detail, but it is worth another look, particularly in terms of Australia's faunal history. One thing most researchers agree on is that central to this extinction was the arrival of a very large asteroid that hit the Earth. Initially, it was suggested that the impact was caused by a piece of rock 10 km in diameter which would have caused worldwide devastation, with gigantic fires across continents; tsunamis several hundred metres high; acid rain; a long nuclear winter, perhaps lasting nine months to one year, resulting from enormous amounts of dust, smoke and other particulates thrown into the atmosphere; and earthquakes close to the impact 1 million times greater than anything recorded in human history.

The selectivity of the K–T extinction, however, has puzzled many and triggered a variety of explanations about how and why such selectivity among animal groups could have occurred in the face of such catastrophic events around the world. Others have also reassessed what the real effects on the Animal Kingdom really were. At first glance it seems to involve just the dinosaurs, who did not look as though they were heading for extinction. These animals had lasted for 160 million years and were the largest, strongest and most ferocious of any land animals ever to exist. They seemed, therefore, almost invincible and had survived previous drastic climatic and environmental changes – for example when global temperature dropped from an average of 22–15°C at the Jurassic–Cretaceous change over *ca.* 144 Ma. But in the last 20 years or so, evidence has mounted that by the end of the Cretaceous they were undergoing population reduction and losing many species. It is now widely believed that they were probably experiencing this decline for perhaps up to 20 million years before they eventually went extinct. That gradual extinction possibly included a contraction of the group so that by the time the asteroid hit at the end of the Cretaceous they were not dispersed across the planet but confined to western North America. Out of 26 end-Cretaceous dinosaur fossil sites worldwide, 20 are in central North America with a few small areas elsewhere. Only 20 genera of dinosaurs seem to have reached the K–T event and 14 of those are restricted to North American localities. The giant meteorite therefore only dealt the final blow to a reduced range of animals that had limited populations in a comparatively small region. That may become a familiar pattern in later chapters.

The Chicxulub crater on the edge of the Yucatan Peninsula caused by the meteoric impact is asymmetrical with a steep southeast side and shallow northeast rim. The angle of the impact is 20–30°, indicating a north-westerly-directed fireball together with the bulk of the debris directed over North America. This event was joined by massive lava eruptions on the other side of the world in western India, known as the Deccan Traps. It has also been suggested that the asteroid impact may have initiated the widespread volcanic activity on the opposite side of the world as the shock wave passed through the planet. The Deccan Traps were part of a series of gigantic eruptions together with others that took place near Antarctica, in the south Atlantic, eastern South America and South Africa. These flood basalt outpourings, measuring 2 million cubic kilometres, occurred between 66 and 65 Ma. The part that this widespread volcanic activity played in the overall extinction process, however, is, as usual, not well understood, but changes to atmospheric composition through CO_2 and fluorine emissions probably caused some climatic variations at that time as they did in the Triassic and Permian. These lines of evidence would at least explain how the vast majority of the dinosaurs could have been eliminated even without implicating the other disastrous effects of the impact. The shape and placement of the impact might also shed some light on what was happening in the mammal and marsupial world at that time, but I will take up those strands of the story below.

One other point worth mentioning in any discussion of terrestrial vertebrate populations at this time is that their distribution and population size depended on the size of continents then. They were all much smaller than they are today because of high sea levels; therefore, the world's land animal populations were overall smaller than they are today and possibly just that bit more susceptible to an extinction event, particularly one so large as this. Also, smaller continents and large tsunamis make for greater destruction of animal populations, particularly those along coastal fringes. Dinosaurs needed very large ranges to cover their food requirements. That limited their populations together with carrying capacities at those times. The comparatively limited amount of fossil remains of these creatures may also testify to smaller populations that popular films might suggest.

So, what about all the other classes of animals that *survived*? The issue is just that: all the other groups of animals *did* survive! Birds lost 75% of their species but this might be expected because of the overwhelming aerial effects that such an explosion and its aftermath must have had not only on the upper but also on the lower atmosphere and their composition. The widespread eradication of trees and other niches favoured by birds (and dinosaurs as well) in a vast conflagration would also have contributed to their losses (and let's face it: eggs break and cook very easily!). Crocodilians and turtles lost 30% and 27% of their species, respectively, which does not seem a great loss in the face of such a catastrophic vision, although it must have been damaging to them in particular regions while watery habitats may have helped save them. Placental mammals and fish only lost around 15%, but marsupials lost the same percentage of species as birds – 75%! So why did marsupials survive at all? How did they manage to hang on? I take that up in Chapter 5.

Why Do Animals Go Extinct?

The object of writing about major extinction events in this chapter is to underscore the natural rhythm of extinction, the myriad reasons for extinctions and the regularity of gigantic upheavals that rock the biological Earth and to show how they are part of the course of world history. What is clear from the previous discussion is that animals go extinct all the time, everywhere, for many reasons and in often peculiar patterns that vary across the Animal Kingdom. Understanding extinction is important because it is a signpost of the loss of creatures that were doing all right for thousands or even millions of years and then disappeared. Extinction can also be an omen of things happening to the Earth that are subtle and do not necessarily flag danger – for example, changes in those four pillars or systems mentioned earlier. What we have seen is that mass extinctions come out of the blue and are caused by a complex network of interacting forces. General extinction, however, is another matter. It shows how nature works through assortment of creatures with respect to environmental and climatic conditions which can cause animals that have existed so successfully for so long to suddenly go extinct. So the question of why those that are perfectly adapted to their environment suddenly disappear can be answered by examining the imposition of changes to their condition. But that is not necessarily as simple as it sounds because those changes are difficult to know in all cases and because the palaeontological and geological record is not always forthcoming or easily interpreted in this regard. Extinction often suggests a disaster for the Animal Kingdom but that is patently not true in the wider view of the history of planetary biology. If there is an overall answer, it must be that it is just a natural process without which we might all be bacteria! One thing we do know is that extinction is not always sudden nor is it caused by a catastrophe. A natural turnover of species goes on all the time; sometimes it speeds up, at other times it slows down. The possibility of the dinosaurs undergoing a population decline over 20 million years until the final nail in their coffin arrived from space is not *really* unexpected no matter what I said previously about them being implacable. And let's face it: animals that lasted at least 160 million years were inevitably going to meet some rare and massive catastrophe that occurs to the Earth over great lengths of time and which is determined by the odds of nature. The biological record shows clearly that almost all groups of animals disappear eventually; so it was with the dinosaurs and probably will be with us.

The process of disappearance without a mass extinction event is one of gradual attrition of individuals which builds among species over time. Some species have fewer animals in their population than others and so they are the first to succumb to this process. Similar processes were probably involved in all extinctions. Yes, cataclysmic upheavals were involved in many but nevertheless faunal populations wax and wane for a whole variety of reasons, some more subtle than others, but most to do with the biogeographic circumstances that blow their way. Animals and plants constantly undergo negative changes, population reduction or contraction and general downturn in their once-favoured status or the environment in which they live. Climate change, isolation, disease – all these are only some of the ingredients that can change an animal's life circumstances, altering for good or bad their chances of survival. Extinctions can grow within species at the local level then proceed to a regional scale and then on to the continental level; at any point along this continuum a notable extinction event can suddenly occur. The accumulation of smaller changes can then multiply to form a much bigger event some time down the track and one that occasionally looks suspicious. Sometimes, however, a catastrophic event coincides with one of those natural downturns which make it look even worse than it is – and the blame is usually put on the mass event. This seems to have been the case for dinosaurs to a certain extent. But what we also have for them is a better view of the fossil and palaeoenvironmental record of the time than we have for any previous mass extinction simply because it was after those that took place much earlier. We have a bigger preservation window in the fossil record at that time than we did before and, thus, more detail and more data. Previous evidence for the earlier extinctions has been lost through erosion, scouring, burying, drowning, tectonic recycling, volcanic activity and sheer destruction over an increasingly longer period of time. This is an exponential phenomenon, so the farther we go back the harder it is to extract detailed evidence from multi-recycled environments and geological landforms to provide a clearer picture of the event. We can place this taphonomic problem next to the time scale squashing phenomenon. Jointly they make unravelling the distant past very difficult.

Apart from the arrival of a giant asteroid or widespread volcanic eruption big enough to affect the whole world – its atmosphere, climate and oceans – extinction is usually not a particularly sudden event. The history of these events shows very clearly that they normally take some time to occur, often hundreds of thousands or even millions of years. Our perception of the time taken is overwhelmingly influenced by, again, the paucity and incomplete nature of the fossil record and by the coarse image that comes to us from the phenomenon of time squashing the farther we go back. It occurs when a 5-million-year time frame seems so much shorter at 400 Ma than it does at 65 Ma. The apparent speed of a particular extinction, therefore, is a matter of time scales. Those usually involved are very long in human terms but are usually a geological blink of an eye. Normally, cataclysmic events like those above are so rare that they perhaps should not be cited as the best examples of extinction *per se*; rather, in many ways small, more unobtrusive extinctions are the norm. Smaller but nevertheless major extinctions seem to occur every 25 million years. But it is those that are even smaller, together with general background extinctions, that make up our planet's gradual biological turnover. Nevertheless, trying to understand why certain groups of animals go extinct when they do, or succumb to an extinction event which other animals survive, is a very tricky undertaking. It is, however, worthwhile because understanding why extinction occurs in all its forms may help us to avoid others or even ours, or at least understand why it will take place.

Explaining survival patterns requires a good fossil record, usually much better than we have – although from my experience we are never satisfied with the fossil record we have! Another aspect of the work is to understand the ecology and biogeography of the animal groups involved in order to tease out what went wrong or, more precisely, what could not be overcome. There are ways of approaching this but nevertheless it is always a difficult or even an impossible task to undertake for fossil animals whose habits and lifestyles are totally unknown to us. So we have to conjure up the sort of world a particular animal required to stay alive. Again, that problem is compounded the farther we go back. Even recent extinctions are difficult to study or understand. It is a sad truth that while we struggle to understand why some species disappear today, even though we are here to record them, they often do so almost undetected until they are gone or almost so; this is exemplified by the fact that we cannot properly explain how some of our Australian species have disappeared during the last 200 years. Moreover, when the wide gamut of extinction forces are surveyed, teasing out the one responsible for any particular event becomes extremely difficult; however, it is almost inevitable that they are not caused by just one factor alone. Simple explanations are alright but can we really see the subtle forces that more often than not are pushing a species to extinction slowly or otherwise? 'Killing mechanisms' that I introduce later in the next chapter as Biological Extinction Drivers (BEDs) can come in obvious violent and catastrophic forms, as we have already seen, or they can be merely a subtle change in the availability of a food resource, in the average annual temperature or rainfall or in the loss of part of the food web. It could also be as simple as a slight alteration in aridity or vegetation structure and composition or a change of sea level.

If we turn to those factors beyond sudden, natural disasters that can cause extinction, we see that life is precarious and depends on a lot of natural and biological balls being kept in the air in order to continue. One very important ball is climate. Animal groups can decline when climate changes, whether it warms up or cools down. Such changes can be subtle or not so subtle and different animals may or may not be comfortable with that change. The degree to which climate changes can be small or large but the result flows on to the particular animal's environment, altering niche composition even slightly or moving that niche elsewhere; this in turn affects food resources and water availability. The root cause of such change can be, as we have seen, massive volcanic or plate tectonic activity detrimental to the climate favoured by the group. But climate change can occur through natural cycles determined by those astronomical factors and, as I mentioned earlier, the Earth has been subject to these for a very long time.

The complex interwoven web of linkages between Earth systems that keep a particular group of animals going add up to a series of extremely delicate balances that tip this way and that. Delicate balances are also easily tipped. Some balances are more easily tipped than others depending on circumstances and influences. The result is a cascade but that may affected some animals more than others. Suddenly, we not only have one creature in trouble: we have others joining it. The planet perches on a delicate balance all the time but the one thing we as humans seem incapable of grasping is that although Mother Earth shakes her skirts and has fits of one kind or another, she will always come back to start a new chapter and she does not need us to help her. The environmental lobby today trumpets *Save the planet*. They don't mean that, of course: they mean *Save humanity*. That cry has an astounding ring of arrogance about it. The climate debate is not really a debate about climate, *per se*: rather, it is a debate about saving our bacon, not that of the planet itself, which we cannot do. What does not seem to be understood by Greens, Blues, Pinks, Reds or any other colour organisation is that there is nothing we can do to save humanity if the Earth decides to bring about its own major biospheric changes – and it can do that: we have all the proof of such changes in the past. They are illustrated above. Even with all our ingenuity and technology, we would be as helpless as all the animals that went extinct in the past if the planet once again went through the same convulsions outlined above.

But the world did not die during those times. No matter how bad the situation was in the past (remember the stagnant oceans of the Permian and Snowball Earth), when most of life succumbed (and so would we have), the planet nevertheless survived. By the way, it has been suggested that if Snowball Earth conditions return we would have a better chance of survival living on Mars. The recent eruption of the Icelandic volcano Eyjafjallajökull in mid-2010 (a mere pimple compared to past eruptions) showed that process most eloquently. As its cloud of ash, dust and gases drifted southeast across Europe, it almost halted the economy of a dozen western nations with airline passengers stranded and goods rotting in storehouses, unable to be shipped through air cargo. Shortages followed and the economic screamers began to search for human scapegoats such as governments and private organisations as share prices began to drop. Share prices! We humans just do not get it! Mother Nature was having a slight snit and we screamed. Imagine the onset of an Ice Age or the eruption of the Toba or Yellowstone volcano! However, if you have a weak constitution, perhaps you shouldn't.

Mammal genera usually last less than 10 million years and individual species no more than 3 million years with primate genera less than 5 million years, while individual species have a far shorter history with a lifespan of 1–2 million years at most. These statistics are confounded, however, by Coelacanths (*Latimeria chalumnae*), horseshoe crabs (*Limulus* sp.), sharks, crocodiles and the Tuatara (*Sphenodon punctatus*), all of which have lasted long past the usual use-by date of animals. To see living fossils like these is amazing but when extinct life 'reappears', as did the Coelacanth and the Wollemi pine (*Wollemia nobilis*), both thought to have gone extinct 90 million years ago, the blood pressure rises and the mind becomes puzzled. We are glad they are here but *why* are they here? Surely they shouldn't be. 'Chance' is probably the best answer. Each of these 'leftovers' found a special niche that, with a lot of luck and exceptional circumstances, including finding a stable environment that did not change very much, allowed them to survive. Another example is the group of cabbage palms (*Livistona mariae*) living in Finke Gorge National Park west of Alice Springs in central Australia. These amazing plants are totally isolated to such an extent that their nearest relatives live over 1000 km away. The ecological circumstances of their niche depend on the porosity of the nearby sandstone ridges that were formed 400 million years ago and which now gradually allow water to seep down through them, ending as a ground water reservoir. This takes place in a comparatively small area but it is enough for the palms' requirements and has maintained their population for 65 thousand years. They are there but they are more than vulnerable because of their limited range and small population: they survive, but only just, just as the Coelacanth does. These are a few factors that determine extinction. Of course, the oceans may yet reveal more species believed to have gone extinct because our knowledge of the two-thirds of the planet's surface, our oceans, is still extremely meagre.

An isolated or very small animal or plant population is vulnerable to extinction but so too are complex behaviour and particular morphologies. Simple marine organisms such as molluscs and bivalves exist for durations of 10 million years. The average rate of marine invertebrate extinction over the last 500 million years is two complete families every million years including those subject to *phyletic extinction*. While some species look as though they went extinct in the fossil record, they actually evolved into other species. This is phyletic extinction, whereby the genes of the previous species actually continue on in a new form. The horse (*Equus*) is a prime example of this process. It began with *Hyracotherium*, a puppy-sized four-toed browser that lived *ca.* 55 Ma – *Eohippus* ('dawn horse') is another name for it. It then proceeded through several more stages and varieties which were geographically separated. These included the Oligocene horse *Mesohippus* (35 Ma) that was a little larger with three toes, then through stages recognised as *Miohippus* and *Parahippus* to *Merychippus* of the Miocene, *ca.* 20 Ma, still with three toes but mainly using a central hoof and still a browser. All these lived in North America but *Hipparion* was a Shetland Pony-sized three-toed version that lived in Asia and Europe in the Pliocene. Meanwhile, *Equus* evolved in North America in the Pliocene. I am oversimplifying somewhat here: I have omitted details and many species that took part in the process and that lived in different places at different times, some overlapping with others. Additionally, the promotion of savannahs across the world at the expense of more forested regions during the early mid-Pliocene promoted selection for the rise of the larger, swifter modern horse which had become a grazing animal. Phyletic extinction is wonderfully exemplified by this once very small, shy, delicate, browsing creature that probably hid in the undergrowth and that then changed over 40 million years into a swift, megafauna-sized animal by 5 Ma, an extremely magnificent creature that many young girls have made their second love (Dad is always first, of course). The horse family has been amazingly adaptive and successful but part of its success was that it became larger. It spread out of North America and into Asia, India and Africa during the Pleistocene only to go extinct in North America during the late Pleistocene. Nonetheless, the tamability and usefulness of horses to humans in the Holocene probably ensured not only their continued existence in the Old World but growth in its populations – something that was not true for many, many other animals of its size. If it had not become domesticated, I have little doubt that there would be few, if any, of them around today. The rabbit did the same thing as the horse, setting out on an extremely successful course that began as *Palaeolagus* of the late Eocene and eventually evolving into a cuddly pet on one side of the planet and an introduced menace in today's Australia on the other.

The mechanisms of speciation and parallel evolution sometimes operate to confound the fossil record but they have also allowed many species to continue on using a more suitable and, more importantly, adaptable guise. So species extinction may look like just that in the fossil record but we have to remember that that record is never complete. It demonstrates fully, however, that extinction is not only a natural part of life on Earth but also one that is difficult to avoid and that will occur to all at some time or other in one form or another. Extinction, whether it is mass or just background, also leaves niches open for others to fill. So the demise of one group opens opportunities for others. In this way, it is a necessary part of the ongoing process of life and remains a factor in the maintenance of species diversity. So, the natural extinction process is not something to lament: rather, it is something to be welcomed.

Well, What Did Extinction Do for Us?

In a nutshell, it paved the way for us to be here – but this book is not about 'us'. Let's go ahead just for a paragraph or two. The road that we have taken to be here is very much like that of a ball in an old pinball machine. To get the best score you need to escape many traps and avoid a lot of obstacles; you also need a great deal of luck as the ball passes vicariously around the table. It needs to bounce in the right place and in the right way. Our distant ancestors passed the five big extinction tests, but only just. The worst time for us was Snowball Earth, but the Ediacaran *Pikaia*, the first vertebrate, survived that, as well as the far greater catastrophe of the Permian event. One consequence of that extinction was particularly fortuitous for humans: it killed all but one group of mammal-like reptiles, the cynodonts and it was from these that future mammals arose. The Permian Extinction and the following J–T event also gave a big boost to the rise of dinosaurs. They dominated and mammals took a back seat in the evolutionary stakes for the next 160 million years, keeping a low profile and a small body size and scampering about very quietly, mainly in the dark. Mammals did not mark time, however: rather, they adapted, diversified and speciated in their own way enough to give rise to proper eutherian mammals, including ancestral primates of the late Cretaceous, as well as the marsupials and monotremes. The K–T extinction did something very important for us. It refined the race, bringing to an end the dinosaur era and opening the way for mammal dominance. They were ready for that chance: they took it and have been top dogs since.

Our big break came with our late Cretaceous palaeoprimate forebears that were scrambling around in the trees before the last group of dinosaurs took their final breath. Their subsequent persistence laid the foundations for a successful lineage that led to groups of anthropoid apes from which one group moved into the spotlight at least 7 million years ago. Therefore, our emergence is directly attributable to the K–T extinctions as well as the background extinction of various less successful members of our wider family. Much more recently *we* have been subject to such processes ask *Homo sapiens neanderthalensis* and the Denisovans or *Homo sapiens altaiensis* or was that phylogentic extinction? Certainly, *Homo floresiensis* vanished only 12,000 years ago and that was extinction. Possibly other earlier hominins such as *Sahelanthropus tchadensis, Kenyanthropus platyops, Orrorin tugenensis, Ardipithecus ramidus, Paranthropus boisei, P. robustus, P. aethiopicus, A. anamensis, A. sediba, A. bahrelghazali, A. africanus, A. afarensis, A garhi, Homo habilis* and a few others that occupied the multi-branching tree that we evolved from all went extinct phyletically. However, we do not have so many branches on our tree; rather, the tree has just been over-taxonomised by the winner sitting on the end of the top branch. However, we probably have as complicated and convoluted family tree as any other species and it will take just as long to work out the exact combination of branches that led to us as it will for that of any other living species of animal – if not longer. If there is one incomplete fossil record, it is certainly ours. But the species that has assumed responsibility for the Animal Kingdom, *Homo sapiens*, has not had a particularly good track record in that stewardship, however. Its act of late is in need of an extensive clean-up before it, too, finds itself part of a mass extinction – possibly by its own hand. We can be assured of one thing, however: if mass extinction does occur, there will be many species that will survive and they will just quickly shuffle into the gap we leave behind and gladly take over.

Rather than leave the 'big break' we mammals got, let's look closer at it and the story of mammal groups that took over from the dinosaurs. Did they suffer extinctions? To find out, let's first take a closer look at those things that cause extinctions – extinction drivers.

2 Extinction Drivers

I now want to focus on extinction processes because many 'things' make animals go extinct besides the vast climatic changes we briefly looked at Chapter 1. So, excluding catastrophic and sudden events, why does an animal that seemingly lasts such a long time suddenly disappear? Although such disappearances are not that simple to explain, neither is the time that it takes for a particular species to go extinct. This is because species extinction is usually not quick or sudden – although the farther back we go, the distance of time (that squashing) makes it increasingly look that way. It is arguably correct to say that extinction occurs when something in the natural surroundings of an extinct organism goes wrong; e*cosystem decay* has been one of the names given to that process. Extinction has been caused by humans, of course, but we all know that. Often, however, what we do not grasp is why it was our fault. Although we hunt animals to extinction and strip, pollute or poison the natural ecosystems that flora and fauna rely on, we are often puzzled as to what it is we have done wrong! This is so because it is sometimes very subtle or seemingly irrelevant changes that can cause modern extinctions. It goes to show how vulnerable organisms are to the extinction process and how easily it can take place. Such a puzzle exists for some of the marsupial extinctions that have occurred in Australia during the last 200 years or so, but I will return to that in a later chapter. Nevertheless, we do alter ecosystems through the mere fact of living on the planet: extinction, if you will, is what we do because we do so much! And many animals have gone extinct through what we do even if we may not have deliberately set out to harm anything; thus we have caused and continue to cause ecosystem decay. From what I can make out, that will only become worse in the future.

There is wide range of factors or *drivers* that cause animals to go extinct, either in the long, medium or short term. They can work subtly or be blatantly obvious. In the past, most were caused by the natural workings of the planet or had their origins in space, either from changes within our Solar System or beyond. Many drivers are links that in themselves may not necessarily bring about extinction individually and they can come in an enormous variety of 'shapes' and constructs. Extinctions can be the result of a succession or combination of drivers interacting collectively or sequentially as one brings another into play. Sometimes the initial force may not have any direct effect but it might trigger a cascade of secondary events that cause an extinction. They can also work in combination with one another, emerging as links in chains whose lengths are dependent on the number of links that are required for the event to happen. Various animals are vulnerable to various chain lengths and combinations of links. Sorting out each one to assess its weight in the process is, therefore, not only difficult but it may also, in many cases, be impossible. With so many things that can act against the continued existence of animals, it is a wonder they survive as long as they do, with some lasting millions of years.

Corridors to Extinction and the Australian Megafauna. DOI: http://dx.doi.org/10.1016/B978-0-12-407790-4.00002-1

I have listed seven major driver categories below in no particular order of importance because that would be difficult to assemble, and pointless – similar to asking
whether strychnine or arsenic is the better poison. It would be nice to write about each
one separately: things are explained so much more simply that way because humans
have a compartmentalised mind that likes to put everything in its proper box. But perhaps with the exception of the Astronomical Drivers, most extinction drivers are not
mutually exclusive: they form an interwoven or co-causative network that works in
different combinations and interplays; so, separating them as though they stand alone
is not the correct way to proceed. However, I will try to some extent to explain them
in that way. Each major driver category is tackled, although it might sound as though
I am repeating myself in some of the details. I am, but only to show how these details
are interrelated and, indeed, precipitate one another with some not possible without
others joining them. The seventh category, *Biogeographic Drivers,* displays the largest
list of factors which are caused not only by other major drivers but also by other natural, biological, environmental and demographic circumstances that can arise either
from major drivers or from other associated factors. Biogeographic drivers, while
looking the most impressive in terms of variety, usually work at the local or regional
level and often in subtle ways. These are also discussed in much greater detail in
Chapter 8 as BEDs; in particular, we will examine how they may have applied in
the Australian Quaternary. The present discussion, however, also shows the range of
forces arrayed against the animal world, and brings a sense of wonder at how life has
progressed and succeeded as long as it has and produced such variety. I will say now
that I do not claim to have covered every possible driver, contributor or reason for
extinction and I welcome the reader to think of more. I am sure there must be some.

1. Astronomical drivers
   a. Meteorite and other bolide collisions
   b. Changes to Earth's orbital forcing: solar orbit, axis angle and precession
   c. Solar fluctuations, flares and sunspots
   d. Insolation changes and irregularities
2. Tectonics
   a. Mountain building: CO_2 absorption, changes to air flow and weather patterns
   b. Continent separation: loss of migratory corridors
   c. Raised seabeds: release of sequestered CO_2; bringing together of competitors
   d. Volcanism: release of volcanic gasses, SO_2, methane and CO_2, atmospheric pollution
   e. And lowering of natural light (lux) levels
3. Volcanism
   a. Widespread lava discharge (basal flows)
   b. Atmospheric aerosol loading
   c. Stratosphere ozone depletion
   d. Sunlight depletion (nuclear winters, rapid environmental changes)
   e. Changes to atmospheric chemistry by volcanic discharge of various chemicals
   f. Alteration of ecological networks
4. Environmental change
   a. Niche shrinkage, loss of niches and habitats
   b. Loss of vegetation

    c. Vegetation changes
    d. Isolation of habitats
    e. Constriction or reduction of habitats
    f. Landscape change: deforestation, aridity, glaciation, etc.

5. Climate change
    a. Ice Ages and rapid climatic irregularities
    b. Albedo changes, changing reflected solar radiation
    c. Changes to land and freshwater sources
    d. Damage to phosphorous and nitrogen cycles
    e. Changes to oceanic chemistry
    f. Oceanic changes (absorption or release of CO_2)
    g. Loss of ice and tundra melt release of sequestered CO_2
    h. Loss of forests (changes to CO_2 regulation)
    i. Changes to grasslands and spread of aridity
    j. Degree of cloud cover

6. Oceanic changes
    a. Changes in ocean chemistry (acidification)
    b. Temperature change (warming and cooling)
    c. Oceanic conveyor loss
    d. Changes to oceanic circulation patterns
    e. Sea level change, release of sequestered CO_2 at low levels, loss and gain of forest and other vegetation and land surfaces, inundation of land

7. Biogeographic drivers
    a. Food chain breakdown
    b. Trophic collapse
    c. Species replacement one by another (invasion and migration)
    d. Intensification of competition
    e. Broad environmental or focussed niche change
    f. Reduced food supply (various reasons)
    g. Failure to compete at same level as competitors
    h. Failure to adapt
    i. Rarity – a small population or small isolated patches, never a common species, a predisposition to go extinct
    j. Limited breading potential and opportunity for replacement of numbers
    k. Genetic limitations through founder effects, isolation, genetic drift and accumulated negative mutations
    l. Geographically restricted (limited food stocks on limited-size geographical areas)
    m. Confinement of a single, small population
    n. Isolation from members of the same species
    o. Failure to migrate against reducing environmental conditions
    p. Overspecialisation (Species with complex, behavioural, physiological and morphological adaptations or those that occupy higher trophic levels, are more likely to go extinct in the face of general or regional environmental downturn and/or climatic change.)
    q. Extinction of prey species
    r. Top predators are more likely to go extinct than omnivores
    s. Large animal species are more prone to extinction than smaller ones (They have greater food requirements and need larger ranges.)
    t. Pathogenic outbreak or new pathogens

Main Extinction Drivers

Astronomical Drivers

We can begin with a brief assessment of extinction drivers by moving in from space. One of the most obvious causes of extinction, and one that can by itself be the sole cause of extinction, is probably the most famous one: meteorite or bolide collision with the Earth. These events have happened frequently in the past but fortunately they seem to have become fewer over time, at least in terms of their size. I say that hopefully, but having said it, there is always a chance for a large meteor or asteroid to approach the Earth from the Asteroid Belt and elsewhere. The best example of this happened in 2006 when Jupiter was hit by several extremely large meteorites. The influence of Jupiter as an attracting agent for space debris has been credited with protecting the Earth from a number of large meteorites during the last few billion years, saving us from obliteration and granting us enough time to evolve complex life. The Jupiter episode did, however, remind us that interplanetary rocks big enough to wipe us out are still hurtling around out there. Hopefully, however, there really are fewer than there were at the birth of our Solar System 4.5 billion years ago. We are constantly bombarded by small space debris, much of which is burned up in our atmosphere before reaching the Earth's surface. Larger pieces of up to 2 km would undoubtedly have localised effects on animal populations living in the impact zone. We know what 10-km-diameter meteorites, such as the Chicxulub event, can do in terms of bringing on a wide variety of worldwide environmental and climatic changes.

Most of our astronomically derived extinction drivers come from closer to home and the majority of those are concerned with our association with and relationship to the Sun – although having said that, most of these are not direct causes of extinction but are contributors. That relationship is succinctly demonstrated by the three Milankovich Cycles mentioned in Chapter 1. The first is related to how our orbit around the Sun (eccentricity or orbital stretching) changes from near circular to an ellipse through 100–400 ka cycles. Orbital shape determines the contrast between winter and summer: the more eccentric or stretched our orbit becomes the colder we become. The second is how the angle of the Earth's axis (obliquity or roll) changes from 21°39′ to 24°36′ in relation to the perpendicular; that cycle takes between 19,000 and 23,000 years. Changes to the angle alter the strength of sunlight (insolation) on high and low latitudes: the greater the angle the cooler those latitudes become. This is offset somewhat by warming of the southern hemisphere, although with less land area in the southern hemisphere heat absorption by the large oceanic masses there reduces the effect. The third cycle is the roll of the Earth (precession or wobble) as the Polar axis swings around a central point like a slowly spinning top and that has a cycle time of 41,000 years. These three astronomical cycles are important in determining the amount of solar heat or insolation that reaches the planet's surface, thus varying our climate. In combination, the three orbital factors play a particularly important role in the timing and length of planetary cooling and the onset and termination of glacials; consequently, they have also been implicated in Quaternary extinctions. I discuss their role, particularly in relation to Australia's megafauna, in later chapters.

The amount of solar insolation reaching the Earth's surface is regulated not only by planetary cycles mentioned earlier but also by cyclic changes in the Sun that change its radiation output and thus the amount of heat that reaches us. It is natural and logical that such changes would have some effect on the Earth's climate but pinning these down to precisely timed regular cyclic episodes or predictive tools has been difficult. There can be little doubt that the output of the Sun has been a significant driver for life on the Earth and that any fluctuations in that output potentially affect life here. Output of solar radiation is affected by the tidal pull of planets in the Solar System. What can be said is that changes to the Sun's ultraviolet range, its enigmatic sunspot activity and solar flares that result from both external planetary influences as well as the Sun's internal nuclear fuel burning mechanisms, do vary insolation and radiation patterns and, therefore, have effects on our planet's climate. The exact effects, however, are controversial and predicting them and their cyclic patterns is difficult – although sunspot activity is believed to be associated with planetary cooling events such as the Maunder Minimum that spanned the 17th and 18th centuries. However, it is unlikely that normal variations in solar activity would be a direct cause of extinction, although they could, in some small way, contribute to them. Normal variation in solar output has probably been only a minor compounding effect of much greater extinctions drivers at work. Nevertheless, extinction is usually made up of varying factors operating together and in sequence. So fluctuations in solar output at the wrong time could have been the metaphorical straw for the camel's back at certain times in the past and may be so again in the future.

This section would not be complete if we did not mention a very large astronomical backdrop for the cause of extinction: the Earth's journey around the galaxy. It is often considered that our 200-million-year voyage around the centre of our galaxy exposes us to a great deal of danger in terms of life on Earth. Mass extinctions have also been linked to a 200-million-year cycle as well as other dangers we can encounter during that extremely long journey. To illustrate, the K–T extinction, the Permian extinction, the Late Ordovician and the Cambrian Extinctions are all roughly 200 million years apart. Those examples are, however, a little misleading because there have been others in between and those are *not* exactly 200 million years apart, but it is worth thinking about. Of course, the cycle itself is not the culprit; rather, the smoking gun (or a series of them) is what is going on intergalactically. One of those guns is the possibility of the Earth passing through clouds of dark interstellar dust that could have changed the climate, probably towards extreme glaciation as experienced in 'Snowball Earth' times; this has, in fact, been one explanation posited for the cause of that event. Another possible danger could have been high radiation bursts from supernovae that could have occurred as we entered areas where star formation was taking place. The galaxy moves vertically like a toy helicopter spinner as it moves through the cosmos, causing a bow shock effect that sends a rain of cosmic rays down onto us. If our planet is sheltered behind the galactic disc, we are fairly safe from these but we do not always stay there. About every 60 million years our Solar System bobs up above the disc because it cycles above and below the galactic plane. Intriguingly, there is a controversial pattern to our planet's fossil record which seems to show a reduction in the diversity of marine animals every 60 million years or so (Rohde & Muller, 2005). We cannot get off this galactic roundabout, but the likelihood of encountering astronomical dangers during our ride is highly probable although extremely long term (we hope) and something we will have to live or die with.

Tectonic, Volcanic and Oceanic Drivers

The enormous movements associated with plate tectonics have played and will continue to play a central role in biological evolution and extinction on the Earth. It is difficult to know where to start with this driver because tectonics produces so many spin-offs on continental shape, mountain building, volcanic, biogeography, zoogeography and climate. The division and joining of continental plates has been a fundamental cornerstone of geographical, environmental and biological assortment on the Earth. The ability of terrestrial fauna to evolve, adapt and go extinct has often been regulated by movement of the planet's plates and the different ecological relationships that have been forged through their fusing and sundering. They have driven life as well as driven extinction. Volcanism, mountain building, the raising of seabeds, the drowning of continents, changes to ocean geography and currents as well as the vast climatic swings resulting from these changes have all played an enormous part in the birth and death of millions of species, in fact, 99.9% of those that have ever lived on the Earth.

Over vast periods of time planetary geography has morphed through myriad shapes and environmental changes, most of which we would not be able to recognise as our planet if we approached it from space. Continents have collided, sea floors have spread and the result has been the disappearance of land through subduction, the sliding of one plate under another, or the building of mountain ranges as continental collisions have uplifted enormous sections of continent kilometres into the air, providing barriers to and compartments for life. Add to this the vast changes in sea levels measured in hundreds of metres that have added and subtracted millions of square kilometres of land joining and dividing continents and islands. Such effects have sorted and resorted animal populations and their gene pools as a major evolutionary mechanism that also drove many to extinction. Merging continents have brought animals together, providing an opportunity for the mixing of genes and exposure to new environments, ecosystems and niches. All these have also caused fundamental challenges for animals. Of course, the shifting of plates and the land forms riding on them is exceptionally slow and measured in millions of years; so, the associated biological and biogeographic changes happen in a subliminal fashion without the obvious recognition of changing circumstances and emerging opportunities, let alone the growth of mountain ranges and emergent continental shelves.

The formation of mountain ranges, however, can do the opposite: rather than bringing animals together they can separate them and provide barriers to movement and migration, sometimes forming funnels pushing species this way and that. Mountain building can also alter weather, atmospheric and wind circulation patterns as well as introducing new ones. These can be major contributors to the growth and persistence of many types of broad ecosystems not only close to the source of change but also at great distances. One example is the formation of the Tibetan Plateau born by the Indian Plate driving beneath Asia, forcing up land beyond the Himalayas that eventually gave rise to the Indian Monsoon. That has benefitted much of southeast Asia and the rainforests that grow there as well as bringing monsoon rains to northern Australia from the northwest. The separation of continents divides animal populations but those in places like Australia, and to some extent South America, forged their own identity and went their own evolutionary pathway although they too suffered extinctions as others did on the conjoined continents.

Plate tectonics also provides another critical element in the process of extinction: CO_2. The natural production of CO_2 is controlled by the effects of continental movements and the volcanic activity often associated with them. The balance of CO_2 production is largely through release and sequestration and this involves three principle processes: volcanic activity, seabed uplift and mountain building. Volcanic activity has now been firmly associated with planetary cooling but this usually follows moderate events and is short term. One of the most famous of these events was the Tambora explosion in Indonesia in 1815 that produced the 'year without a summer' in Europe in 1816 as the injection of masses of sulphur dioxide droplets, ash and other gases into the atmosphere reduced solar insolation, cooling the planet by 0.5–1.5°C. Major eruptions of the past, however, such as those of the Deccan and Siberian Traps, caused just the opposite effect, that of global warming. Indeed, the extreme volcanic events of *ca.* 650 Ma actually freed the world from a 'snowball' state by massive injections of CO_2, as well as other gases, into the atmosphere. It is clear that the Permian, late Triassic and K–T extinctions were all driven by volcanic events, although meteorite impact was also a firm contributor to the latter extinction. While explosions or pyroclastic and basaltic lava flows from massive eruptions may not directly cause mass killing of animals, they can drive them from large areas, trap and destroy regional groups and certainly burn, bury and spoil millions of square kilometres of grazing land and food stocks as well as change whole ecosystems for thousands of years. The real danger from these eruptions, however, comes from the effects of their gaseous injections into the atmosphere which can reach 60 km high and spread around the globe. Gases include sulphur dioxide (SO_2), methane (CH_4), chlorine (Cl), fluorine (F) and the infamous CO_2. Injection of this vicious cocktail into the atmosphere causes many, often contradictory, effects for the environment but when it happens on a massive scale the result is inevitably extinction for many aquatic and land animals as the world heats, climate changes and ecosystems collapse as we saw in the Permian.

The chain reaction of this process is roughly as follows. Atmospheric aerosol loading not only poisons the atmosphere but it also depletes sunlight, reducing the amount of solar insolation the Earth receives and bringing on 'nuclear winter'. This is the term often applied to describe severe atmospheric pollution from exceptionally large blasts. Initially, it reduces sunlight penetration of the atmosphere and changes diurnal variation to such an extent that day and night blend into one. It also brings plummeting temperatures. Sulphur dioxide causes short-term cooling, glaciation, sea-level drop and marine and terrestrial extinctions. Chlorine and fluorine emissions bring acid rain with increased continental weathering and terrestrial extinctions. With higher CO_2 levels and volcanic ash injected into the atmosphere, the planet gradually begins to warm up. So while there is an initial cooling period, this is only brief and disappears as the heat of the Earth builds up within its thick atmospheric ash and aerosol overcoat. Long-term warming also heats the oceans and reduces the oceanic temperature differential between upper and lower levels of the deep ocean that keeps ocean currents such as the ocean conveyor moving. That can slow oceanic circulation to a point where the conveyor stops and oceans can begin to stagnate.

The scenes described above are truly catastrophic occurrences and hardly bear thinking about, but they are also rare and usually take millions of years to go through the cycle. Humans would be the most vulnerable of species if these sorts of events took place now – and that is always a possibility. Imagine: there was always the day before the Siberian or Deccan traps began to erupt. That could be tomorrow. But we have not finished. Volcanic ash fall would cover landscapes far beyond the area buried beneath hundreds of metres of lava flow, which was 2000 km deep in the case of these eruptions. One statistic worth noting is that while Mount St Helens produced 1 km^3 of lava, the Deccan Traps (the smaller of the two examples) produced at least 517,000 times that amount. This is a very sobering thought. These events lasted millions of years but even one lasting a few years or perhaps several decades, a microsecond of geological time, would be enough to cause massive loss of life, perhaps almost all humanity. It would probably destroy the world as we have constructed it, taking us back to the Stone Age, but the mere onset of the next Ice Age might do that anyway.

The theme of global warming and climate change through CO_2 buildup in the atmosphere runs through many extinction events. The reason is the accompanying environmental changes that can be comparatively abrupt. These are secondary effects and need the primary driver to initiate them; nevertheless, CO_2 seems to be the key to understanding a lot of the story of evolution in terms of climate change and global warming from carbon dioxide buildup in the atmosphere. Mountain building can counter that buildup to a certain extent by CO_2 absorption through chemical and physical weathering. Another contradictory consequence occurred at the end-Permian volcanic event: while acid rain increased weathering, it also absorbed more CO_2, but obviously not enough. Carbon dioxide dissolves in rainwater and helps the weathering process by decomposing rock mineral to form bicarbonates. The solubility of bicarbonates allows them to be conveyed through stream and river systems to the oceans where the CO_2 they contain becomes sequestered on the sea floor. Carbon dioxide is then locked away for millions of years, thereby reducing the amount in the atmosphere. This process is useful in reducing CO_2 emissions during super-volcanic events which have transcontinental effects, but it is slow. While CO_2 can be removed from the atmosphere, its steady release can occur from uplifted seabed. Tectonic activity not only opens oceans but also closes them and examples of raised seabed can be seen in the Andes, Himalayas and almost all large mountain chains. These seabeds were raised as continents were uplifted, usually at their margins. Typically this occurred when India approached Asia, its plate both squashing and raising up the bed of the ancient Tethys Sea. On these occasions, seabed-sequestered CO_2 was released into the atmosphere, contributing to global warming. Thus plate tectonics even without volcanic activity can raise CO_2 levels in the atmosphere. So the swings and roundabouts of tectonic forces push the Earth's climate one way and another, often far more drastically than we contemplate with our present climate change. On the receiving end, however, is the biology of the planet. Life adapts or perishes in the face of these changes as natural forces hone species that are adaptively suitable to continue and weed out, sometimes drastically, those that are not. Once again the cry of *'save the planet'* seems so futile when the events of the Deccan and Siberian Traps and slithering continental plates are contemplated. I have said elsewhere that if the Earth has survived this long, it can survive anything short of a collision with a 5000-km-diameter asteroid. The only thing is that *we* would not survive. 'So what?', I hear the rest of the Animal Kingdom chorus ....

Climatic and Environmental Drivers

Species distribution and migration are in constant turmoil, at least in geological time frames. Various drivers are responsible for this including plate tectonics which brings together and pulls apart continents, providing avenues of species exchange, migration, escape or separation. Going hand in hand with this process are the climatic and biotic changes imposed on the shifting continents as they move into different latitudes. That in itself is a driving force, albeit slow, that alters faunal composition and sorts out those that can best adapt to new conditions from those that are adaptively stagnant. Overlaid on this are major climatic shifts including global warming and Ice Ages as well as the sea level changes that accompany them and that alter the shape and size of continents. Further, there are smaller or regional climatic or environmental shifts that originate from locally derived factors that affect one particular continent either as a whole or in part. Climate change alters biotic regimes such that rainforests come and go and savannahs appear, bringing fresh environments for those that can take the opportunity to adapt. Deserts replace savannahs and then become overgrown with shrub and forests. Incessant environmental turnover is a constant force for change which then pushes adaptive responses among animal families as well as dragging those that are pre-adapted to certain niches as they move from one place to another. I will talk about these processes again below.

The key themes running through the previous section are the environmental consequences of climate change, plate tectonics and volcanism that drive extinction. Such extreme events as those described, however, are not necessary to drive natural background extinction. Changes in the natural processes that control climate are continuous but are not necessarily extreme. During the last 60 million years there were wide swings in world temperatures that radically changed all Earth's ecosystems and brought fluctuating sea levels, changing the size of continental landmasses with the emergence and drowning of continental shelves as well as joining and separating continents. These not only triggered biological responses in the form of new species but also brought about the demise of others, while the mixing and separation of animals initiated cladistic change as a biological response that produced new species.

The emergence of continental shelves also brought expansion of terrestrial ecosystems while offshore shallow marine environments suffered. These processes became increasingly important as the world moved into the glacial Quaternary and they will be discussed in more detail in Chapters 7 and 8.

Climate change can be either slow or rapid and its rate of change also determines how quickly environments change in response. Change from one ecosystem to another can go both ways, for example, from rainforests to tundra and desert or the reverse, and there are many examples of such radical changes in the Tertiary as the world swept through the Eocene Optimum on to the cool Oligocene, then back to a warmer early Miocene, then cooled off again in the Pliocene as the planet plunged towards the icy Quaternary. It is clear that Tertiary animal groups went through spectacular changes in their particular compositions and migratory patterns as they tried to fit their adaptations into a constantly changing world by migrating and maintaining their populations within familiar ecosystems. It is also obvious that there were two types of continental form during most of the Tertiary: first was that of the northern hemisphere, where the Old and New Worlds were, in effect, one continent for most of the time with long connections between North America, Asia and Europe. Even Africa joined that club half way through the Tertiary. The result for animal populations living there was a greater range of environments to choose from, with adaptive opportunities for species variety as well as space for maintaining them. Competition was probably also greater and the merging of Asia with Europe and Africa with Eurasia provided catalysts for this process. All these factors played a central role in the success of animals in the northern hemisphere. The second was the southern hemisphere continents of Australia and South America, which provided no such opportunities for their animals, with the former being the most conservative of all the continents during this time. South America was isolated for much of the Tertiary but it did enjoy on/off linkages with North America, while Australia was completely isolated from at least 45 Ma onwards. It is worth thinking about what Australia would have been like biologically if it had broken away from Gondwana earlier, before marsupials had reached it. Would it have been a continent of monotremes, with the Echidna (*Tachyglossus aculaetus*) and Platypus (*Ornithorhynchus anatinus)* as our only quadrupeds?

Enviro-climatic change constantly selects for animal groups most suited to those changes. Climate change both shrinks and expands niches, thus providing a variety of suitable homes for an equally varied fauna. This depends on temperature, precipitation, humidity, evaporation, solar radiation, cloud cover, wind patterns and so on. The system is similar to an amplifier's graphic equaliser where certain bands of audio frequencies can be adjusted by a series of sliders on the front of the system. Similarly, if we imagine each of the above ecological determinants (as well as many more) as a slider and if we could move them to different points on a scale in different combinations, we could then change ecosystem conditions this way and that in response to the different combinations we selected. Moving individual sliders can produce a wide variety of environmental conditions. With all sliders down, for example, we have ice and cold everywhere; with them all up, we have a hot planet and total desert. But by positioning each one at some point along its sliding scale we might have temperate or tropical rainforest or open woodland, closed woodland, thicket scrub, tundra or savannah, with differing grades among all these. Adjusting perhaps two or three sliders very gently might change thick scrub into forest or savannah pushing into previously dense forested landscape. Our drivers push the sliders: a nudge here and a tweak there and some animals are suddenly disadvantaged while others have their day.

Biogeographic Extinction Drivers

The principles of biogeography have been with us for over 150 years and it is now considered by many that the co-discoverer of evolution and natural selection, Alfred Russel Wallace (1823–1913), was its founding father. A prolific collector of beetles, like Darwin, Wallace became a free-spirited traveller, unlike Darwin. Also unlike Darwin, he set out with the question of the origin of species on his mind and was not on a finite journey but one of open-ended discovery. His initial travels in the Amazon and his later journeys across the Malay Archipelago allowed him to amass large collections of beetles, butterflies, birds and animals. Again, unlike Darwin, he was a comparatively poor adventurer and sent collections back to England to pay – barely – for his expeditions. He seemed not to worry too much about making ends meet but lived frugally and natively and concentrated on his work, notes and collecting. But like Darwin, Wallace intimately studied the natural world he was passing through and that galvanised, for him, a recognition of a natural force behind the emergence of species rather than a supernatural one. Many, many nights spent in jungle huts and dilapidated lodgings, in heat, humidity, days of monsoonal rain, and suffering sandflies, leeches and mosquitoes resulted in tropical ulcers and malarial stupours but that did not deter him from studying his collections, writing his notes, reading the few books he had and, quite obviously, thinking about all of the above. It was through his sheer doggedness and fierce drive that he amassed the knowledge to construct a picture of why different insects, birds and other animal species existed on different islands with apparently little ecological difference between them. Why do some animals live in certain places and others do not? he asked and that is a fundamental question of biogeography. How can one group of birds and animals be so different from another while being separated by only 20 km of ocean? The now famous Wallace Line, an invisible biogeographic border, has become the principal delineation between placental mammals and marsupials as well as bird and butterfly species. The principles that determine where animals live are now encompassed securely within the tenets of biogeography of which Wallace laid the foundations. But those same tenets are double edged because while they explain why animals live where they *do*, they can also explain why they *don't* live in certain places. Indeed, they can be used to see why animals can also go extinct. I have chosen to call those tenets *Biogeographic Extinction Drivers*, or BEDs. These will be used in the following chapters in discussion of megafauna extinctions.

Biogeographic study is an extremely powerful tool for understanding the biological status of the planet. It is used also to determine how shifts in ecology and environmental circumstances can alter, either in a limited way or profoundly, the chances of animals surviving. Thus the process of extinction has always been very much a part of biogeographic study. BEDs are the bread-and-butter mechanisms that cause population reduction and contribute to extinction. The ecology of animals and their placement in the landscape governs their chances of persisting or going extinct; this has been amply demonstrated above when considering the macro-planetary changes that take place. However, I am not actually going to discuss BEDs here (I take that up in detail in Chapter 8) – but what are they? They are, in fact, the micro-environmental, -behavioural and -biological changes that occur in response to regular episodes of climate change or other natural changes that disrupt the natural course of events and that allow animals to thrive. They are more often than not delicately interrelated with subtle or not-so-subtle consequences for one another, a delicate web that, like a lace curtain, keeps everything in equilibrium for faunal populations. But when a thread is cut in a lace curtain, it begins to unravel and BEDS come into play. They cascade when one disappears or changes that causes many others that depend on them to change in response, and this, in turn, causes the disruption of others. If you like, a big hole appears in the curtain and its spreads. The list of BEDs provided at the beginning of the chapter demonstrates what I mean: they can cause extinctions on a local or regional scale, Although probably not complete it is worth reviewing these in relation to Australia's Quaternary megafauna extinctions, dealt with in Chapter 8.

Now, eliminating cataclysmic mass extinctions, the puzzle remains as to what caused more minor extinctions before humans arrived on the scene? In Chapter 3 we shall see that there was constant mammal turnover during the Tertiary. Sometimes it was slow; at other times it speeded up. All that was needed for the most part was a tweak here or a small turn there in the form of a rise or fall in ambient temperature, an environmental change, a loss of habitat, a reduction or rise in rainfall, the formation of desert or the rise of a mountain chain. This was caused by the natural sequence of planetary climate change and its environmental consequences. What we see that far back are gross changes in terms of world temperature, changing sea levels and the coming and going of ice sheets, rainforests, deserts and so on. They are every bit as catastrophic in their way as impacts by very large meteorites; it is just that their impact was not as instantaneous as that of a massive explosion. Rather, these processes are drawn out, usually over a great length of time. They can be thought of in a similar way to the slow release of nuclear energy in a power station reactor as opposed to a nuclear bomb explosion. They can be easily hidden in the undergrowth of time as they work decade to decade, century to century; for example, the changes that take place between the beginning of the drying out of a rainforest, for example, and the time that a savannah or even a desert forms in its place. The many stages involved in this sort of process each have an effect on an animal as a vegetation mosaic on which that animal depends alters from one composition to another or its geographic distribution changes through the rise of a mountain chain. Incremental changes add up to the movement and distribution of animals and the favouring or otherwise of others. The resulting waltz of creatures around the ecological dance floor may not always be recognised in fossil deposits but it has certainly taken place, and the study of biogeography over the last 150 years has shown how this happens to modern groups of animals all too often. But it does not only take place now the waltz is very ancient and the notes played in the tune are as constant and loud now as in the past. Ediacara fauna, Dimetrodons, Hadrosaurs and Trilobites must all have been where they were and did what they did best because they were subject to the very same tune of biogeographic principles operating at those particular times and they are the same ones that still operate today. But those same principles eventually worked against all of them as their opposite effects came into place and they became extinction drivers.

3 After the Dinosaurs

Starting Again

The time after the dinosaur extinctions often appears as an anticlimax to the enormous forces that were unleashed on the Earth across the K–T boundary and the resulting loss of the magnificent giants. The story that followed the dinosaurs is far too complex for the brief mention it receives in this book, and presenting that story is not what I really want to do anyway. But this is a good time to show how environmental and landscape changes affected the world's fauna and shaped its future. Principally, however, it is a good time to concentrate on mammals, particularly marsupial mammals. The Tertiary is the lead-up to the extinction event that will take centre stage in later chapters: the megafauna extinctions. Looking at the background to those requires a long lead-up through a long corridor back into the Tertiary. It is for this reason that it is worth looking at some of the history of mammals and marsupials during that time, in terms of what makes things evolve as well as what makes them disappear.

The Chicxulub bolide together with the massive volcanic activity of the Deccan Traps brought a finale to an extinction process among the dinosaurs that had begun millions of years before. What that extinction did was to open up opportunities for mammals in terms of diversification and this would lead to their eventual dominance. The loss of the dinosaurs left some gaping faunal holes among world's environments that needed to be filled and mammals obliged. But first they had to pick themselves up from being small and rather insignificant animals and become worthy of their future dominant role. They had to become variable in size and develop capabilities that gradually allowed them to adapt to all the world's environments. They already had long experience of the world around them but they had also played the underdog for a long time, with limits on their adaptive skills and diversification and how big they could grow without drawing too much attention to themselves. They must have been bursting with drive and potential to do this and the Tertiary Period was their time to take over. They would now become giants too; very different from their predecessors but, in their own way, magnificent examples of how animals evolve, adapt and fill the vacant niches and environments that were left behind for them to occupy.

Before proceeding it is worth knowing the basic timeline nomenclature that makes up the Tertiary which may be unfamiliar to some and which will be used freely here. The *Tertiary Period* began 65 Ma (million years ago) and ended 1.8 Ma. It is divided into two sub-periods: the *Palaeogene* (65–23 Ma) and *Neogene* (23–1.8 Ma). Each of these is divided into *epochs*: the *Palaeocene* (65–55 Ma), *Eocene* (55–34 Ma) and *Oligocene* (34–24 Ma) forming the *Palaeogene*, and the *Miocene* (24–5 Ma) and *Pliocene* (5–2.6 Ma) forming the *Neogene* (Table 3.1, my

Corridors to Extinction and the Australian Megafauna. DOI: http://dx.doi.org/10.1016/B978-0-12-407790-4.00003-3

Table 3.1 Dates of the Tertiary Eras

Palaeogene			Neogene		
Palaeocene	Eocene	Oligocene	Miocene	Pliocene	Quaternary
65–55 Ma	*55–34 Ma*	*34–24 Ma*	*24–5 Ma*	*5–2.6 Ma*	*2.6 Ma–Present*

italics). The Quaternary (1.8 Ma – today) followed the Neogene and is divided into the Pleistocene (1.8 Ma–10 ka) and Holocene (10 ka – today). I will sometimes refer to the Pleistocene but much more regularly deal in Quaternary terms. I hope you feel better for knowing that, although it has been known to bring on mass yawning in a lecture theatre. Unfortunately, these time segments are essential navigational tools that enable us to trace the turmoil of faunal development, change, replacement and extinction that took place during the last 65 million years. The Tertiary seems a long period probably because we know more about its detail, although never enough. Once again, we know more because the closer time scales are to us the more detail is available. It will also bring us to the near present, represented by the *Quaternary*, which is the time period of the main focus of this book.

Palaeogene Extinctions

The Palaeocene was an extremely important time for life on the planet because it was the first unfettered step in the evolution and radiation of mammals in their own right without the dinosaurs looking over their shoulders (some might say repressing them), and it gave rise to modern animals, including us. It was an active world with continents moving in all directions and climate fluctuating wildly. In the previous chapter we saw plate tectonic shifts that contributed to past extinctions; the situation was no different in the Tertiary and one of the most important shifts in this period was the final break-up of the megacontinent Gondwana.

Biological assortment is not only the prerogative of genetics: genes are spread out and brought together by geological processes like tectonic plate movement and sea level change, both of which provide the opportunity for the bringing together and sundering of gene pools. Chance is important in gene exchange, and in the Animal Kingdom chance depends on many things including continental movement and environmental and climatic changes; these can cause the geographic isolation of genes as well as their mingling and remixing as they are brought together again at a later time. They also determine the approximation of species, or how they are brought together, which tests their ability to interbreed, brings vigour and affects their chances of speciation. It also brings the possibility for extinction through competition and replacement of one group of animals by better-adapted or fitter ones. Even if their biological chances of interbreeding are good, animal A is never going to meet animal B if they live on opposite sides of the world, separated not only by distance but also by a large expanse of ocean: genes cannot assort at that distance. In biogeography animal groups are referred to as the South American, Australian or African faunas, which reflect their respective isolation and unique life histories as well as the fact that there was no chance at all for species to meet and integrate on one level or another. Past isolation provided the opportunity for faunas to take on their own characteristics and special identities over millions of years; this was enough to make the name of each animal synonymous with a particular place or continent. However, while plate tectonics can bring faunas together, it takes an inordinately long time to do so, particularly if continents are widely separated. The Australian fauna is a prime example of that. However, the Tertiary provided a continental geography that enabled the mixing of many faunas and we will see later how this took place.

The Tertiary saw vast changes in mammal populations around the world that encouraged diversity through their intercontinental movement. As I have said before, this sounds simple enough, but it is important to remember that we are dealing here with a period of millions of years: the bringing together of animal groups is only made possible by the joining of regions or continents through one geological mechanism or another. That is achieved through the animated three-dimensional jigsaw of plate tectonics shifting continental plates around at between 2 and 8 cm annually, or changes in sea level that either joined continents and islands that were once separated by land bridges or, conversely, sank such connections. As with genetic assortment, continental assortment comes about in various ways and produces equally varied results. When continental plates collide they form mountain ranges or one colliding edge can slip or subduct under the other to disappear into the Earth's mantle for recycling. Merging continents bring animal groups together and facilitate migrations across freshly expanded horizons. Mountain building can form barriers isolating one group from another or it can redirect, push and funnel animals along narrow corridors or passages such as between mountain ranges or along isthmuses or archipelagos. In wide open spaces on savannahs or plains, animals can split up as they spread out and become separated. They can also meet others in competition. That is when chance moves in and being in a particular place at a particular time starts to play a large part in how animals will fare in the future: they might thrive or they might go extinct under the dominance of others.

Extinction can take place when continents merge, bringing together previously isolated animal groups; this can lead to competition and possible replacement or outcompeting of one group by another that shares a similar diet and/or niche. The arrival of a new predator is another consequence that can alter the balance in such cases. The subsequent loss of one or more herbivores can also cause the demise of the carnivores that preyed on them, regardless of whether they were browsers or grazers. A new browser or grazer can then be a new selecting agent for plants and shrubs, and this in turn can have consequences for plant communities, altering their composition or survival chances. This then feeds back to assort the kind of animal which will inhabit that area and other creatures further along the chain may or may not benefit from these changes, and so on. Of course, such processes do not happen overnight, and alterations to animal populations resulting from them can be slow. They are also difficult to detect in a sparse and very long palaeontological record that is often difficult to date precisely. The point to be made here is that extinction also comes about by chance and the long-term consequences of these forces are not always obvious. Separation of continents puts expanses of ocean between once sympatric animal groups, beginning the process of division and possible isolation of genera and families and the loss of individual benefits that emerged from their previous association. Each will then follow its own path of survival or oblivion depending on its adaptive capabilities and environmental and competitive chances. Something similar to this happened among the early primates that led to the establishment of the New and Old World monkeys.

Tertiary Geography

To exemplify the above processes, it is worth looking at how they worked and what form they took during the Tertiary. First, we have to look at what the Earth was doing – in other words, what was happening to continental geography that would lay the foundation of the biogeographic stage. In the Palaeogene, the relic Gondwana supercontinent, consisting now only of Australia, Antarctica and South America, was in its final stages of break-up. After 600 million years of being part of Gondwana and 25 million years of rifting that began *ca.* 80 Ma, Australia and Antarctica were almost separated. That had followed the previous loss of Africa ~130 Ma and India ~105 Ma, both taking around 30 million years to make the split. Australia's break with Antarctica began in the west with a rift valley that gradually unzipped over 10 million years to a remnant join at Tasmania as we peeled off clockwise from our southern neighbour. The final separation was completed by the middle Eocene (*ca.* 45 Ma).

At the same time, India was moving across the old Tethys Sea at a variable speed of between 7 and 20 cm per year, very fast for tectonic plate movement. It was on its way to a collision with southern Asia, leaving the newly formed Indian Ocean behind it. It squeezed the Tethys before it like a bulldozer blade, and pushed the Tethys seabed along with it. The crust was driven down and became subducted beneath Asia's continental crust, causing the uplift that eventually formed the Himalayan Mountains. A similar subduction was taking place on the other side of the world as the Pacific Plate pushing beneath North America's west coast and the Rocky Mountains began to form. An extension of the same process was taking place right along the west coast of South America, resulting in the uplift of the Andes. At the same time, Greenland was separating from North America and Europe and became isolated on either side by a Y-shaped fissure that formed at the northern end of the mid-Atlantic ridge. By the mid-Eocene, Africa was approaching Europe, a movement that would eventually cut off the western end of the Tethys Sea, eliminating it altogether and forming several large inland lakes. The Tethys had been a very large ocean since the Triassic and an important force in the distribution and direction of world ocean currents. It had also been dominant in the transport of heat and nutrients throughout the world's oceans, but now it was disappearing, although at the same time giving way to the new Indian Ocean. It is always difficult to believe how such a large and important feature can be squashed out of existence by the movement of continents, but the power of the Earth makes us remember *we* can *save the planet*.

The formation of the North African Atlas Mountains was also a consequence of the Afro-European collision. Africa's relentless northward push continued during the Neogene, eventually raising the Alps and altering Europe's geography. These ancient Alps would form part of a mountain chain that included the Pyrenees, Italian Pennines, Carpathians and Caucasus, all of which would join up with the Himalayas in a long mountain chain encircling southern Eurasia. Massive volcanic events accompanied Palaeogene tectonic movements. Prominent among these were those which preceded India in its northward passage, spreading out across the narrowing Tethys Sea. Another volcanic outbreak accompanied the activities of the mid-Atlantic Ridge. Named the North Atlantic Igneous Province, it formed Iceland *ca.* 50 Ma as well as a number of very large volcanoes and basalt plateaus in the North Atlantic region, some 1.5 km thick. Atmospheric CO_2 began to rise from disturbed carbon-rich sediments on the Atlantic seabed, and methane and sulphur dioxide were released by the massive volcanic activity. Those greenhouse gases were joined by carbon dioxide that had been sequestered on the now-uplifted Tethys seabed and was now released into the atmosphere. This combination brought on massive global warming and the formation of a 'greenhouse' world. By the early Eocene the planet's climate had reached the warmest it would ever be in the Tertiary, with annual average temperatures 9–12°C higher than those of today, even warmer than at the time of the Permian extinctions 200 million years before (Figure 3.1). Ocean temperatures also rose to between 4°C and 9°C, which brought on an extinction of foraminiferans. The EO had begun. It initiated the spread of rainforest into high latitudes with deciduous forests at the Poles. Trees there lost their leaves in winter not because of cold temperatures but in response to the long Polar night. As an example of the different world of the early Palaeogene, Britain at that time had a climate more like present-day Malaysia.

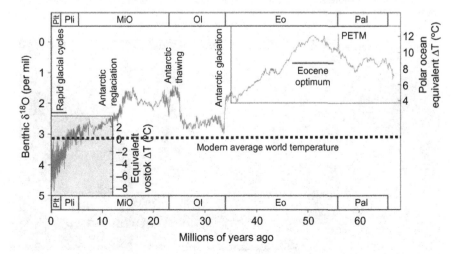

Figure 3.1 Average world temperature fluctuations during the Tertiary showing prominently the very warm EO around 50 Ma and world cooling that began around 12–15 Ma.

The map of the world was changing; it always had been, mind you, but now it was taking on the appearance that we recognise today and was helping shape the future of the rise of mammals. There were, however, still enough differences to make it stand out, particularly continental shapes. An obvious example was the region between Europe and Africa, which was filled with islands floating in the remaining western portion of the Tethys which still drowned parts of southern Eurasia. Naturally, the biogeographic status of animals in the Palaeogene reflected the geographical differences in the way continents were related to one another and the very warm planet they lived on. The spread of warmer environments to higher latitudes also provided greater range sizes for animals which encouraged diversity and speciation and encouraged many species to live in high latitudes. Animals could also travel between Europe and North America by a North Atlantic Route, a permanent land bridge between the two continents that existed via Greenland (Figure 3.2). Animal migration at that time was enhanced by the EO, which reached its peak *ca.* 50 Ma, but that migration route was terminated when Greenland was isolated on both sides at the northern end of the Y-shaped mid-Atlantic Ridge. In contrast, the Ural Trough or Turgai Strait that had divided Asia from Europe since 160 Ma prevented fauna moving west almost to the end of the Palaeogene, when it closed in the Oligocene *ca.* 29 Ma. They could, however, move east via the Alaskan–Siberian or Bering Bridge, into North America and possibly on to Europe, taking the long North Atlantic Route before Greenland became isolated. The Bering Bridge was an intermittent feature but lasted on and off till *ca.* 23 Ma. When the Ural Trough

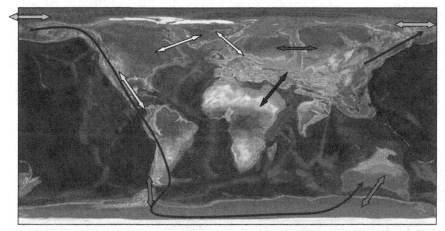

Figure 3.2 Tertiary mammal migration routes.

closed at the end of the Oligocene, both Asian and European mammal faunas had formed their own respective compositions and could now mix. Closing of the Turgai Strait brought Europe and Asia together but that was not the only closing-up event. Another dry-land collision had previously occurred between Africa and Eurasia at the start of the Oligocene, forming the Afro-Asia Bridge and bringing together African and Eurasian faunas. The African plate had actually collided with Europe much earlier, beginning the *Alpine Uplift* (Orogeny), but the opportunity for Africa's animals to run ashore had to wait till later.

The Panamic Bridge between North and South America existed during the Palaeocene but fractured before the beginning of the Eocene and did not form up again till the mid-Pliocene (*ca.* 5 Ma); this produced a long period of isolation for faunas on that continent. However, evidence from excavations carried out during recent widening of the Panama Canal suggests that the Panamic Bridge might have formed much earlier – around 18 million years earlier! This suggests that the bridge actually began with the emergence of a series of volcanic islands, which later formed an archipelago. The archipelago rose as part of the tectonic processes of the Caribbean Plate which became a plug sealing the join between North and South America. That connection was important in producing the world we know today. Separation of the Atlantic from the Pacific produced the Gulf Stream that warms Europe in winter, and with the westerly trade winds from the Atlantic, more rain dropped in the Pacific, making it less saline than the Atlantic. But most important for this story, whenever that join took place, it allowed the transfer of animals between the two continents: creatures including sloths, marsupials, porcupines, anteaters and armadillos headed north, while pigs, dogs, sabre-toothed cats, camelids, horses and elephants moved south. Animals were again on the move, mixing, competing and replacing one another, and, as always, some went extinct in the process.

By the end of the Palaeogene Asia, Africa and Europe had joined together, allowing animals to roam free across an east–west expanse of over half the globe. That brought widespread assortment to a large number of the world's faunas. Each group had been adapting and diversifying in its own realm but now there was a fresh chance of dispersal to new areas where newly emerging ecosystems enhanced further diversification and increased opportunities for adaptation and speciation. The palaeogenetic 'washing machine' that followed was a driving force that invigorated the world's mammals. The EO had eminently driven the take-off of mammal faunas, providing many different opportunities for them to show how diverse and adaptive they could be. But now climate change was again driving adaptation as the world began to cool from the Eocene into the Oligocene, bringing an 8°C drop in average world temperatures. That began a glaciation at the Poles, with the loss of large areas of forest and grazing land and contraction of rainforest to lower latitudes, a process that undoubtedly affected animal groups dependent on those ecosystems. Now they had to move or perish. One other change was that with fewer forests, there were fewer animals – in other words, smaller populations. One major change was the formation and spread of grasslands in lower latitudes, favouring grazing over browsing animals. Then there were animals not in the direct line of genetic assortment – the isolated populations like those of North and South America. They would have to wait some time, till the Early Miocene and then Pliocene, before they could participate in free exchange through Central America. But at this time both groups were isolated from the world. Of course, at the beginning of the Oligocene the most isolated fauna of all was that which slowly floated north on the Australian continent, the marsupials. Every day they were becoming more isolated but their story will have to wait till Chapter 5.

Animals of the Palaeogene World

Rainforest ecology provides a magnificent setting for diversification and speciation among myriad creatures spread through its various high-rise niches, although it does not encourage bigness. So, the hot world of the Palaeogene gave mammals just what they needed after the dinosaurs: a chance to get on their feet. This was the time of widespread rainforest growth and their complex stratification offered support and safety for various groups through its many levels. Small animals in particular could grow and thrive in the comparative safety of jungle branches and canopies rather than be exposed on open plains. From the complex ecosystems of their undergrowth, streams and swamps to upper canopies, rainforests encouraged speciation, specialisation and diversity, providing an enormous boost to mammal variety. It gave them a chance to see what they were made of and put them on the road to dominance through growth in size, stealth and hunting capability. Small was the watchword: there were no large carnivores or herbivores at that time and large bodies would not evolve till the Eocene, but the stage was set. Most essential was the plentiful breathing space that allowed them to evolve. Extinction did not seem to stalk these creatures at this time but it was never far away.

There were no real browsers or grazers because open woodland and shrub lands became rarer with the EO onset and rarer still at its peak. Grasslands were small and would stay like that for almost 20 million years till the world cooled. Evolution continued its march, however, and one notable example was a hoofed carnivore (a mesonychid condylarth) that, almost unbelievably, exchanged a life on land for one in the ocean, beginning a line that would lead to modern whales (cetaceans) and sea cows (sirenia) and reversing the old habit of life moving from the ocean to the land. While this might sound as though all was well, however, the following 50 million years was not without its turnover of animal families and extinction events, albeit none extensive, dramatic or sudden.

One consequence of extinction mentioned in the previous chapter is habitat vacancy through species loss. The comparatively sudden demise of various types of dinosaur left open a range of ecological niches that were sooner or later filled by the surviving mammals. So the speciation and radiation that was beginning to take place among Palaeocene mammals was also driven by niche availability and newly vacated landscapes. This process also brought about *novel* specialisations among animals in the form of anteaters, grass grazers and those that gnawed. Peculiar anatomical arrangements later emerged, particularly among the new larger species, as though nature was once again experimenting with size and introducing different models till the right one was found. The Palaeogene fossil record is littered with weird and wonderful examples of this process, some indicating creatures whose life-styles and habits we know nothing or almost nothing about; this was also happening among Australia's marsupials on their isolated raft. Indeed, the mammal groups of the Palaeocene are often referred to as primitive but really some just consisted of odd designs. They were probably as adapted to their world as modern animals are to theirs, but in these early days of mammal dominance it did not really matter. It is also worth remembering that even if a creature lived only between, say, 58 and 60 million years, that is still a considerable period of time – although it looks fleetingly brief at this distance. It may also have been well adapted to its surroundings and to a special niche that only lasted for that particular time.

Unusual appearances flourished in the mid-late Palaeogene and many exam-ples included large bodies which were emerging by the end of the Eocene and early Oligocene. Two examples are the multi-horned and short-tusked *Eobasileus*, a plains-dwelling uintathere from North America, and the North African *Arsinoitherium*, with its gigantic, front-facing, twin horns, both rhino-sized ani-mals (Figure 3.3). North America almost cornered the market on these sorts of crea-tures with other examples such as *Brontops* and *Brontotherium* ('thunder beast') of the early Oligocene (Figure 3.4). All were very large and collectively referred to as titanotheres. Many reached 2.5 m at the shoulder, but *Brontotherium* was the larg-est and, although it looked like a rhino, at around 5 tonnes it was closer to elephant size, although shorter. Asia also had *Embolotherium*, which was also bigger than the largest modern rhino and sported an upright, shield-like nasal horn. The brontoth-eres were very successful and widespread and must have been impressive to see, particularly as a charging herd, but with their rather small brains they were prob-ably not very bright. Ironically the modern form of grazing rhinos that the titanothere most resembles would not make an appearance till the larger forms had disappeared;

Figure 3.3 One of the large early Oligocene brontotheres *Arsinoitherium* of North Africa with its massive pair of bilateral horns.
Source: Photograph by Steve Webb.

Figure 3.4 The North American titanothere, *Brontops robustus*, with its twin, laterally positioned nasal horns.
Source: Photograph by Steve Webb.

Figure 3.5 The small, early proboscidean *Moeriotherium* of late-Eocene Africa.
Source: Photograph by Steve Webb.

this was also true of elephants. Both were a product of the Neogene, although the African ancestral forms were smaller late-Palaeogene animals, like the pig-sized *Moeritherium*, which lived in the late-Eocene, and the larger 2.5-km-tall *Phiomia* that followed in the early Oligocene (Figure 3.5). The earliest member of the true rhinoceroses, the hornless *Trigonis*, also appeared in the early Oligocene. This animal was much smaller than modern rhinos and lived in North America and Europe. The big Palaeogene browsers, however, were about to go extinct as shrub lands and forests contracted ahead of spreading grasslands, in response to a cooling planet.

One animal that lived in the Oligocene that did fill dinosaurian boots was *Paraceratherium bugtiense* of Pakistan and India, also known as *Indricotherium* or *Baluchitherium*. It dwarfed the largest titanothere and had the distinction of being the largest land mammal ever. Its statistics are stunning for a recent terrestrial

Figure 3.6 *Andrewsarchus mongoliensis*, a rhino-sized, wolf-like carnivore of the late-Eocene.

animal, standing 6 m at the shoulder, 7.5 m at the head, 8 m long and weighing in at between 10 and 20 tonnes, the weight of *three* elephants. It was a browser and, in fact, a good equivalent for the smaller brontosaurs. *Paraceratherium* was another highly unlikely evolutionary story because at present all the evidence seems to suggest that it evolved from the lightly built, 1.5 m long, fast-running *Hyracodon*; but it outlived the titanotheres, disappearing at the end of the Oligocene *ca.* 23 Ma. *Andrewsarchus mongoliensis* (Asia) of the late-Eocene was another spectacular animal that could also occupy a previous dinosaur niche, that of *Tyrannosaurus rex* (Figure 3.6). This was a big niche to fill but *Andrewsarchus* was probably up to it. It was a fearful, gigantic wolf-like carnivore the size of a rhino with massive 1-m-long jaws, equipped with a dentition that could tackle almost any animal it came across – although a bull *Paraceratherium* might have stretched its capabilities. But as with *T. Rex*, arguments still continue as to whether it was a predator or largely a scavenger; like *T. Rex*, no doubt it did both.

The Palaeogene set the stage for the development of variety among mammals, with enough breadth to provide radiation of larger varieties with a range of adaptations that could confront a rapidly changing world. Change occurred as the Earth became colder, grasses spread, savannas emerged, rainforests retreated and sea levels dropped. The last stage of the Palaeogene, the Oligocene, heralded a different world from that which had gone before and required a new direction in adaptation which threw switches in the mammal world that began to produce templates for animals that are more familiar today. They could live on a much cooler planet and were arguably more readily adaptable. What we do know is that during the Palaeogene animals came and went, large fierce types and smaller quiet individuals alike. Turnover continued and individuals and groups of species disappeared in what is termed *background extinction*, where animals come and go with no apparent reason – usually because the details are lost in the fossil record. Assessing rates of extinction at that time is difficult because there were no humans to record the speed at which species and genera were actually disappearing. But it is worth looking at the transitions that took place between the Eocene and Oligocene as it was then that mammals were faced with their first big environmental challenge.

Eocene–Oligocene Boundary: the End of an 'Era'

The Eocene–Oligocene transition is marked essentially by the Earth moving to a much cooler and more temperate climate. The late-Cretaceous/early-Tertiary 'hothouse' that had reached a zenith in the early Eocene now cooled to a level not seen since the end of the Jurassic. With termination of the EO, global temperatures dropped uniformly although they still remained 4–5°C above today's levels. That temperature drop culminated in a rapid cooling event *ca.* 34 Ma. Another 4°C drop then occurred very quickly, marking the Eocene–Oligocene boundary and bringing temperatures almost to the level of those today. It took place as the final two pieces of Gondwana separated: South America separated from Antarctica *ca.* 35 Ma, opening the Drake Passage. Continental division in the high southern latitudes caused a major change in oceanic currents, particularly in the Australian–Antarctic region, resulting in the formation of the world's largest current: the *Antarctic Circum-Polar Current* (ACC). The ACC has also been called the Clipper Current for its ability to speed up the sailing time of the old clipper ships going from west to east across the southern Indian Ocean; they were also aided by the Roaring Forties and Screaming Sixties latitudes. It dwarfs the Gulf Stream with a flow equivalent to 600 Amazons and puts up a barrier to warm water reaching Antarctica, thus helping to prevent the continent from defrosting. Before, the process was in reverse, with warm ocean currents transporting heat down to Australia and Antarctica, which helped keep Antarctica ice free and encouraged high levels of rainfall in Australia. This, in turn, maintained temperate rainforests across most of both continents. But now, circulating only around Antarctica, the ACC became very cold, deflecting warmer currents, cooling down the Antarctic and promoting ice formation, lowering sea levels and reducing rainfall in southern parts of Australia. Sea ice began to form and the cold, dense water sank, precipitating the extinction of deep-dwelling molluscs and foraminiferans as well as some plants and marking the end of the ancient whales, the Archaeoceti.

One other element that might have contributed to global cooling at that time was a series of meteor impacts. One occurred in Siberia *ca.* 35 Ma, forming the 100-km-diameter Propigai crater. Two others struck North America around the same time (35.7 Ma): one formed the 85-km-diameter Chesapeake Bay crater (85 km across), and a second, the 22-km-diameter Toms Canyon Crater, on the continental shelf off New Jersey. Again, we are confounded by the length of time between these events, the various interrelationships that may have existed between them and their consequences for North American and world faunal populations. What we do know is that a rapid 4°C planetary cooling took place at or close to the impacts and that continued in the Oligocene, encouraging Antarctic glaciation, dropping sea levels and bringing aridity to some parts of the globe as world rainfall decreased.

In Europe the *Great Break*, or *Grande Coupure*, of the Eocene–Oligocene boundary marked a turnover in mammal populations which, of course, meant extinctions. Some of the faunal groups representing larger animals previously dominant in Europe, including perissodactyls, odd-toed animals, such as early horses, and artiodactyls or even-toed and cloven-hoofed varieties, went extinct. These included six families: Anoplotheriidae, Xiphodontidae, Choeropotamidae, Cebochoeridae, Dichobunidae and the Amphimerycidae. They were replaced by the modern rhinoceros family (Rhinocerotidae) and three families of artiodactyls, the Entelodontidae (pigs), Anthracotheriidae (ruminants) and Gelocidae (hippos); this also marked the end of Europe hosting members of the early primate families Omomyidae and Adapidae. Again we see the extinction followed by replacement. These changes coincide with both the rapid cooling event *ca.* 33.5 Ma and the later arrival of Asian mammals that moved west into Europe following the closure of the Turgai Strait *ca.* 29 Ma. So the debate arises: Was it a cooler planetary climate, including ice formation in Antarctica and the accompanying drop in sea levels and vegetation changes that accompanied climate change, that caused the disappearance of European mammals, or competition from incoming Asian animals or a combination of some or all these? There may have been another cause that is not clear but whatever it was, it was not anthropogenic. The onset of aridity in Asia may have precipitated the westward dispersal of animals from there but that is not directly correlated with the Great Break, which occurred slightly earlier. Rather than a replacement of animal families by entirely different animals, it was an event that saw the replacement of one series of families by similar families that may have been better adapted, or more able to adapt, to changing conditions. The cooling argument is undermined somewhat by the lack of an equivalent event in North America, although North America did lose some animals around that time.

These events were not part of a sudden mass extinction but a gradual one somewhat more severe than that of the usual background type, and it varied in the degree of animal disappearances over large geographic regions across the world. It was also an opportunity for the replacement of animal types due to comparatively rapid environmental change and the rapid adaptive qualities that mammals needed to keep up with the changes. However, if we had been around at the Eocene–Oligocene boundary we would probably have had no idea how fast the faunal changes were taking place because of the comparatively long time scales involved. Any change that we did observe might have been put down to a trend due to a much shorter mechanism or perhaps associated with environmental change across local or regional niches. The 'sudden' changes taking place saw two genera (*Palaeotherium* and *Anoplotherium*) as well as two of the six artiodactyl families (Xiphodontidae and Amphimerycidae), disappear completely. These changes were not abrupt, however, but probably took place over 6 million years. Nevertheless, in terms of the number of species involved it was a comparatively large extinction because it involved a great turnover of animals – certainly the biggest since the dinosaur extinctions – and we are only dealing with mammals.

The Isolated Continents

But what of animals living on the two isolated continents of South America and Australia? During the Tertiary both continents supported their respective faunal populations, which diversified in their own way according to their needs and without outside influence from incoming animal populations and new species. Australia had only marsupials but South America had both marsupials and placental mammals. The latter included the more primitive edentates, the ancestral forms of sloths, armadillos and anteaters, but South America did not experience the extinctions that occurred elsewhere at the end of the Eocene. It was separated from its northern neighbour by fluctuating sea levels that left only an intermittent island chain between; that isolation differentiated it from North America, where animals could move in and out via two intermittent land bridges with Europe and Asia. The ancestors of all the South American fauna had moved into the continent at the Cretaceous–Tertiary boundary, after which South America became isolated. Condylarths entered the continent, eventually giving rise to the litopterns, short-trunked, hoofed ungulates, the most famous of which is probably *Macrauchenia*, a rather Disneyland-looking animal that went extinct in the middle Pleistocene (Figure 3.7). Some of the last animals to arrive in South America before its isolation were the protitheres, early monotremes. Technically, it is not quite true to say that South America was isolated after the K–T boundary. It certainly was from the north but not from the south: it maintained its links with Australia via Antarctica till 45 Ma and Antarctica till *ca.* 35 Ma. However, protitheres seem to have been the last animal group to enter the continent before the final break between Australia and South America occurred. They had probably begun their journey from Australia, where the oldest fossil

Figure 3.7 A tapir's head on a camel-like body; the odd looking *Macrauchenia* of South America sporting its short trunk.

evidence for them has been found, dating back to the early Cretaceous, long before the time when Australia broke away from the last to Gondwanan continents. The fact that larger placental mammals did not move in the opposite direction to prototheres and enter Australia seems to suggest that colder, higher latitudes at the bottom of South America may have been a deterrent, although at that time it was not as cold as it would become later.

Other South American animals included the large tapir-like *Astrapotherium*, various large-bodied hoofed ungulates (*Toxodons* and *Trigonostylops*), endemic giant sloths (*Glossotheres* and *Megatheres*) whose ancestral forms (the *Promegatheres*) began in the Miocene, and a strange marine sloth called *Thalassocnus*, a seaweed and seagrass grazer, a sort of hairy dugong with legs. Others included *Pyrotherium*, a small elephant-like ungulate, and Scarrittia, a rhino-like hoofed mammal. Similarities between many South American mammals and those living on other continents (highlighted by the hyphenated '-like' used above) suggests some parallel evolution took place, with many of its species taking on roles similar to those of Old World mammals, something that also took place among Australia's isolated marsupials. Another peculiarity that occurred in South America was that its placentals were being hunted by marsupial carnivores like *Borhyaena* and *Thylacosmilus*; this may seem odd to those of us brought up on quite opposite ideas that marsupials were always hunted, or at least outcompeted, by mammals. Ancestral New World primates did not arrive till much later, during the Pliocene.

With the lack of arrivals from the outside, South American faunal changes could only come from endemic evolutionary factors and universal world events that affected that continent. Its animals diversified into a variety of niches stretching from the sub-Polar regions of Patagonia to north of the Equator. Such a broad geographic range presented a wide choice of habitats for the marsupial-placental population to occupy, although it is possible that the bulk of the population lived in more northerly parts of the continent, which might explain the small number of placentals living in the south ready to move into Australia.

For most of the Tertiary, South America was home to the flightless 'terror birds' (Phorusrhacidae) (Figure 3.8). They were powerfully built and ranged from 1 to 3 m tall, with the largest having a deep, terrifying beak nearly 1 m long, often tipped with a nasty hook, with which they were able to consume smaller mammals in one gulp. Opinions differ about whether they were pack hunters or loners but their fast-running abilities were probably not developed just to scavenge. Like *Andrewsarchus*, the biggest of these birds filled a dinosaur niche, that of velociraptors; they were fast and dominant predators, afraid of nothing and without enemies (except perhaps *Andrewsarchus*), but as far as we know that did not live in South America and was confined to East Asia and Mongolia. The earliest members of the Phorusrhacidae lived in Europe and North America in the early Palaeogene, but became extinct. They seem to emerge again in the late-Eocene in South America where they lived for a very long time, from 3 to 36 Ma. Obviously some migrated to North America from Europe then south, crossing the Panama isthmus. The question remains: Why did they go extinct in Europe only to remain for over 30 million years in South America? Perhaps they filled a niche that no other animal could fill. At least one species

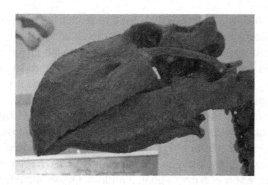

Figure 3.8 The powerful beak of a 'terror bird' (Phorusrhacidae), a fast, top predator that filled the niche of velociraptors and occupied top predator during the Palaeocene. *Source*: Photograph by Steve Webb.

returned to North America during the *Great American Interchange* in the Pliocene, when a number of marsupial and mammal groups also moved north after the Panama Bridge formed during the Pliocene. That bird was or gave rise to *Titanis*, a 2-m individual that, it has been suggested, could reach speeds of 65 km, a truly frightening predator – but even it went extinct in the early Pleistocene. It is worth noting that a number of very successful species eventually went extinct in the Pleistocene, the Age of Ice Ages, and this will become relevant in later discussions.

What can we take from the story that emerges here? Animals appear and disappear; some last for short periods of time, others for longer ones. It matters not whether they are large or small, terrifyingly strong or rather puny; whether they live a precarious existence or are totally adapted to their environment; whether they are adaptively gregarious or specialised – extinction comes to all of them in the end. Eventually a combination of circumstances produced by geotectonic factors, climate and environmental change, as well as a range of other forces, confounds the strongest and most prolific of animal groups. The conveyor of inexorable world change never stops and continually throws up new challenges for the Animal Kingdom, including new combinations of habitat that require new adaptations, migration and a healthy element of chance and luck in order for them to survive. We must also not forget that we are mammals viewing the vulnerability of other mammals. In that role we tend to see others as, perhaps, inheriting an *inevitability* towards extinction while ignoring our own inevitable end. One reason we do this is probably that, generally speaking, we cannot control the normal planetary causes of extinctions no matter how sophisticated our technology and abilities become, so we ignore them. But the Earth, she will win.

It seems clear that the central reason for the faunal changes, exchanges and extinctions of the Eocene–Oligocene transition and the *Grand Coupure* was the chain reaction of planetary geological change in the form of plate tectonics, followed by a climatic see-saw, its accompanying environmental changes and the consequences of those that occurred to various degrees on all the continents. Central to the environmental changes were world cooling then warming and cooling again.

Comparatively rapid climate change was important in these processes. As the world cooled, rainforests and other forests retreated, leaving more land open for the spread of grasslands and grassland communities. Grasslands had a higher albedo, transpired less water and grew in soils that weathered faster than those in which woody vegetation grew; thus, organic carbon accumulated better in grassland soils, all of which contributed to a cooling planet. Thousands of square kilometres of rainforest disappeared as it contracted in the face of lower temperatures, particularly in high latitudes, which probably offset CO_2 sequestration somewhat, another feedback for a cooler planet.

Major areas for the environmental changes that followed occurred in places like Europe, central and northern Asia and North America where grasslands began to expand. Dwindling forest saw the reduction in the suites of animals that occupied them. Grasslands then began to favour broad species diversity among grazing animals while at the same time rainforest contraction mitigated against those species inhabiting them; this in itself brought about alterations in faunal variety and distribution across major parts of the world. Smaller rainforests meant smaller rainforest animal populations, making them more vulnerable to extinction and producing a smaller variety of mammals across the world. Smaller populations brought about viability problems for some species that probably meant extinction. But these changes took time and they were subtle over the short and medium term. If we had been observers of these changes we, in our short human generation time frame, would probably comment that certain species were never common or had always been rare and therefore vulnerable to small or almost insignificant environmental or climatic changes. But those situations had actually built up over many hundreds if not thousands of years. Tracing the course of extinction even over thousands of years is difficult because of the long time frame compared to our lifetimes. In geological time scales, such extinctions are often seen as rapid if not sudden. For example, how can we know about or assess the gradual disappearance of a particular species over 1000 years if it lived 20 million years ago?

What we see in the Palaeogene is the continual replacement of one group of animals with another on all continents, independent of one another, and this takes place on a regular basis. From this distance we can see these extinctions as the natural progression of evolution, even though it meant the loss of hundreds of species. We can also see it as a natural reaction to changes in the world's climate and environments. They are not disasters but rather the Animal Kingdom reacting to altered circumstances, over which it has no control, as best it can; that process continued into the Neogene.

The Neogene Extinctions

The Palaeogene contained an interwoven tapestry of factors and forces that were aligned against mammal populations not only during that time but also continue to operate against them. It was important to outline those interrelationships, i.e. the role of migration and the chances for adaptive radiation that open and close for animals as the planet moves through a wide range of geological, geographical and climatic changes. The study of changes to animal dispersal and population placement through time usually comes under the heading of palaeobiogeography or palaeozoogeography, and we will delve further into these in later chapters as we move into the mid-late Quaternary. The interrelationships among factors that send animals extinct become a little more apparent during the Neogene, although we know a little more about them the more closely we approach the present because more information is available from fossil and geological records. We can now move on to understand better the pulses of extinction that continued in the Miocene and Pliocene in particular, but the details of the extinction drivers covered in the previous section will not be repeated here.

The goal is to concentrate on the timing of faunal disappearances, the reasons for them and what animals were doing to adapt to a climatically changing planet. In tracing the comings and goings of mammal genera around the world during the Tertiary and into the early stages of the Quaternary, I refer the readers to the following five tables (Tables 3.2–3.6). They highlight many of the genera that went extinct on the five major continents of North and South America, Europe, Asia and Africa throughout that time. The lists are not exhaustive but I believe they provide a reasonable picture of the great turnover of groups around the world during that time and the co-occurrence of many with major climatic changes. The resultant losses are graphically illustrated further in Figures 3.9 and 3.10 and Table 3.7. They show the Miocene extinction hiatus as the world entered a new long-term ice age with temperatures reduced to a level where many animals reached their adaptive limits as the warmer Earth slipped away. That trend continued into the Pliocene as planetary cooling continued.

Miocene Environmental Switching and Extinction

The beginning of the Neogene is not marked by any particular event but global warming was occurring once again during this period with world temperatures climbing 3–4°C higher. That took the planet back to conditions similar to those of the late-Eocene. The Antarctic ice cap that had developed through the Oligocene thawed, seas rose and the world returned to a partial hothouse, although temperatures did not reach those of the EO. Tropical forests expanded once again but not nearly to the same extent as they had before, and they tended to be restricted to mid-latitudes. This time warming also brought planetary drying rather than a much higher rainfall. Western parts of North and South America and parts of east Africa received less rainfall, which impacted on their animal populations. These factors, together with the development of tougher scrub plants, open woodlands and even deserts, brought new challenges for animals including fresh selection pressures that demanded not only adaptive changes but also behavioural ones for those who aspired to live in more arid conditions. These acted in three ways: animals either moved or adapted, and some

Table 3.2 Mammal Extinction in North America During the Tertiary and Early-Mid Quaternary

N = 65	Paleocene	Eocene	Oligocene	Miocene	Pliocene	Quaternary
Chriacus	▬					
Diacodexus		▬				
Trogosus		▬				
Heptodon		▬				
Coryphodon		▬				
Eotitanops		▬				
Uintatherium		▬				
Hyrachyus		▬				
Stylinodon		▬				
Dolichorinus			▬			
Eobasilius			▬			
Protylopus			▬			
Arsinoitherium			▬			
Brontops			▬			
Brontotherium			▬			
Embolotherium			▬			
Trigonius			▬			
Elomeryx			▬			
Eusmilus			▬			
Marycoidodon			▬			
Poebrotherium			▬			
Hyaenodon			▬			
Metamynodon			▬			
Hyracodon			▬			
Archaeotherium			▬			
Nimravus			▬			
Amphicyon				▬		
Cynodesmus				▬		
Miotapirus				▬		
Oxydactylus				▬		
Promerycochoerus				▬		
Stenomylus				▬		
Syndyoceras				▬		
Oreodontids			▬			
Brachycrus				▬		
Moropus				▬		
Desmostylus				▬		
Dinohyus				▬		
Hemicyon				▬		
Aepycamelus				▬		
Barbourofelis				▬		
Ilingoceros				▬		
Amebelodon				▬		
Protoceratids			▬			
Aphelops				▬		
Astrohippus				▬		
Blastomeryx				▬		
Teleoceras				▬		
Procamelus				▬		
Synthetoceras				▬		
Gomphotherium				▬		
Borophagines			▬			
Dinohippus					▬	
Cranioceras					▬	
Hippotherium				▬		

(*Continued*)

Table 3.2 (Continued)

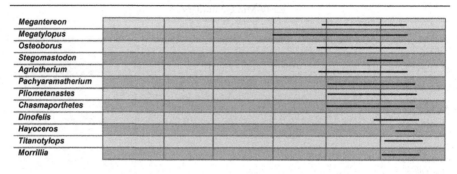

Megantereon					
Megatylopus					
Osteoborus					
Stegomastodon					
Agriotherium					
Pachyaramatherium					
Pliometanastes					
Chasmaporthetes					
Dinofelis					
Hayoceros					
Titanotylops					
Morrillia					

Table 3.3 Mammal Extinctions in South America During the Tertiary and Early-Mid Quaternary

N = 24	Paleocene	Eocene	Oligocene	Miocene	Pliocene	Quaternary
Trigonostylops	—					
Thomashuxleya		—				
Notostylops		—				
Pyrotherium		—				
Scarrittia			—			
Borhyaena (M)				—		
Diadiaphorous				—		
Thesodon				—		
Adinotherium				—		
Peltephilus			——	—		
Astrapotherium			——	—		
Homalodotherium				—		
Telicomys					—	
Thylacosmilus (M)					—	
Trigodon				——	—	
Thalassocnus					—	
Chapalmalania					—	
Stegomastodon						—
Macrauchenia						—
Pachyaramatherium					——	—
Glossotherium					——	—
Toxodon					——	—

M - Marsupial

Table 3.4 Mammal Extinctions in Asia During the Tertiary and Early-Mid Quaternary

N = 32	Paleocene	Eocene	Oligocene	Miocene	Pliocene	Quaternary
Ankalagon	—					
Yangtanglestes	—					
Diacodexus		—				
Coryphodon		—				
Eotitanops		—				
Sarkastodon		—				
Andrewsarchus		—				
Mongolestes		—				
Phiomia		—				
Paraceratherium			—			
Metamynodon			—			
Hyaenodon			—			
Archaeotherium			—			
Metaschizotherium				—		
Hemicyon				—		
Desmostylus				—		
Percocruta				—		
Platybelodon				—		
Thalassictis				—		
Chalicotherium			—	—		
Gomphotherium				—		
Giraffokeryx				—		
Hippotherium				—		
Merycopotamus				—		
Ancylotherium					—	
Bramatherium					—	
Deinotherium				—		
Agriotherium					—	
Anancus					—	
Megantereon					—	
Chasmaporthetes					—	
Dinofelis						—
Elasmotherium						—

were able to do that, but they also pushed others to the edge of their capabilities and they went extinct.

Regional environmental variations dominated, affecting local faunas and biotic communities. Just imagine: the Palaeogene finished with a cooler Earth, lowered sea levels, acceleration of grassland expansion and rainforest contraction. These conditions had been taking animals in one selective direction for some considerable time, but now, with a return to warmer conditions, selective pressures were acting in the opposite direction, as well as in some new directions. Such switching is really the stuff of rapid evolutionary change and requires very strong adaptive abilities: this really sorts out those that can cope from those that cannot. In itself it becomes an extinction driver. Fauna faced the challenge of adaptation in many areas as conditions swung back and forth, and only those able to stay with these challenges

Table 3.5 Mammal Extinctions in Europe During the Tertiary and Early-Mid Quaternary

N = 36	Paleocene	Eocene	Oligocene	Miocene	Pliocene	Quaternary
Diacodexus	—					
Coryphodon	——					
Eurotamandua		—				
Anoplotherium		—				
Trigonius		—				
Elomeryx		——				
Eusmilus			—			
Hyaenodon		——	——			
Nimravus			——			
Amphicyon			——			
Cainotherium			——			
Potamotherium				—		
Metaschizotherium				—		
Hemicyon				——		
Percocruta				—		
Platybelodon				—		
Thalassictis				—		
Chal(c)iotherium				——		
Gomphotherium				——		
Giraffokeryx				—		
Ictitherium				—		
Hoplomeryx				—		
Chilotherium				——		
Hippotherium				——		
Ancylotherium					——	
Kvabebihyrax					—	
Deinotherium				——		
Agriotherium					—	
Anancus				——		
Megantereon				——		
Elasmotherium						—
M. meriodionalis						—
Dinofelis						—

remained. The challenges, however, were not the same for all but varied temporally, spatially and across families, genera and species, and they were probably much harder for larger animals to cope with than smaller types. One of the selection pressures that emerged at this time took the form of changes in dental morphology, particularly among herbivores, as vegetation responses to the drying brought about broad development of grasses that were about to become dominant over millions of square kilometres of the Earth's surface.

Miocene mountain building took place on three continents with the continued rise of the Andes and Himalayas and the formation of the Cascades in North America, although the latter were largely volcanic in origin rather than purely tectonic. Mountain building disrupted and altered wind and weather patterns around the world; this, in turn, brought regionally sweeping environmental changes. The

Table 3.6 Mammal Extinction in Africa During the Tertiary and Early-Mid Quaternary

N = 21	Paleocene	Eocene	Oligocene	Miocene	Pliocene	Quaternary
Moeritherium		—				
Arsinoitherium			—			
Bothriogenys			—			
Phiomia			—			
Prolbytherium				—		
Kanuites				——		
Kenypotomus				—		
Percocruta				—		
Platybelodon				—		
Chalicotherium				——		
Ictitherium				——	—	
Giraffokeryx				——	—	
Hippotherium				——	—	
Ancylotherium					——	—
Archaeopotomus					——	
Agriotherium					——	—
Megantereon					——	——
Chasmaporthetes					——	—
Deinotherium					——	—
Metridiocheoerus						—
Dinofelis					—	—

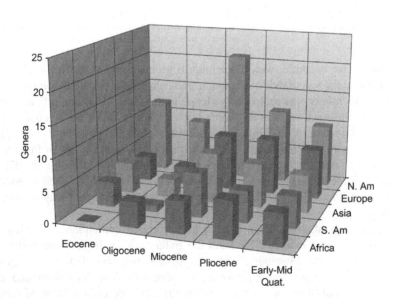

Figure 3.9 Extinction of mammal genera on five continents during the Tertiary and early-mid Quaternary (refer Table 3.6 also).

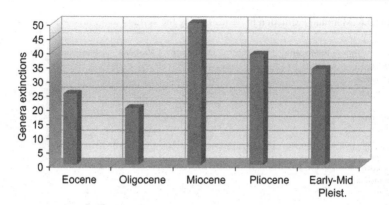

Figure 3.10 Tertiary–Quaternary mammal world extinction trend 65 Ma–300 ka.

Table 3.7 Mammal Genera Extinction Timing on Five Continents from the Eocene to Early-Mid Quaternary

	Eocene	Oligocene	Miocene	Pliocene	Early-Mid Quaternary	Total
North America	12	9	21	12	10	64
South America	4	1	7	5	5	22
Asia	5	3	8	7	6	29
Europe	4	3	9	9	8	33
Africa	–	4	5	6	5	20
Total	25	20	50	39	34	168

rising Andes, for example, were beginning to produce a rain shadow to their east which replaced the woodlands of Patagonia and some of the Brazilian rainforest with savannah, which later was replaced by steppe in the south. The Indian Plate continued pushing into Asia and was beginning to have another effect on weather in addition to the production of the Himalayan wall that was building before it. Throughout the Miocene and into the Pliocene, continued subduction of the Indian Plate under the Asian Plate was forcing up terrain north of the Himalayas, forming the Tibetan Plateau. The process had begun *ca.* 50 Ma and the range had been raised to 9000 km by 20 Ma. But it had little effect on weather till *ca.* 8 Ma, although it may have begun *ca.* 15 Ma.

The formation of the Tibetan Plateau had profound consequences for regional weather patterns that extended far from the Indian region. It gave rise to the formation of the Indian Monsoon, a weather system that would affect the growth and promotion of rainforest habitats across southeast Asia, Indonesia and even Australia. It would, therefore, set up opportunities for the development of rainforest faunas throughout the region. Australia was, of course, gradually approaching southeast Asia from the south and it was around this time that its plate collided with that of southeast Asia. That collision brought about the formation of Papua New

Guinea (PNG) north of the Australian mainland and the raising of the New Guinea Highlands. The result was an increasingly high mountain range which now reaches around 4500 km and stretches like a backbone down the centre of PNG. Its formation cast a rain shadow across northern Australia, one of several factors that eventually contributed to making Australia the driest inhabited continent on Earth.

The Miocene heralded a world that was also undergoing a shift towards more pronounced and contrasting seasonality. By the early Miocene there was little left of the old Tethys Sea. That had largely been squashed out of existence by the crushing northward movement of Africa in the west and the Indian Plate in the east. Today only the Black, Caspian and Aral Seas remain as remnants of that once great ocean. A series of lakes (Mediterranean Lakes) were left where the Tethys had lain between Africa and Europe and these would eventually join up to form the Mediterranean Sea. This meant the disappearance of a vast ecological network based around littoral as well as broader coastal environments. But what was more important about the Miocene, particularly in terms of mammal populations, was the series of climatic changes that took place during that time.

The Miocene began with high temperatures; it then cooled and then temperatures rose again during the middle Miocene. Around 20 Ma, world temperature dropped again, ending the warm phase that had melted the Antarctic ice sheet. Early Miocene World temperatures dropped by a couple of degrees and that condition remained fairly steady till another warm period began towards the end of that time. But these are extremely long-term climatic shifts; within them were embedded equally severe temperature changes that were rapid, wildly fluctuating and short term, and which show up like noise along the general temperature line (Figure 3.1). These were extremely short-lived compared to the overall trend, although they lasted tens or even hundreds of thousands of years, but we still know so very little about the effects these rapid climate changes had on the environment and subsequently animal populations. We can begin to appreciate the nuances of survival when we examine the cumulative changes that animal groups faced in any one epoch and, as usual, these become somewhat clearer as we approach the present because of time scale expansion. For example, such repercussions would have affected a few animals here but many more there so that no one change can be extrapolated to all populations and geographical regions. The variability in adaptive plasticity among some groups varied leaving specialists vulnerable and generalists with a better chance of survival. Such is the way of natural systems.

The warm world of the late-middle Miocene saw the greatest diversity of mammals on most continents, but then planetary temperatures began to drop again. This time, however, they did not plateau as they had before, instead, they just continued downward although continuing to fluctuate on either side of the general cooling trend. The building and melting of ice sheets and glaciers associated with temperature change raised and lowered sea levels repeatedly and brought broad ecological changes around the world. It was these that really affected the biogeography and demographics of animal populations; this process would really sort out who could survive and who could not. Mammalian modernisation was at a premium during the Neogene and it took place in fairly brief amounts of time. For example, the low sea levels of the early Miocene brought opportunities for extensive faunal interchange among North America, Eurasia and Africa and faunal groups underwent massive turnovers, although most took place towards the end of the epoch. Major mammal turnover occurred in Europe and Asia when a number of terrestrial and aquatic species disappeared at the beginning (16 Ma) and end (11.5 Ma) of the middle Miocene, placing, as it were, bookends to the middle Miocene extinction event.

World temperatures began to decrease ca. 14 Ma, corresponding with increased cooling of deep water around Antarctica and major growth of the East Antarctic ice sheet. This process may have been exacerbated by widespread volcanic activity along East Africa's Rift valley. These conditions were reflected in Europe where a temperature fall together with drying in a number of habitats brought about a decrease in rhino diversity and a size increase in the three-toed anchitherine horses there. That adaptive size increase was, however, in vain because this long-lived line that had originated in North America then dispersed to Europe and Asia, eventually going extinct at the end of the Miocene. The hipparionid horse went a similar way with a late Miocene radiation which then reduced in diversity during the early Pliocene and succumbed to eventual extinction in the Early Pleistocene. Woodland browsing ruminants had their greatest diversity during the middle Miocene but were extinct by the end of the epoch. Probably the most important factor that affected Miocene fauna was climatic deterioration in the late Miocene. At that time the world entered a new ice age, with temperatures going into a steady decline after 13.5 Ma, but this time it did not go in reverse and has lasted till today. Probably it was partly driven by chemical weathering of rock as mountain building continued around the world, a process which removed carbon dioxide from the atmosphere and, through water transport, sequestered it in ocean beds. By the time of the cooling onset, temperatures had dropped 6°C to a level sitting just above those of today. It was a major turning point in planetary climate that would see it cool to a level not experienced since the Permian Extinctions 250 million years earlier. Cooling heralded drying around the world and an accompanying reduction in the diversity of mammal species. It was also one of the most severe selective pressures that the Animal Kingdom had faced and would see many species gradually disappear while others continued to adapt to a cooling planet.

The large browsers of North America began to disappear as the extensive midcontinental savannah in turn gave way to steppe. At that time mixed feeders and even grazers, including peccaries, horses, rhinos, antilocaprids, and gomphotheres, became extinct together with large browsing camelids and browsing horses, victims of expanding grasslands. The archaic herbivores of the Palaeogene were replaced by modern ungulate groups such as pigs, grazing horses, rhinos, hippos, deer and cattle and this is where the earlier changes in dental morphology among some grazers came to the fore. Moreover, while the expanding grasslands initiated the development and expansion of grazing animals, it also increased numbers and varieties of carnivores that preyed on the different emerging herbivore classes. Therefore, speed became important to herbivores living on open plains and savannahs as their visibility to predators increased. With plenty of carnivores inhabiting the same niche, body shape for speed became a premium, a process that involved both prey and predator. This is not really a 'chicken-and-egg' situation, however, because it would seem that the prey, in this case grazers, would necessarily need to adapt to faster movement which would then be matched by their predators as a catch-up process, although it was probably not at all as simple as it sounds.

The Miocene also saw the last of the giant pig-like entelodonts and anthracotheres disappear while the bear-dog amphicyonids were restricted to larger forms but they also went extinct in the late Miocene. Environmental changes responding to a steadily cooling Earth were constantly sorting out faunal populations around the world. Animals tried to respond and adapt to the new conditions and many succeeded for some time; others lasted and some succumbed almost immediately. One response was size increase in many species which was a reaction to dietary changes emanating from the reduction in nutritional content of plants and the ever-spreading grasslands and savannahs. These were replacing many areas of shrub lands and forest but the replacement was not necessarily uniform. Ecosystem change took place in fits and starts, often sporadically and with geographical variability. Nevertheless, the temperature trend was down, and global temperatures continued to drop as the Miocene gave way to the Pliocene.

Growing Antarctic ice sheets locked up oceanic water, lowering seas to 50 km below previous levels, adding to world cooling and severing contact between the Atlantic Ocean and Mediterranean Sea. The latter dried out and became an enormous halite evaporite basin, a process that culminated in the *Messinian crisis* between 6 and 5.3 Ma. When the Mediterranean filled once again it was due to a gigantic 1000-km-high waterfall over the 200-km-long channel that took the waters over the Camarinal Sill through the Straits of Gibraltar. This filling lasted 100 years, an event that has been described as making Niagara Falls look like a dripping tap: it has been suggested that 90% of all the water arrived in the Mediterranean basin in just two years, with a flow 1000 times that of the Amazon! It was around the time the Mediterranean dried out that the North American prairies began to expand, together with the African savannahs, and grasslands were expanding more than ever before. General planetary cooling in the late Miocene caused extensive extinctions in Europe, Asia and North America. Those changes continued to keep step with the continued downward plunge of world temperatures as the new ice age began to grip the planet and the Miocene gave way to the Pliocene.

During the last 10 million years of the Miocene, six major extinctions devastated North America's terrestrial mammal fauna, bringing on a major transformation in the character and composition of the group. North America's browsers and others adapted to woodland habitats gradually disappeared by the late Miocene and were replaced by browsing and grazing animals adapted to savannah environments. Changes in atmospheric CO_2 levels as well as a shift from C_3 to C_4 grasses may have accompanied the broader changes from woodland to prairie during the slow onset of aridity that took place from 15 Ma onward. Larger herbivorous mammals such as horses, rhinos, deer, antelope, cattle, pigs, camels and elephants chose a wide range of browsing and grazing targets with some specialisation (Janis et al., 2000). But that made them vulnerable to the slow but inexorable changes that were taking place around the general trend of world aridity. Extinctions among these groups took place while new niches opened up for those animals adapting to the new conditions. Climate change for faunal populations living on the isolated South and North American continents also brought about extinctions and reduced populations. Larger animals were more susceptible to these forces because of the quantity of food and living space they required compared to smaller varieties. Consequently, they are often the first to go extinct in that habitat if they cannot move on to a larger one. But that does not always work: similar-sized species will already occupy niches into which the migrants move and consequently the overall numbers of animals will equilibrate to the carrying capacity with a net loss of large animals. I take this up again when I discuss the Australian megafauna in later chapters.

What the Tertiary as a whole teaches, however, is that mammal populations were as vulnerable to extinctions as any other animal group that had preceded them. What we term *natural population turnover* or *change* really means *extinction* but it is the rate of turnover, the situation in which it occurs and the circumstances in which it takes place that seems to define or, more specifically, divide them. While looking at the long-term coming and going of animal genera and individual species may look like *turnover*, it is, nevertheless, extinction, the same as that resulting from the impact of a massive meteorite or immense volcanic activity. The difference is that one results from natural forces that act long term through selection, while the other is an accidental occurrence, which may still be natural, but is sudden and often devastating, seeing off a particular group of animals before their time. As far as extinctions were concerned, the Neogene continued in the same way the Palaeogene had left off, with biogeographic change and species turnover in many faunal populations; Appendix 1 is a summary of many of the genera that went extinct during the Tertiary. Extinctions took place and animals were replaced in response to climatic and environmental change as well as to the movement of the Earth's plates that brought not only continents but also animals together as well as dividing them. Turnover and change in large animal populations has always taken place and the many megafauna extinctions that took place in the Tertiary continued until recently. The one thing that was new in the second half of the Tertiary was the advent of a cooler climate, one that was not static but continued to become colder until a second phenomenon was superimposed on this trend: the Ice Ages.

Pliocene Extinctions

Although Antarctic glaciation had been an on-again, off-again process throughout the late Palaeogene and early Neogene, a permanent ice cap had formed by the late Miocene and it was expanding. Continuing global cooling now brought about the major environmental change of the Pliocene; glaciation began to take place in the Arctic *ca.* 2.5 Ma as average world temperatures reached close those of today. By the Plio-Pleistocene boundary (1.8 Ma), that cooling had translated into arid and semi-arid landscapes and emerging desert around various parts of the globe. All these major environmental changes were a response to a steadily cooling climate and they brought significant pressures on terrestrial mammals to adapt to the new conditions. One aspect of these changes worth remembering is the speed with which they occurred. Although extremely slow in terms of our short life history, the Pliocene saw speeded-up climate and environmental change as temperatures continued to fall in a way probably not seen since Snowball Earth time. Indeed, the Pliocene itself was only 3 million years long –quite a contrast from the 19 million years of the previous epoch, the Miocene – but average world temperatures dropped almost as much during that short time as they had in the last two-thirds of the Miocene. But tectonics and volcanics were about to catch up with the Animal Kingdom once again. In the late Miocene, the Pana Ma arc collided with the South American continent and bridged the two. Some recent palaeontological work has suggested an earlier bridging *ca.* 23 Ma (mentioned earlier in this chapter), but until more detail is provided about this development I have decided to remain with the previous data which indicates a Pliocene bridging event.

As part of the mountain building along the western Americas, rapid uplift accompanied by volcanic activity occurred in Central America. It began in the late Miocene and continued into the Pliocene, resulting in the formation of the Pana Ma land bridge joining North to South America again after a 50-million-year separation. The Central American closure was at first a series of islands followed by a direct thread of land. It heralded an event that saw a major swap of fauna between the two continents *ca.* 2.5 Ma, a process referred to as the *Great American Biotic Interchange*. One of the first species to enter North America from the south was *Titanis*, the flightless 'terror bird', just before the Panama Bridge formed around 5 Ma. Later, other southerners moved north including the three giant ground sloths (*Eremotherium*, *Paramylodon* and *Megalonyx*), the large relatives of armadillos (*Holmesina* and *Glyptotherium*) (Figure 3.11) and the giant capybara *Neochoerus*. The giant sloth

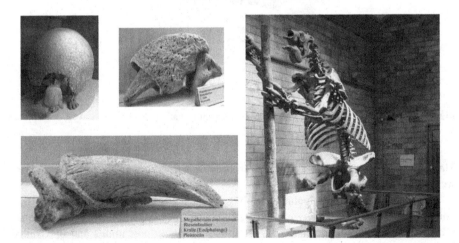

Figure 3.11 The armoured *Glyptodon* (left) and the giant ground sloth *Megatherium* (right) two migrants to North America. One of Megatherium's massive ripping hind foot claws (bottom) which were retractable into a bony sheath (top centre).
Source: Photographs by Steve Webb.

Figure 3.12 Front and side views of the placental scimitar cat *Smilodon populator* the North American invader of South America and possibly one reason for the demise of the marsupial scimitar *Thylacosmilus atrox* (see *Figure 3.13*).
Source: Photographs by Steve Webb.

Figure 3.13 *Thylacosmilus atrox*, the marsupial scimitar 'cat'.
Source: Photograph by Alexei Kouprianov.

Megatherium also made the journey north because their remains are found widely in western North America.

Those travelling south included wolves, bears (*Tremarctos* and *Arctodus*), the gomphothere elephant *Cuvieronius*, tapirs, horses, peccaries (*Platygonus*), llamas (*Palaeola Ma* and *Hemiauchenia*), armadillos, anteaters, porcupines and opossums The placental sabre cat, *Smilodon populator*, also moved down into South America from North America (Figure 3.12). Its most well-known armour was its long ferocious-looking canine teeth. But a similar animal had evolved in South America, *Thylacosmilus atrox*, only it was a marsupial (Figure 3.13). Clearly, convergent evolution had taken place in these two species, given their similar body type and size as well as the incredibly long upper incisors they displayed. *Thylacosmilus* had the longest teeth and they were tapered to two edges front and back; it also lacked incisor teeth. A marked difference was that it had evolved a downward-protruding guard on its mandible parallel to the jaw to protect the tooth (or protect those it did not wish to harm) from the long dagger-like teeth. Perhaps this was similar to a cutting board more suited to carving up meat or scavenging, whereas Smilodon's daggers were for a fighting purpose. The jaw sheath feature of *Thylacosmilus* was very similar to that of another earlier sabre cat from North America, *Barbourofelis* sp. that emerged in the Middle and went extinct at the end of the Miocene, long before *Thylacosmilus*. There is a little irony here in that the arrival of *Smilodon* marked the beginning of the end for its South American counterpart: *Thylacosmilus* went extinct by the early Pleistocene, possibly outcompeted by its placental look-alike, although the fossil remains of this animal are not common and that may indicate it was not widely distributed or very common or both. Oh, one other thing: although *Thylacolsmilus* is often termed a sabre cat, it was no relation to the cat family being closer to kangaroos than cats. This is a similar story to that of the koala 'bear' which, while imitating a small bear at times, has no taxonomic relationship to them.

Many more species moved to South America than moved north including horses, pigs, wolves, elephants and mastodons, sabre cats, jaguars, deer, bears, rhinos, camels and other smaller varieties including cats and dogs. Both migrating collections were largely successful and initially mingled well in their respective new homes. Success was more limited for those that moved north, with six families going extinct by 1 Ma, as opposed to two families that went south. Some species in both regions later went extinct in their old homeland while surviving in their new one. Of all the marsupials that moved north, only the possum (*Didelphis virginiana*) survived; later it moved as far north as Canada. In contrast, the South American immigrants proliferated but many endemic South American animal groups disappeared at this time, hence the 'outcompeting' idea hitched to the dominant placentals. They included the litopterns (horse-like animals the size of gazelles), the notoungulates, a sort of giant rodent, and the weird xenarthrans, also known as edentates, such as the club-tailed, armoured glyptodonts, as well as the wolverine-like marsupial borhyenids.

While these extinctions happened at the same time as the new Panama Bridge opened, argument continues about whether they were caused by the arrival of the northern immigrants across it. They were, after all, placental mammals which are often blamed for eliminating many of the South American marsupials after their arrival, and we are well aware of such possibilities in Australia. On the other hand, some argue whether sections of the South American fauna were already on their way to becoming extinct before this, a common argument in discussion of extinctions (remember the dinosaurs?) and often the point at which natural turnover and accidental extinction overlap. The most favoured theory at present is that the South American extinction was probably caused by a range of events. That would make it a more complex issue than simply the southerners being outcompeted by the new arrivals, which is another argument with a familiar ring when it comes to extinction. Indeed, it seems to have been the case that many of the new arrivals did not have to compete. They found niches previously left unoccupied by endemic populations.

So, many arrivals did not take anyone's spot by force: they merely occupied vacant homes. Nevertheless, the outcome was a net increase in the amount of South American species, although the range of endemic forms was now reduced. It is important to remember the latter had been isolated from North America and, consequently, the rest of the world for 50 million years, although they had a pathway south all the way to Australia for the first 20 million years of the Tertiary and some smaller species did travel in that direction (see Chapter 5).

It seems that the larger South American fauna could not wait to leave, entering the northern continent as soon as a bridge was formed, although it may have been too late for others with reduced populations or those living in southern parts of the continent. Perhaps *Thylacosmilus* was one of those; it seems to have also had a small population living in southern regions that we might call *endangered* today. North American animals had been part of a mammalian gene flow coming from both directions, Europe and Asia, almost throughout the Tertiary and pulses of species turnover were part of that continent's biogeographic make-up for much of that time. So North American faunas probably had a greater genetic vigour that took them in different directions and which inevitably provided them with better adaptive capabilities than South American populations. That may have also played a part in their successful invasion of the southern continent. South American faunas had a more conservative gene pool with little or no input from outside for millions of years. They had evolved differently and many looked different from those in the north; thus they had somewhat different lifestyles and travelled along different ecological and adaptive pathways than the southbound migrants.

Even though they supplemented the variety of South American families by their arrival, the diversity of all North American groups was severely reduced by the late Pliocene. A net trend towards replacement of large herbivores by smaller varieties occurred together with the climatically induced extinctions of old native ungulate groups and rodents rising to become generically the most diverse herbivores in North America and were another group that entered South America. It must be said that the equatorial and tropical north of South America and southern North America had largely different environments than those of today and they were integrally mixed. At least that is particularly so for northern South America. That sounds a little confusing but there was much less rainforest in northern South America and more savannah at the time of the interchange than today. The presence of savannah is where the similarity with the North American landscape overlaps and a savannah landscape naturally holds fewer species than rainforests.

Although the Pliocene was much shorter than preceding Epochs, particularly the preceding Miocene, a lot happened during its 3 million years. In that time the world continued on a path of cooling; vegetation changes included the continued spread of grasslands on all major continents, while in high latitudes tundra and taiga developed, particularly in Eurasia and Siberia. High-latitude landscapes were taking on the familiar appearance of pampas, steppe and open treeless prairie while mid-latitudes saw the development of temperate woodlands. The squashing of faunas mentioned before increased, particularly for tropically adapted species, as rainforests

continued to contract to a belt around the Equator and gave way to subtropical woodland or grassland, which meant an overall reduction in tropically adapted faunas.
World cooling with ice formation lowered sea levels, exposing continental shelves
and giving rise to land mass expansion and development of ecosystems in tune with
their respective latitudinal potentials. For example, rainforest emerged on some
exposed shelves while others developed savannahs or mixed ecosystems. So while
rainforest contracted generally, that loss was offset by its appearance on emerging
coasts in the tropics on which animals were also finding homes.

These environmental changes were selective drivers of change among mammal
populations, challenging adaptive abilities, sorting out those who were best adapted
to changing conditions, pushing animals here and there and weeding out those who
could not cope in an ever-cooling world. Again the fittest continued to thrive and
one example of this was the three-toed horse *Hipparion* that crossed from North
America to Asia via the Bering Bridge. The wide plains it found there suited it
so well it became common in the Old World. To survive in the face of spreading
grasslands, many herbivores not only had to become speedier but they also needed
high crowned teeth that enabled them to cope with the high silica-content grasses
and the dental attrition that came from that diet. Such teeth were a partial deterrent
against the high attrition placed on dentition by the silica phytoliths embedded on
the grasses. Grazers grew larger with larger guts to enable them to consume higher
volumes of low-nutrient grasses. Indeed, it was the continued spread and dominance
of grasslands, particularly in Asia and North America, that saw the loss of all horse
varieties but one. The survivor was a grazer that was becoming speedier in the face
of the expanding variety of carnivores that haunted the wide open grasslands across
which its herds roamed. As mentioned earlier, large wide open spaces were a particular selective force for change, selecting for speedier animals with good eyesight both
in predator and prey. Such selection did not just encourage the emergence of new
species; it also honed those already in existence. For example, the evolved hooves of
the horse were not only good running shoes but they also made quite good weapons,
especially on the end of lengthening legs with powerful muscles.

A range of extinctions took place over time as some older lineages disappeared
in the face of the increasing cool of the Pliocene. Animal interchange between continents continued but that took place among species that were capable of moving within
high-latitude regions whose environments were changing radically. A variety of antelopes were living in the Old World but they had not adapted so well to the changing
conditions. They were one of many mammals that had been common in the earlier
Miocene but disappeared by the mid-late Pliocene. Now larger bodies appeared
among the herbivores. Increased body size would culminate in the Pleistocene as a
reaction to accommodating a higher fodder intake to cover the required nutrient
extraction from reduction in plant nutrients – another effect of a continuously drying world. So, what we see in the Pliocene is the gradual attrition of animal varieties
as a consequence of the onset of cooling across the planet. But what was happening in the most isolated continent, Australia, during the Tertiary while the rest of the
planet underwent its tumultuous faunal turnovers, exchanges and extinctions? I have

deliberately avoided talking about the Australian experience, saving it instead for special consideration in Chapter 5.

Where to Now?

Apart from a basic documentation of the Tertiary extinctions, why have I written this chapter? There are many reasons. In particular, I wanted to show some of the extinction mechanisms or drivers that I spoke about in the previous chapter in action. Extinction is like an irregular heartbeat: it comes in random pulses of varying amplitude or intensity with a seemingly unpredictable frequency. The turnover of species is sometimes slow in a background extinction mode, while at other times there is a larger pulse in a short time. But on closer inspection, the natural changes in the balance of climate and environment, tectonic movement and other natural systems are always implicated and they are the random factors in this apparently random process. If all these could be kept constant perhaps there would be no extinctions! The world does not operate like that, however, but it does operate as a system. The story of the Tertiary shows those systems at work. It has all the right ingredients for extinctions and the reason for documenting them the way I have is to introduce them as well as others into the equation so as to proceed to the next phase of the book.

Another reason for writing this chapter is to highlight the important role that faunal biogeography plays in the demise of various groups. That also highlights the fact that while many animals exist for a long time, extinction comes to them all in the end whether they have survived many past natural changes or not. Even the ability to adapt to changing circumstances is no guarantee of continued existence in the long term. Sometimes those so affected cannot adapt as fast or as far as is required by circumstantial forces. Climatic change and variability, tectonic movement and volcanic upheaval and the environmental consequences that come from these act in various combinations and degrees both spatially and temporally to condemn even the adaptively best-prepared species. This chapter also introduced the principles of animal populations moving and migrating as a result of geographical changes large and small, the replacing of one group with another and the rise of one group over another. Adaptive forcing squeezed out some groups while letting others in, showing the various abilities of animals to adapt or perish or even perish after they had adapted to change in their natural surroundings but which they could not overcome. I now want to continue with almost all these issues in more detail as we move into the next chapter of the extinction story in the Quaternary.

The fifth pillar of world systems, the cryosphere, now takes pride of place with the other four. Ice was about to increase on the planet in a way not seen since the Permian. This would alter world landscapes in a very different way than had ever before been experienced by mammals. That was brought about by the Quaternary Ice Ages which were the culmination of the continuous process of world cooling that began 12 million years earlier. So, the Earth continued to get colder at the end of the Pliocene, forcing world temperatures ever lower. Dropping temperatures presented a particular set of stresses in terms of selection and they required special adaptations on the part of the world's animal populations. But the Quaternary Ice Ages that were superimposed as fluctuations on the general cooling trend brought another selective process: that was a series of rapid reversals in climate and environmental conditions that cycled through at least 100 times. As the world swung to and fro between these extremes, so an even greater adaptation was required, one that would allow animals to live in those extremes. We now turn to Australia and begin with a brief look at its biological and geological heritage. That sets the background for the following chapter, where we look at the story of Australia's marsupials. We will examine how they coped with and adapted to the Tertiary changes that we have just looked at around the world and how that set them up for their great challenge in the Quaternary.

4 Australia: From Dreamtime to Desert

Aboriginal people have a term for all time: whether it is the distant past, the present or the future, they call it the 'Dreaming'. A prominent feature of the Dreaming is that it is a continuum of timeless time: no dates, Epochs or Eons, just ageless time that emerges from a time when the whole world was dark. It tracks to the present, then continues into the future, forever. The Dreaming is, in fact, much more than that because it was the time when all things were formed and placed in and on the Earth, including people. Aboriginal people believe they have the oldest 'Dreaming' in the world, extending back to a time when the natural and human world was brought to life by special Ancestral Spirit beings with enormous powers. These giant beings were in different forms, some animal, others human, and still others half human half animal. They came down from the sky, rose from the ground or arrived on the dark, lifeless shores of Australia and then they set about, in one way or another, giving life and form to the continent. They travelled the barren Earth, having adventures and occasionally fighting, creating everything as they went: the land, the people, their language, ceremonies, lore, songs and all other aspects of their culture. They taught people how to look after and respect the land and one another and they taught them the stories that explain what the land is all about. So, from those ancient and time-less beginnings Aboriginal people emerged '...from the rocks'. That is why many Aboriginal people believe they are part of the Australian landscape and the Earth itself more than anyone else. Their land is their mother and it is their greatest love. The three words quoted above were spoken to me years ago when, as an enthusias-tic young biological anthropologist mesmerised by human evolution, I discussed the origin of Aboriginal people with a group of Aboriginal Australians. I was quickly told, however, that they came from the rocks of Australia, not as I did, in a boat from across the sea. They were not migrants: I was. They were part of the land because the spirit of each of them came from and was forever part of the land. Although I did not see it then, I now see that in many ways they are right; they are from the rocks. They have been here for so long that their culture, unique to Australia, seems as though it did, indeed, emerge from this ancient continent.

The Dreaming explains things simply through story and song and the transmis-sion of culture through oral traditions. The natural systems and mechanisms that drive the Earth are every bit as epic as the songs and stories of the Dreaming but how they work is far more complex. I have spoken about the Earth's natural systems in previous chapters but understanding and appreciating how these systems work and interact with one another is not easy; nor do we know everything about those inter-actions. The subtle and not so subtle interplay between the threads that link them is sometimes difficult to follow and nobody would claim that we fully understand how

Corridors to Extinction and the Australian Megafauna. DOI: http://dx.doi.org/10.1016/B978-0-12-407790-4.00004-5

they work. To follow how natural systems work requires some mental gymnastics; probably a better analogy is juggling. That juggling is required to keep the many natural effects that emanate from the myriad changed states that can occur in the four spheres in our mind at the same time, thus enabling us to appreciate their relationships to one another. In our present society we are taught to think in a compartmentalised manner which does not help in this regard: history in there, geography over there; this is ecology and that is genealogy. Everything is given a label and filed under a category. We think in boxes and that does not allow many of us to easily bring things together to see relationships and integrate functions. In many ways, and with the many things we need to learn about, a good filing system is essential. But we did not always think that way. The hunter–gatherer looked at the world in cross section, seeing many of its many parts in one scene, and that is the teaching that has emerged from the Dreaming. It is certainly the case in Aboriginal culture, where people living close to the environment appreciate the many things that make up the world they live in and which intimately affect their lives as they scan the world around them. But they put all those many things in one box, not many different ones. The hunter–gatherer never did and still doesn't look at them individually. They bring the world together holistically by cutting across the vision of all the things that they rely on and affect one another and literally mapping themselves into it. Without seeing 'all' there is no view of the world.

Australia: A Palaeohistoric Glimpse

As all Aboriginal people naturally understand, Australia is very old, arguably the oldest continent on the Earth. It is a vast geological wrecking yard and, as such, a fitting testament to its age. However, because it is so old it is also a very tired continent. It has played both mother and midwife in the birth of other continents and countries. During the last 4 billion years Australia has travelled over 100,000 km, visited the North Pole twice and is on its way back for a third visit if it does not subduct under the Pacific Plate to its north. It has grown from embryonic crust that eventually formed two enormous granite blocks, the Yilgarn and Pilbara blocks that now make up the West Australian Craton, to be the world's largest island continent with its own unique flora, fauna and human culture. Some of the major steps in the evolution of the Earth and life itself are documented in Australia. It is a place where much of the oldest evidence for the different stages of life and the world's geological events can be found. Australia has many places where people can walk on ancient landscapes not changed in hundreds of millions of years. Stand on one of the Devil's Marbles in the Northern Territory and you stand on 1.5 billion-year-old ground. Stand on the crumbling, iron-rich cliffs of the Pilbara and you are on a 3.5–4 billion-year-old landscape. The stubs of worn-down hills of the Peterman Ranges and the almost endless plains of crushed rock and ancient soils that surround them are all that is left of enormous Himalayan-sized mountain ranges that ran across the centre of the continent. Those mighty mountains were ground to dust by time, water and glaciers kilometres thick that moved across the entire continent.

It is no surprise that because of its continental stability, discoveries of early life are constantly being made which continue to push back the time at which simple as well as complex life forms first emerged. Therefore, Australia plays a special role in our understanding of the planet's geological and biological history and the many life forms that have evolved on it. We are the driest habitable continent on the Earth yet we float on fresh water that lies under one-third of the continent in the form of the Great Artesian Basin. It is the most isolated continent but has received more than its share of floral, faunal and human visitors. It has been submerged under seas, frozen over, torn apart, crushed and cloaked in rainforest. Now, it is almost covered in desert and semi-arid rangelands.

At the beginning of the third millennium and after more than 220 years of non-indigenous occupation, most Australians know comparatively little about the origins and history of the continent on which they sail at 6 cm annually. Our population desperately clings to the coasts to be near 'life-giving' water, although it might come up and bite us in the near future. Arguments over the future of the Australian population range from *'keep it like it is'* to *'let it grow to 50 million or more'*. The latter would not be a problem if the Australian environment was like that of the United States, but it is not. We are not spread right across our continent like Americans – and for good reason. We occupy only about 25% of the 7,500,000 km^2 available albeit sparsely because 75% of our island continent is classified as arid or semi-arid even in an interglacial. And even much of our coastline is arid to the edge of the ocean. In other places such as the vast Kimberley of the northwest, the Arnhem Land seacoast and the Gulf of Carpentaria the coastline consists of mangroves, remote wilderness, difficult rocky terrain, sand flies, mosquitoes and crocodiles. Our deserts are very remote regions traversed by a few dirt roads and even smaller, barely open tracks, many of which are not open to the public. Much of this is Aboriginal land on which ceremony and 'business' are still conducted and requires permits to pass through. So the deserts remain as secret and as largely pristine as they ever were. Our population has concentrated on the 'fertile boomerang' that extends from Cairns in the northeast to Adelaide in the south, with some nestling in the southwest corner of Western Australia, and that is a pattern that will become familiar in later chapters. Only modern mining has opened up to any degree remote areas such as the Pilbara and the western Desert regions.

We are suspicious of the inland, almost as though we know it is not 'our country' and we don't really belong there. If anything, we are less likely to live in Australia's interior now than we were 100 years ago. I would like a dollar for every person I have spoken to about my research activities in remote Australia who has never seen the remote outback or a desert of any kind or has not even been to one of the many tourist icons that are distributed throughout Central Australia. Our non-Aboriginal population strongly holds a 'coastal cling' mentality, probably derived from a subconscious feeling inherited from the earliest European settlers: we will be safe only in an environment that reminds most of us of our personal or ancestral origins in temperate, watery, leafy-green homelands. Indeed, it has probably been a subconscious yearning for 'the old country' that has prevented many settling migrants from accepting Australia for what she is and who have tried to make her what she isn't. Even with a greater heterogeneous population and the multicultural society that has emerged here during the last 60 years that view has not changed and few have braved the hot, dusty interior. By and large, the migrant influx into Australia post-World War II only reinforced the urbanisation of our big cities and that trend continues with the migration of the last couple of decades. Both old and new arrivals have largely been and continue to be reluctant to *venture* into the interior, fearing that the 'bush' starts at Toowoomba, the Blue Mountains or north of the Murray River. How many people can say that they have been in the real bush where there are no shopping centres, petrol stations, doctors or obvious water for hundreds of kilometres? Ah, there's one over there, in the corner....

Over the range (the Great Divide) is the 'eternal bush' consisting of heat, flies, dust, sand, sore eyes, no showers and so very little of that lovely water that you have to dig your own toilet: the unlucky part of the lucky country, some might wrongly say. On the positive side, the 'coastal cling' has been a blessing for the inland. Since the Australian landscape is so old it cannot take the cut and thrust of habitation, development and attempts to make it like Europe or anywhere else. As it is the *driest inhabited continent*, most efforts to do this have largely come to nothing: the ruins of small mud brick homesteads, rusting remnants of machinery and the splinters and tin of long-forgotten mining leases are testimony to those who have tried. Nobody said Australia was not tough on humans; thus only the toughest humans, the Aborigines, lived there permanently. Moreover, our ancient nutrient- and mineral-poor soils and the lack of surface water in the centre have been barriers that confounded grazing, farming and settlement across wide areas – although no serious attempt at setting up a town in Australia's deserts has ever been undertaken. Places like Alice Springs and Tennant Creek were expansions of a couple of buildings associated with telegraph repeater station on the overland telegraph line. The early search for gold and other precious minerals helped these places get off the ground but they took a long time to grow and even now they are not large. On the other hand, those attempts at changing the environment to suit us have had negative consequences for the little water we do have and the flora and fauna that have evolved here over millions of years on meagre rations.

With the exception of the formation of the Great Divide and two phases of volcanism along the eastern edge of the continent between 14–23 Ma and 50–80 Ma, there has been little geological activity across the rest of the continent for hundreds of millions of years. The last mountain building occurred 350 million years ago. Volcanism and mountain building recycle, form and rejuvenate soils; thus most of our soils are old and tired, a trade-off for being largely tectonically inactive, making us possibly the *quietest continent*. Australia is a rugged and very experienced continent, but it is like an old person: it has seen much and travelled widely and there is a wealth of knowledge to be learned if we spare the time to listen and study her, but she is also very delicate and vulnerable. Like an old person, our continent must be treated properly and looked after. Its special nature requires considered treatment, and if we give it that treatment we will live better on it and learn much not only about its history but also that of the planet itself. However, time is nothing to

Australia and as we try and make her into a place she is not, like a stubborn elder who is forced to do something she does not want to do, she bides her time and watches us suffer the consequences. She will give us enough rope and I maintain the hope that we will not entangle one end around our necks.

Australia has a unique, long and often tumultuous history reaching back over 4.4 billion years; it is arguably the *oldest continent*. We know this from SHRIMP (Sensitive High Resolution Ion Micro-Probe) dating of decayed radioactive isotopes of uranium detected in zircon crystals found in the Jack Hills of Western Australia, part of the Yilgarn Block. Standing on Australia's very small shoreline at that time we would have seen a vastly different world around us with

- pink skies;
- a 4–6 h day;
- a much larger moon because it was much closer;
- no ozone layer;
- poisonous volcanic gases everywhere;
- vast quantities of ultraviolet radiation hitting the few small land surfaces;
- an extremely hot volcanic environment;
- no flora and fauna (in fact, no life), and
- showers of meteoritic rock of all sizes striking the Earth regularly.

A very inhospitable place, but these are only some of the differences. On present evidence simple bacterial life forms would not emerge for at least another 500 million years and when they did they would be unimpressive, slimy, blue–green algae or cyanobacteria. Life might have begun here on the Earth or come in on one of those pieces of meteorite, but let's not go there. Bacteria rarely fossilise, and the older it is the less likely it did. Indeed, it would be difficult to prove life began 3.5 billion years ago if this simple life form had not built rocky colonies in the form of stromatolites, layered rocky structures that seem to last forever. The oldest occur in the Strelly Pool Formation of the Pilbara, Western Australia – the second piece of Australia to emerge *ca.* 3.8 Ba. Bacteria also came in tough-nut forms collectively called 'extremophiles', scientifically labelled archaebacteria. They evolved to cope with the wild, merciless conditions of early Earth which were extreme, violent, chemically toxic and generally hostile to life. That first Era in Earth's history has been named the Hadean, primarily because of its supposed similarity to the human imaginings of Hell. As their name implies, extremophiles lived in extreme environments and today one species can be found living around the rims of undersea geothermal vents ('black smokers'). They are called, appropriately, Thermoproteus and as hyperthermophiles they can probably survive living in 140°C. Other extremophiles can withstand highly toxic, chemical environments such as sulphuric acid (sulfolobus) or have the ability to live in deadly methane (methanogens) or pure salt (halophiles) or even inside rock. All were ideally equipped for an environment that would be intolerable to us or any other multicellular life, vertebrate or vegetable. It is these 'beasts' that some hope may have left their shadowy imprint on rock scarps and in long-dried river beds across the Martian landscape as a sign that our two planets did, for sometime at least, co-evolve 3–4 billion years ago and that life did not begin only once.

Figure 4.1 Buckled BIF in a Pilbara gorge northwestern Australia.
Source: Photograph by Steve Webb.

Australia has made some very special contributions to the museum of the Earth history in the form of some of the earliest evidence for land formation and the emergence of various life forms and as a record, in and of itself, of the climatic and environmental milestones the planet has passed through. It was during the time of the cyanobacterial empires that it began its 100,000-km journey around the world on its tectonic plate at the breakneck speed of roughly 6 cm per year, give or take the odd speed-up or slowdown. The Australian Plate has visited the North Pole twice as it just as slowly spun around on its axis. In the meantime, Australia has been sliced, torn apart, scoured, squashed, drowned and bombarded by large and small meteorites. A billion years of rusting seas eventually produced our 'red centre' with its iron oxide-stained quartz grains, and it has laid down banded iron formations (BIFs) and iron ore deposits up to 2 km thick in the Pilbara that now help drive our economy (Figure 4.1). We have the cyanobacteria to thank for this because they usually lived in shallow, sheltered and probably warmer ocean waters, and during their photosynthetic process they used water's hydrogen molecule for energy while releasing the oxygen molecule that was poisonous for them. Originally, most of the oxygen remained in sea, becoming attached to vast amounts of free iron molecules left by the volcanic activity of early Earth. The result was oxidisation or rusting that then sank to the bottom of the shallow ocean. Later, as the amount of free iron molecules was reduced, more oxygen was released from the oceans which began to build our atmosphere. That process took about 2 billion years, two-fifths the age of the Earth.

At the time of Snowball Earth *ca.* 700–600 Ma Australia received gigantic glaciers 4 km thick which rolled across the continent producing piles of debris half the height of Mount Everest. They scraped the top of the continent, which sank half a kilometre under their weight. After two extreme glacial episodes the Earth was released from its frozen state, probably by massive volcanic activity that built up the now much maligned CO_2 in the atmosphere to greater than 2000 times that of today; this gradually warmed the Earth and the ice slowly retreated.

Besides having arguably the oldest evidence for life on the planet in the form of tell-tale stromatolites, Australia also has some of the earliest evidence for complex life in the form of the Ediacara Fauna. These enigmatic fossils were originally discovered by Reg Sprigg, a geologist who trained under the famous Antarctic explorer Sir Douglas Mawson. This enigmatic group, named from their site of discovery in the Ediacara Hills of South Australia, probably evolved during the last stages of the snowball event ca. 630 Ma, although the exact process of how this happened is still hazy. Whatever that process was and regardless of when it occurred, it produced a range of basic multicellular animals consisting of sea pens, jellyfish, anemones and flat worms. These were not particularly complex creatures compared to, say, a tiger but they were anatomical mega-structures compared to bacteria, and they evolved sometime before 600 million years ago. They could also be easily seen with the naked eye (if eyes had been present at the time), whereas earlier life was microscopic. The Australian evidence for this early life joins that of other fossil localities such as the famous Burgess Shales of Alberta in Canada, which are slightly later, and Mistaken Point in Newfoundland, somewhat earlier, as well as a variety of other locations in Europe and Russia. While there remain important questions to be answered in the history of life's development at this early time, these three different assemblages make up a basic stepwise continuum in the process of how the first complex life forms evolved and what they looked like. All of which suggests that their evolution from simple organisms probably took place in different parts of the world and that it was not confined to nor did it originate in one small part of the globe even though all the present-day places where they are found were in different geographical positions on the Earth's surface from those they occupy today.

An event that closely followed the end of Snowball Earth was the arrival of jostling continental neighbours that began to cuddle up to the small and very unimpressive-looking Australia of that time and that heralded the beginning of the super-continent Gondwana. Large pieces of what was to become Africa and Antarctica began to nestle against Australia's south, sending shock waves across the continent. The process was extremely slow but the shocks that resulted from the coming together of these large continental blocks gradually squeezed up massive Central Australian mountain ranges (the Peterman Orogeny). They grew as high as the Himalayas and they formed twice between 600 and 350 million years ago as the continued activity of the building Gondwana consolidated as a megacontinent. Around 650 million years ago, the top half of the continent slid 500 km eastwards, dividing a mineral-rich area into two: the Mount Isa and Broken Hill mineral provinces. During that time Australia constantly changed its shape in other ways as a series of giant seaways sliced the continent in different directions, separating it into sections of various shapes and sizes. The largest of these, the Larapinta Seaway, extended from the northwest into Central Australia and at certain times exited out through the south of the continent into what is now the Great Australian Bight. The ancient seaways opened and closed like vast lock gates. The entry of the

Figure 4.2 The ancient 100 m high coral cliffs at Windjana Gorge in the southern Kimberley, once part of a 1000 km Devonian-reef system.
Source: Photograph by Steve Webb.

sea into northwestern Central Australia around 380 Ma enabled the formation of a 1000-km-long barrier reef along what is now the southern edge of the Kimberley ranges. Large sections of it can now be seen in Windjana Gorge, standing 100 m above ground and 400 km from the sea (Figure 4.2). The gorge is part of the Napier, Lawson and Ningbing Ranges, a complex of sharp, jagged limestone formations skirting the southern and eastern Kimberley. To the south is more evidence of the amazing ecology of the reef at that time in the form of the life that swam in it, but before going further I want to go back 100 million years. What has been related so far has taken many years of research by many scientists doing long field work in sometimes arduous and uncomfortable conditions – I know: I have done some of it. I have not referred to them thus far because they are too numerous to mention but a few of these intrepid souls do need mentioning because of the significant discoveries they have made, and because I know some of them.

Australia has not always been the most *isolated continent*. It has been a speck of land, a series of islands, an archipelago and several large, individual land masses that have blended and parted several times. Australia's fabled 'inland sea' for which early explorers searched in vain did exist – only the explorers were tens of millions of years too late. There have been many inland seas and as evidence of this, one of the oldest true fish (*Arandaspis*) was found by Alex Richie close to the geographic centre of the continent at Mt Watt. It was a jawless filter feeder with an external skeleton and was partly named after the Aranda indigenous Language Group on whose traditional country it was found. It swam in the Larapinta Seaway 480 million years ago which flags the rapid evolution of multicellular life that took place in the previous 100 million years. Many early fish species that lived 100 million years after *Arandaspis* have been found exquisitely captured in three dimensional within the abundant limestone nodules of the Gogo Formation of the southern Kimberley. The resulting details of these fossils are quite unlike the one-dimensional flattened fish fossils most commonly found around the world. The amazing depth of the details they contain have taught us an enormous amount about early fish diversity, biology and evolution. Consequently, the descriptions and observations made from these beautifully preserved members of the early fish families living 380 million years ago have allowed some important bricks of fish evolution to be put into place at that important time when life was emerging from water onto the land. Specifically, the pectoral fin skeleton of the Devonian *Gogonasus* fish has shown it to have been developing a transitional limb form that signals the evolution of the terapod arm that would take these animals from the water towards the movement onto land. The brilliant work of John Long and his colleagues in releasing these fossils and their story from their stony coffins with acetic acid and describing them is testament not only to their skills and persistent research over many field seasons but also to the nurturing that was provided to fish by the ancient barrier reef (Long 1995, 2006; Long et al., 2006). A barrier reef is the aquatic equivalent to a rainforest. In the same way as a rainforest, it provides a magnificent array of shelters and niches for small fish and myriad species can shelter and thrive in its environment. The many refuges allow fish to adapt, speciate, specialise and evolve and provide places for them to hide from predators. But it is also a great place for predators with a wide variety on the menu. The shelter and beneficence that the Gogo reef provided for such a long time allowed the growth of fish variation and the anatomical beginnings of the transition of fish to land-crawling tetrapods which is clearly shown among its fossil treasures.

Much later, oceanic drowning of northeastern Australia allowed marine dinosaurs, such as mosasaurs, ichthyosaurs and plesiosaurs, to swim in western Queensland 100 million years ago. Pioneer researchers such as Mary Wade from the Queensland Museum first brought this Cretaceous life to the eyes of the public but since then others such as Alex Cook and Scott Hocknell, as well as natural outback palaeontologists like David and Judy Elliot and Stuart and Robyn Mackenzie, are continuing her legacy and are now forging the next chapter: the discovery of Australia's terrestrial dinosaurs, particularly some of the largest sauropods in the world. In particular, that legacy will take the form of research centres set in the outback that will excavate, research and display the vast palaeontological riches from Queensland's Early-Mid Cretaceous. In southern Australia small dinosaurs from Apollo Bay, excavated by Tom Rich and his colleagues, as well as other sites along the Victorian coast, show that these remarkable animals lived in a very cool but ice-free polar region because the South Pole then was where the New South Wales outback town of Bourke is today. Although the Earth was ice free at this time, it was dark for six months of the year. It is likely that because the Earth's average temperature was around 8°C hotter then, these Australian dinosaurs were able to stay warm enough to survive the 'Polar' winter. But it is likely that they would have needed to be self-warming (endothermic) rather than relying on sun bathing (ectothermic) to keep warm. Another possibility is that during winter they may have migrated to other parts of Gondwana that were not so dark for so long. Africa broke away from Gondwana 130 million years ago after a 30-million-year process of separation and India followed *ca.* 105 million years ago, leaving its long-time anchoring point at the tip of southwestern Australia. So they were places that these 95 Ma dinosaurs could not migrate to, leaving only Antarctica and South America as refuges. Australia was the next to depart from Gondwana *ca.* 45 million years ago and South America then followed 35 million years ago.

Australia has wandered north across the vast southern ocean for 45 million years and continues to move towards Japan at 6 cm annually. But its departure from Antarctica was much more significant than just the separation of two pieces of Gondwana. It had a special job: carrying a cargo of marsupial passengers that for one reason or another luckily found that they had the place to themselves after arriving here *ca.* 55 million years ago. So far as we can tell, the marsupial story seems to begin in China at least 125 million years ago. They occur in North America and Europe between 90 and 100 Ma and reached South America 20 million years later *ca.* 70 Ma. To reach Australia, however, they had to cross Antarctica but it was not the continent we know today. At that time it was joined to South America in the west and Australia in the east and was covered in a temperate rainforest that stretched unbroken from eastern Australia to Bolivia. Marsupial movement to Australia was probably comparatively easy because with no obvious barriers and rainforest all the way, they could not miss it as long as they kept moving eastward: it lay directly opposite South America on the other side of Antarctica. When they arrived, the world was enjoying the 'EO', when there was no ice in Antarctica. Like boarding a departing cruise ship, the marsupials arrived just in time to hitch a ride on the good ship Australia as it broke away from Antarctica. Casting off had been a long process beginning 30 million years before as a giant rift valley began to form between Australia and Antarctica. A giant tear was also occurring under the Tasman and Coral Seas off the east coast, which allowed the continent to move away in that direction. The enormous stresses emanating from the separation pushed up along eastern Australia, forming the Great Divide.

Isolation now allowed the marsupial refugees to slowly evolve into many hundreds of species that began to adapt to the very wet, cool, rainforest conditions covering the continent. But they were safe, although from what we are not quite sure. As I mentioned in the previous chapter, it has been suggested that placental mammals outcompeted marsupials when and wherever they came into contact with each other. The latter either succumbed to the dominant placentals or they migrated away. Placental animals do have some advantages over marsupials. Generally, they are probably more skilful and adept at moving into niches; they are more cunning, have larger litters and are faster breeders than marsupials, all factors that can make them the dominant group. That could explain why the Old World does not have them today and North America has only one. South America, however, has over a dozen species and this does not totally comply with the theory, although it is basically a logical one given present evidence. It might also explain why Australia, at the end of the Gondwana continental chain and attached to eastern Antarctica, found itself with these animals while they disappeared on most other continents. A couple of placentals may have been accidentally stranded in Australia but they quickly disappeared by 50 million years ago (see Chapter 5). The impact that placentals have on marsupials has been borne out starkly in Australia since their introduction by colonial immigrants. The introduction of cats, dogs, foxes and other placental animals since then has had dire consequences for our native species. The marsupial dog, the Thylacine (*Thylacinus cynocephalus*), seems to have disappeared on the mainland at around the same time as the arrival of the Dingo (*Canis familiaris*) ca. 4000 years ago. The Tasmanian Devil (*Sarcophilus harrisii*) did the same thing but that is a more difficult case to pin down because it seemed to go extinct in stages with the last ones probably living in the Northern Territory perhaps less than a thousand years ago. Certainly, the introduction of the European fox (*Vulpes vulpes*) has greatly impacted our native Tiger cat (*Dasyurus* sp.), with fox populations pushing them to the verge of extinction. Cats also have had an impact, particularly on small marsupials, native lizards and birds. But the marsupial migration to Australia paid off: it was as if they were made for this continent. Finally, they had a place of their own, ideal for their future preservation and survival – or at least the best they were ever going to have.

The marsupials were safe but they were wet, very wet. As much as 20 m of rain fell annually in parts of Australia in the early Tertiary, a time of high world temperatures, high sea levels and high rainfall across the planet. Australia's climate was, however, cool to cold because it lay in high latitudes but the conditions were ideal for the growth of vast temperate 'palaeoaustral' rainforest which grew thickly right across the continent. The continental plate was largely tectonically quiet except on its northern frontier, but Australia's position in the middle of the plate kept it reasonably untroubled by those disturbances as the raft drifted slowly north, meeting the Pacific and Eurasian plates head-on. Thus, during the 20-million-year period from 45 to 25 Ma Australia underwent a very stable phase in its history because of its placid conditions. Those conditions were ideal for the evolution, adaptation and dispersal of its cargo of marsupials. They had time, space and safety to do this in a way they had never been able to do previously. They could evolve on their own terms and in their own time. The palaeoaustral rainforests with their dominant *Araucaria* and *Nothofagus* tree species offered an almost infinite variety of micro-environments for marsupials to choose from, from top emergent canopy to bottom creek line, swamp and undergrowth. The marsupial world was just about as idyllic as it could be, so much so that this period could be called the *age of marsupials*. The population blossomed with many becoming highly specialised, appearing and just as quickly disappearing again. Species have been found that are still not well understood although they must have been adapted to specific conditions that arose during this time, developing specialities and lifestyles in the niches of their choice that were appropriate at that time.

But how do we know these details? The fossil treasure trove of the Riversleigh fossil site in north Queensland has allowed researchers, like Mike Archer, who have worked there for more than 30 years to put together a magnificent picture of the evolutionary explosion of marsupials that lived in the temperate rainforests that covered the region. He and his principal co-workers Henk Godhelp and Sue Hand have detailed the life of the early rainforest marsupials in their book *Riversleigh* (1991). The animals they have discovered are representatives of the earliest known marsupials and Riversleigh is one of those extra special places where life from the late Oligocene to the Quaternary can be traced almost without a break. But the work has taken years of painstaking effort with not a little heavy lifting of limestone blocks containing the valuable and wonderfully preserved fossil treasures. Work has been accomplished with many volunteers and equally many baths of diluted hydrochloric acid followed by detailed preparation and painstaking assembly of fractured, and in some cases shattered, fossil remains – some of only very small fragments. Even more amazing is that the extracted fossils themselves represent an almost unbroken sequence for the last 25 million years. That has been a special gift for palaeontologists and something rare in the world of palaeontology. It has, therefore, provided a unique opportunity to understand how our marsupials evolved and adapted over a large part of the Tertiary when vast changes took place in Australia. In particular, it was the time when Australia swapped its rainforests for deserts and became the *driest inhabited continent* in the world. Riversleigh has now been designated a World Heritage area not only because of its fossil horde and the time span it covers but also because it will continue being worked for decades to come, due to the fact that it covers a huge area that will no doubt reveal many more secrets of the evolution of Australia's marsupials. That future biological information has also been a cornerstone of Riversleigh's in its successful World Heritage nomination. Riversleigh's record, therefore, charts the change that made Australia what it is today: as it dried out, swapping one sort of animal for another, that record showed that extinctions occurred all the time as those environmental changes rolled forward.

An Introduction to Ice Ages and Deserts

Several factors encouraged the drying of Australia. One was the buckling of the lighter elements in the northern leading edge of the Australasian Plate as it collided with the Pacific Plate. That process gave rise to the formation of the New Guinea Highland backbone that runs from Papua New Guinea west through Irian Jaya, a process that began *ca.* 24 Ma. The raising of these mountains resulted in a rain shadow across northern Australia, reducing the amount of rain entering from the north. Another factor involved Australia's northward drift, which took it into drier latitudes positioned between 20° and 30° in both hemispheres. But the major factor in the drying of the Australian continent was the onset of the Ice Age that began 12 million years ago, which also triggered the beginning of glaciation of Antarctica. At the beginning of the Tertiary the EO (Figure 4.2) enabled the giant boa *Titanoboa cerrejonensis*, measuring 12–14 m and weighing over 1 tonne, to evolve and function with the exceptionally warm climate, enabling its large ectothermic body to warm up enough to move around. Over the next 16 million years Earth slowly cooled to

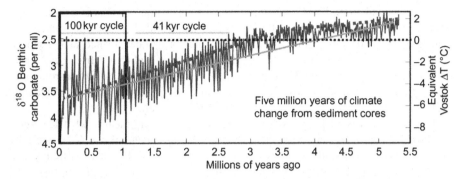

Figure 4.3 Pliocene–Quaternary glacial cycle build-up showing an average downward trend (light, thin line) in world temperature that began with the greater Ice Age beginning around 12 Ma. Note the trend line division around 3.5 Ma (heavy dotted line) shows a shallow slope before that time but a marked downward trend after it which kept below the present average after 3.0 Ma (straight black dotted line). Glacial cycles became established then consolidated in an increasing cooler, deep amplitude cycle pattern in the early Quaternary. That trend increased during the last 1 million years (box) with interglacial excursions to modern temperatures occurring as short interglacials during that time.
Source: Main graph after Lisiecki and Raymo (2005).

5°C higher than present; then *ca.* 34 million years ago temperatures dropped suddenly to 1°C, almost those of today; finally, *ca.* 25 million years ago, they rose again by 5°C. During that time Antarctica experienced its first glaciation, which ended with deglaciation *ca.* 24 million years ago, and large animals like *Titanoboa* and *Andrewsarchus* disappeared. There were a series of smaller temperature fluctuations before Antarctica once again began to reglaciate *ca.* 12 Ma as temperatures dropped below 2°C. Temperatures continued to fall during the rest of the Tertiary and into the Quaternary as the Earth cooled; Figure 2.2 shows a gross view of that event. A closer look, however, shows how ragged the temperature trend line becomes as it approaches the Quaternary. Important for our later discussion is the exaggeration of its raggedness and increasing fuzziness, or noise, as it approaches the Quaternary and how that becomes even more exaggerated as we enter that period. Figure 4.3 shows that more clearly as an expansion of the area within the box. The fuzziness present during the last 2 million years can now be seen for what it is: vast fluctuations in temperature that form a regular pattern over that time. Those are glacial cycles or Ice Ages. They are superimposed on the greater glacial cycle that began 10 million years earlier and continues today. The fuzz is caused by increasing cycle frequency and amplitude that took place during the Quaternary. They don't look like much on the broad scale of Tertiary climate change, but they have played a major part in human and animal history during the last 2 million years. The trend line across 5 million years shown on the graph marks the slow downward creep in world temperatures, but if we divide that line at *ca.* 3.5 Ma, just above the point where the temperature drop crosses the modern temperature boundary, and make two trend lines, the second from 3.5 Ma onward shows a steeper plunge. This is a significant time in terms of environmental change around the world. It is the time when the Atacama Desert began to form, when a permanent ice sheet formed at the North Pole. In addition, there was a significant development of aridity in Africa *ca.* 2.8 Ma and another *ca.* 1.7 Ma. The 3 Ma date is also significant in the debate surrounding the mid-Pliocene faunal turnover that has been noted in South America (Vizcaino et al., 2004) and Europe (Meloro et al., 2008; Palombo et al., 2009).

The vast majority of people have, at best, a very sketchy idea of what Ice Ages are, how long they last and what their presence means for the planet. They are further perplexed, as my students are, when I say that an Ice Age began 12 million years ago and we are still in it, but another began 40,000 years ago! Their confusion is understandable. The Quaternary Ice Ages or glacials are part of a cyclic phenomenon determined by astronomical factors that alter the shape of the Earth's orbit around the Sun, the angle of its axis and how it spins and wobbles on that axis. These orbital phenomena are the Milankovich Cycles, named after the Serbian mathematician who discovered them. They are regular and as far as we know they are continuing; therefore, they should condemn us to further glacials. At least 100 glacial cycles have occurred during the last 2 million years, moving the Earth from a cold (glacial) to a warm (interglacial) state and back again through glacial–interglacial cycling. For the layman, 'Ice Ages' conjure up all sorts of images: Neanderthals, snow and ice everywhere, grey skies, woolly mammoths and equally woolly rhinos, and exceptional cold with even more snow. Some of these pictures are correct but they do not fit all parts of the globe. For example, the Earth not only cools at these times but it also dries because there is less evaporation. Water is also locked up in expanding ice sheets; consequently, sea levels drop. We receive over 99% of our rain from the oceans, so with less oceanic water there is less rainfall. With the reduction in snow and rainfall, deserts spread while glaciers grow, with the result that habitable regions in temperate and subtropical regions shrink and become sandwiched between the two more extreme environments. The build-up of ice and snow has another effect: it increases the Earth's albedo so more of the Sun's rays are reflected back out into space with less heat reaching the Earth's surface, compounding the effects of the cold. During the Quaternary most glacial construction took place in the northern hemisphere, where ice pushed south covering millions of square kilometres of land, often building to a couple of kilometres thick. Ice cover was variable and unpredictable, however, so that in places like Siberia and northern Russia ice formation was not uniform with many areas below the Arctic Circle remaining ice free. In Western Europe ice spread out across the North Sea covering Britain almost to the south coast. The result of glacial formation was the loss of land for almost all animals together with their ranges and food supplies. Vegetation changed, disappeared or shrank and habitats did the same, so the animals and people that once lived there moved away to better pastures or perished.

Climate change has not been and never will be stable. That is perfectly clear from the records and perhaps we should modify the old phrase, *'The only things we can be certain about are death and taxes,'* to include *'and climate change'*. Short-, medium- and long-term climate fluctuations are as inevitable as death. The Vikings went to Greenland and farmed in the 900s AD; they also found grapes growing in Newfoundland, or Vinland as they called it. They were able to go to Greenland, begin their settlements and farm because the world was in the Medieval Warm Period, but they soon retreated when the Little Ice Age began *ca.* 1450 AD. They were a tough, hardy and capable lot but even with the relatively small climatic changes of the Little Ice Age they had to retreat in the face of both Arctic Indian movement south in response to the rising cold and their own inability to live in the colder winters and summers. But the Little Ice Age was only a small example of climate change towards cold. Our evidence for constant climate change is made starkly clear from the study of ice cores, speleothems (stalagmites), palaeontological evidence, satellite imagery, geomorphological work around the world and the use of a wide variety of geochronological dating techniques that have been developed to complement these studies. If there is one thing the work shows us it is that however warm we become in the short term, the interglacial in which we now bask will end. We will then enter a new Ice Age and that may not be in the too distant future. Today, climate change and a warming planet are for many reasons the topic of choice, but these changes are minute compared to past changes and are incomparably trivial compared to what humanity will face when we enter a new Ice Age. The planet became warmer than it is now during the last interglacial *ca.* 125–130 ka, when temperatures suddenly jumped up to 5°C higher than those of today and sea levels rose between 4 and 6 m above present levels as a warmer world partially melted the West Antarctic Ice sheet and the Greenland Ice Cap. But the cycles were not uniform; they did not consist of nice smooth U-shaped curves showing 'now we are warm' and 'now we are not'. They fluctuated, sometimes wildly, and warm interglacials were on many occasions not very warm. I will take this further in Chapter 6.

Glacial–interglacial cycling during the last 2 million years represents climate change in the extreme, each phase bringing vast changes to world environments. But this is not just a case of good-change/bad-change: rather, these cycles drove a set of forces that must have impacted on the direction of evolutionary development among all animals, no less than in Australia. Glacial cycles were, therefore, another force for change and adaptive selection as well as another extinction driver. For example, before the Quaternary glacials began, the set of evolutionary selection processes were different from those after they were in full swing, certainly in some parts of the world, and as the cycles grew more intense so did the selective pressures that they generated. That switch must have been a major selective force for rapid adaptation among animals, pushing variety, behaviour and adaptive strengths in all sorts of directions and producing rapid changes in animal types. It would have selected particularly for those species who could adapt to rapid change in their respective environments: the increasingly long glacial cycles, the abrupt enviroclimatic swings associated with glacial cycling and the increasing amplitude of those swings. These conditions must have militated against some species as they found themselves increasingly unable to adapt fast enough through their particular specialisation to an altering niche or habitat and being in the wrong geographical place at the wrong time. An inability to alter behaviour or living strategies or move away from increasingly changing environments and shrinking niches through, for example, migration, would also underpin the forces acting against them reducing their ability to thrive. These issues took place not only with the passing of each cycle but also as the Quaternary progressed the amplitude of each cyclical swing became deeper and more challenging.

Glacials radically changed vast areas of the planet and these are of concern because what we see as 'Australia' today is not really Australia at all. The real Australia, meaning the one that has been most represented during the last one or 2 million years, is very different. That was a glacial Australia which lasted for a very long time: ~90% of the time. The 'normal' Australia as we know it is actually 'interglacial Australia' that exists for the other 10% of the time, much less during some of the past interglacials. The present one began around 17,000 years ago as the Earth emerged from the last glacial but, as I have said, all interglacials are short. During our comparatively short interglacial we have domesticated animals, taken up farming and cultivation, built pyramids, formed city-states and empires, and invented writing, machines of all kinds, mega-financial structures, trans-world and electronic communications systems, developed an extremely sophisticated electronic world we now rely on heavily and gone out into space. But we have also grown an enormous world population that now exceeds 7 billion or 80 million annually and growing. People living during and before all previous Ice Ages lived the life of a hunter–gatherer, without written records, experiencing a simple, uncomplicated, largely hand-to-mouth existence without diets or processed foods and always ready to take advantage of any opportunity that came their way. Survival was the overwhelming issue facing humanity all day (and night) and every day. People certainly lived long enough and successfully enough to grow a future population which has bourgeoned. They used a comparatively simple technology made only from stone, bone, sinew, wood, fur and vegetable products. They were tough, resilient and faced day-to-day danger in a way we cannot really imagine and could not undertake because they possessed skills, cunning, toughness and ruthlessness which we have lost, and lived by wits that today are not continually sharpened by the danger from animals, weather, impending starvation and terrain. They lived in the open, at best a cave with an open entrance. They relied on the four great senses of taste, hearing, smell and sight, all highly tuned – the same ones we have blunted through noise, pollution, processed foods, disuse and looking constantly at screens of one sort or another. Consequently, we are not as tuned into the world around us as they were. Our ancestors hunted or gathered every day because, with few exceptions, food was not stored in sufficient quantities to last and even if it was, it could be raided by large and nasty animal scavengers. Life expectancy was generally not long and their populations were small. Many starved; some were taken by wild animals or trampled, gored and mutilated in various ways during hunting. Others froze and many perished during childbirth or following childhood and through injury, the lack of food and water or localised skirmishes with others. We might quote Thomas Hobbs' well known phrase that life then without proper government was 'nasty, brutish and short' for many, (Hobbs 1651:89). This view has been condemned with post-modernist derision as a totally false impression of our 'hippy-like' ancestors at one with a world they 'looked after'. The fact is life was hard, but humans were hard and to survive they did anything that was required. But they were adapted to the world in a way we cannot even imagine, because we live in various cocoons of one form or another, mainly an electronic one that isolates most of us from actually experiencing nature and our environment beyond going to the local beach, park or supermarket – the rangelands of the modern hunter–gatherer. The old hunters also knew nothing else but their glacial world, as we know nothing else but an interglacial-electronic one. We could not be more removed from one another.

Today, we contemplate temperature changes of 1°C or 2°C and sea level rises of a metre or two in the next 100 years, and we quake at how such changes will affect our lives so radically. Imagine rapid sea level changes of 140 km and average world temperature swings of 10°C below those of today. These occurred during the last glaciation and the one before that and the one before that, but what would these do to our present environment, its animals and our society? (We are also due for a reversal of the world's magnetic polarity from north to south. It last happened 760,000 years ago. What might that do to our eWorld or iWorld is anyone's guess, but we won't go there.) While we talk of global warming, it is inevitable that we will enter another glacial sooner or later and if it is anything like the last, then Canada, half of the United States and the whole of northern Europe will disappear under ice sheets like the Scandinavian Cordilleran and Laurentide sheets, which are 2–3 km thick. The Sahara Desert will extend 700 km south into tropical Africa. The Namibian Desert will send its sandy tendrils into the Congo, Alaska will join Siberia and Britain's channel tunnel will become defunct as the United Kingdom and Ireland literally join the European Union by land bridge. The massive loss of grazing and cultivatable land worldwide will mean hunger on a scale that cannot be contemplated. The loss of habitable land will turn the world into a massive human migration trying to reach warmer and/or more fertile areas. Unfortunately, those areas will not sustain any population increase and hundreds of millions, perhaps billions, of people will die. The planet will have time to recuperate, breathe again and restore itself: it will be saved, but it will not be humans who save it....

Although the horrific picture painted above suggests a scene from a riveting Hollywood movie, at sometime in the future it is going to happen. The story that I want to continue in this book is set during such times in the past but not too distant past, when the world moved through multiple Ice Ages during the mid-late Quaternary – and this time the story I want to tell only affects animals. It deals with only one effect of glacial climate change, animal extinction in Australia, but there were many others taking place around the planet. Australia, being the 'lucky country', did not suffer many of the extremes felt in other regions of the world, particularly the northern hemisphere, such as massive glaciation, but there were climatic and environmental changes that must have had severe detrimental effects on its fauna and flora, so much so that they incrementally changed the composition of Australia's animal populations and the direction that they took.

5 The Australian Tertiary and the First Marsupial Extinctions

Introduction

This chapter sketches out the story of Australia's marsupials across the Tertiary. 'Sketch' is the operative word because although many details are known, large gaps in our knowledge exist – indeed, half of the Tertiary is almost unknown. The main aim here is to show something of the way in which the marsupial population lived, changed and went extinct against a changing climatic and environmental background that swept through the Australian Tertiary, as outlined in previous chapters. It also provides an important background to later discussions concerning the megafauna. Specifically, the Neogene was a particularly important time for our fauna. During that time the Australian continent underwent one of the greatest environmental changes in its history, at least as far as fauna were concerned. It turned from a place of widespread luxuriant temperate rainforests to a desert, from one of the wettest continents to the driest and, of course, it was increasingly becoming the most isolated. Australia's Tertiary history, however, is largely a mystery, because we know comparatively little about the first half, the Palaeogene. All that can be said is that on present evidence, marsupials arrived on the continent 55 million years ago and they may not have been alone. They lived here at least 10 million years before Australia finally separated from Antarctica and became its own continent. We also know that after the breakaway, Australia's marsupials diversified and adapted to their new home, but exactly what form these adaptations took and how animals spread throughout their new island home is a mystery. They must have liked the place because they were a great success, but that tells us little about how they thrived in their formative years and what forms they took because those successes are hidden in an extraordinarily long period that extends from 55 to 25 Ma. At present our view of this first 30 million years is fogged over by an almost total lack of fossil deposits from that time. Nevertheless, for the purposes of this book the ability to assemble a decent picture of the Neogene allows us to see the evolutionary and biogeographic preamble to the emergence of the species that we now call the megafauna; that in turn enables us also to compare their final stages against a background of the 25 million years that preceded their extinction. One thing we do know is that marsupials thrived on their new island home but they were not immune to extinction.

Corridors to Extinction and the Australian Megafauna. DOI: http://dx.doi.org/10.1016/B978-0-12-407790-4.00005-7

Marsupials Go to Australia

One story that lies in the shadow of the dinosaur empire and their extinction is the movement of marsupials around the world and their pattern of migration that brought them to Australia. All took place under the gigantic noses of the dinosaurs. Initially thought to have developed in Australia, marsupial fossils now show that these animals originated elsewhere. The oldest found so far is *Sinodelphys szalayi*, a squirrel-sized, near complete fossil found in the shales of Liaoning Province in northwestern China and dated to 125 Ma. Marsupial fossils from North America (didelphids) have been dated to *ca.* 100–90 Ma, while the South American evidence shows they reached there by 70 Ma. Therefore, although not unaccustomed to long journeys, marsupials certainly had one to face when moving to Australia. Conveniently for them, there existed a 'Gondwana highway' that enabled them to cross from North to South America, move on to Antarctica and from there enter Australia. Therefore, because South America, Antarctica and Australia were the remaining part of Gondwana, the marsupials had a choice of three continents to move around in. The geographical positions of South America and Australia placed them directly opposite each other on either side of Antarctica and I like to think that the marsupial journey took them in a straight line across that continent and into Australia, although this was almost certainly not the case. An important point to remember about their journey is that their transfer from north to South America was probably at its height during the time of the Chicxulub explosion. It may be that they suffered their estimated 75% loss of species from this event just because the bulk of the population was in the wrong place at the wrong time, i.e. in southern North America and Central and northern South America. This also goes for any placental mammals that were in the area at the time. Perhaps this also contributed to giving the southward marsupial migrations an even break by affecting placental migrations in the same way at the same time.

The exact timing of the north–south continental crossing is not well understood because of the intermittent linking of the Panama Bridge in the late Cretaceous. Moreover, marsupial migrations probably took place over a long period of time on either side of the dinosaur extinctions. They were probably accompanied by placental mammals on their migration from North to South America, although we have little idea concerning proportions of the two groups and the species involved. One possibility is that the marsupial migration south involved members of most species remaining in the world at that time because following this movement, marsupials comprised only a minor element of the North American fauna till going extinct in the Middle Miocene. It was not until South America rejoined the North that they moved back north in the late Pliocene *ca.* 2 Ma, only to go extinct once again (see Chapter 3). At present the marsupial mystery is enhanced by the paucity of information available from the fossil record. It tells us little, for example, about the southward migrations but it seems the North American source population were perhaps the last representatives of the group beyond there. Marsupials may have been confined largely to the Americas by the time of the K–T event, with fewer species in South America and fewer still beyond there. Also, if the marsupial component in

South American populations was smaller than those of placental mammals, the effect of the K–T impact could have been felt proportionally more by them. At the same time, their geographic position placed the bulk of the population further south of the ensuing devastation throughout Central and possibly northern South America. That may have given them breathing space to move ahead of animals caught up in the conflagration. Therefore, depending on their population dynamics and the pattern of devastation caused by the K–T event, they may have been given an advantage in the migration stakes. Both marsupials and placental mammals could have been affected but those with the smallest population fared worse. But the marsupials had one very timely piece of luck and one probably not experienced by a group of animals before. They reached a continent that was about to depart for isolation and that guaranteed their survival. Australia was their ship of safety and once aboard, they thrived.

It is worth noting that the other group of animals Australia is famed for, monotremes, left Australia for South America and reached there on present evidence *ca.* 62 Ma. That evidence is in the form of the early platypus, *Monotrematum sudamericanum*, found in Argentina and similar to Australia's early but modern-looking *Obdurodon dicksoni*; the former subsequently went extinct in South America. It should also be noted that *Monotrematum* was found in Patagonia, so on that very slight evidence it could be suggested that while a monotreme did escape from Australia, it may not have moved very far beyond Antarctica. We know monotremes evolved in Australia because we have the oldest evidence for them, *Steropodon galmani*, which has been found in the New South Wales opal field of Lightning Ridge, dated to 110 Ma. Another specimen, *Teinolophos trusleri*, consists of a partial left jaw with one molar and is also from the early Cretaceous. Slightly older Australian evidence in the form of *Kollikodon ritchiei* may put the Order back to 115 Ma but the placental or marsupial status of this fossil is still under investigation. I have alluded to the competition between placental and marsupial groups but this has also been extended to the monotremes; the latter two groups always come off worse when placentals arrive on the scene. So the South American monotreme extinction may have been caused by mammal competition but that argument also continues. There may be a number of other reasons for the disappearance of monotremes in South America, although only one species has been discovered so far, and that in itself points to a tricky status for an ongoing, thriving population of any description. One thing seems certain, however: it is probably unlikely that the monotreme extinction was caused by competition with marsupials because these two groups have accommodated each other for at least the last 55 million years in Australia. On the other hand, they may also have been affected by the fallout from the K–T impact to the north, which meant Australia was a *safer* haven, positioned almost on the opposite side of the world, and left largely untouched by this event as far as we can tell. At present the evidence strongly suggests that the monotreme population must have been very small and possibly contained only one species. That in itself would have predisposed them to extinction, particularly as a new species on the continent meeting its first placental mammals.

The fossil record suggests that placental mammals were more abundant than marsupials in the late Cretaceous, some much larger than the mouse-sized individuals that we usually associate with mammals of the time. They could be found on most continents, albeit each with its own variable assortment of species. At one time it was believed that placentals did not reach Australia when marsupials arrived. We now know they possibly did but the evidence for them is slight and controversial. Their story here largely rests with one tiny jaw of *Ausktribosphenos nyctos*, from the early Cretaceous, two teeth representing two individual species (*Tingamarra porterorum*, late Palaeocene, and *Yingabalanara richardsoni*, early Miocene) and a fragment of cranium (*Chronozoon australe*, early Pliocene). The difficulty of analysis and interpretation of these fossils is such that all that can be safely said is that mammals *probably* reached Australia, those that did were small, there were not very many, and they all soon went extinct. Indeed, the earliest, the tiny, shrew-like *A. nyctos*, ca 115 Ma was contemporary with the oldest monotremes yet found as well as dinosaurs of course, but the subclass status of its tiny jaw (it is either a placental or a type of monotreme) is also unresolved. On the other hand, Australia was joined to a Gondwana that only had Africa missing at the time and so, with the long placental record going back at least 225 Ma, there is no reason to think placentals did not live in Australia at that time: they lived in other parts of Gondwana!

The earliest evidence for marsupials found so far comes from the Tingamarra Formation on the Palaeocene–Eocene boundary *ca*. 10 million years before Australia's final breakaway from Antarctica. It is also the time of the Eocene Optimum and the warmest Tertiary temperatures. Murgon contains a diverse vertebrate fauna and it is from there that evidence has emerged for a placental condylarth mammal, *Tingamarra porterorum*, which must have entered Australia in the Palaeocene or even the Cretaceous. In a way this is a shame because it spoils the myth that only marsupials ever reached Australia. In fact, it is probably more realistic to believe both placentals and marsupials arrived here, rather than believe *only* marsupials made it, but it also seems logical to suggest marsupials must have made up by far the largest component of that migration. Well, what is the evidence for the placental *Tingamarra* animal? Only a single lower molar tooth 3 mm long, I am afraid, which is not very much even in palaeontology's often depauperate world of fossils. As with interpretations made on any tiny piece of skeletal evidence, scepticism and controversy arise among the scientific community, and it has about this evidence. That is as it should be, however; a rigorous examination of evidence and a lively debate is always required for small fragmentary finds, and one unwritten law of scientific enquiry is that the size of the evidence is inversely proportional to the amount of heated argument it usually generates. Consequently, there is controversy over the allocation of *Tingamarra* as a placental mammal. It has been proposed that it might be a marsupial that underwent convergent evolution that conferred placental mammal characteristics on its dentition. Whatever the reasons for its convergence (if that is what it was), its numbers were too small to sustain an ongoing population, or environmental conditions did not suit it, or the local marsupials muscled it out or outwitted it in the competition stakes, because *Tingamarra* soon went extinct – or did it? Marsupial fossils found on the Antarctic Peninsula close to South America resemble much earlier South American types. Two, however, are very different, suggesting they had a long time to diverge and thus had been living in Antarctica for some considerable time. Similarly, the earliest specimens from the 55 Ma Tingamarra Formation at Murgon in southeastern Queensland are unlike South American or Antarctic marsupials, suggesting that they might have lived in Australia much longer than 55 Ma, enabling them to develop local characteristics. They also lived a long way from the Antarctic Peninsula. Both pieces of data suggest, therefore, that marsupials had penetrated the Antarctic–Australian region possibly as early as the earliest evidence for them in South America. In other words, they had completed their migration from North America, probably crossing Australia's borders long before the K–T event. Therefore, any disturbance of this process by the impact was by populations of marsupials endemic to South America and possibly a few placental types rather than those that were making their way here or already here. This also suggests that a kind of founder effect may have taken place in Australia whereby the genetic stocks already here were cut off for some considerable time from newer arrivals that were severely affected by the Chicxulub event. Thus, there may have been little new genetic input here before Australia finally broke away from Antarctica; there certainly was none after it.

On present evidence it seems most likely that marsupials moved from China to Australia via the Bering Bridge and North America. They may have moved into Europe at the same time. The Turgai Strait divided Europe from Asia, blocking a westward marsupial movement from China until its closure in the late Oligocene, far too late for marsupials to come that way; they probably came via North America and the Bering Bridge. They then became extinct in the Old World and in North America while flourishing in Australia. Monotremes became extinct in South America while also flourishing in Australia but their migratory population may have been very small, predisposing them to a hard time surviving. Marsupials returned to North America much later when the Panama corridor closed in the Pliocene. Placental animals thrived but they had only 15% species loss in the K–T event as opposed to the 75% marsupial losses. The main problem facing those wishing to put together a better and/or more accurate picture of these events is the long periods of fossil-empty time, specifically the 90–70 Ma gap in North America and the 55–25 Ma and 115–55 Ma gaps in Australia. The South American record for those time frames is also very sparse. The overall emptiness in the story of the marsupial and placental migrations in this part of the world is frustrating but typical of the many frustrations palaeontology always faces in terms of the fickle nature of the fossil record.

Australia's Earliest Mammals

To my mind, Australia's continental biology did not really start till we became an independent island continent *ca.* 45 Ma. It marks the real beginning of the Australian continent after being just a region of Gondwana for *ca.* 600 million years. Breaking away also marked the independence of the fauna from external migrations and the beginning of its development as a symbol of Australia's special biological heritage. There really is no other continent with a comparable palaeobiogeographic or palaeozoogeographic story that compares to that of Australia. While other continents exported and imported species in various frequencies and at various times during the Tertiary, Australia was isolated from such movements as it floated farther and farther away from its nearest neighbour. It was also moving from a cool to cold, wet climate towards a warmer, drier one. India had done something similar, moving north across the Equator; additionally, its plate had joined that of Australia's so we were both distant companions on our northern journey. Africa too was moving north but its northern half was drifting away from the Equator while the south moved towards it with one part always straddling the Equator. Africa collided with Eurasia in the mid-Tertiary firmly enough to swap fauna, something Australia's isolation always prevented. Indeed, it was the new continental linkages brought about by the formation of various Tertiary land bridges that allowed animals to migrate right across the Old World but not to here. This brought about various genetic assortments as they moved from Asia to Europe in both east-to-west and west-to-east directions, and from North to South America as well as from Africa to Eurasia. In contrast, Australia's animals were totally confined to one continent, the smallest. This was an important palaeozoogeographic difference from other continents and one that eventually would impact on the history of Australia's marsupial populations. Their only migrations were internal: genes moved neither into nor out of the continent, only within and around it. Moreover, internal migration was almost certainly only precipitated by environmental changes and this was later flagged by the parallel evolution that took place among marsupial populations. Australia's animals settled into similar ecosystems and niches to those of placentals in other parts of the world. That produced animal forms similar to those outside; trends resulted in marsupial dogs, moles, deer (kangaroos), anteaters, bears (koalas), flying squirrels (gliders) and many more.

Australia's gene pool was fixed; this was a situation that probably saved marsupials initially but would have an impact in the very long term. Founder effect operating in the original gene pool no doubt helped produce the unique array of species that had arisen by the end of the Oligocene. That process was guided by ecological shifts which brought various botanic mosaics, opening opportunities for many species with habitats growing and disappearing as Australia slowly moved northwards. Isolation in itself tells the story of the preservation of the vast majority of marsupial species in the world, although the original population of migrants probably did not represent a very big range of types. But it has also provided a unique opportunity for a whole subclass of mammals to evolve almost as though they had the planet to themselves. The comparison of Australia to an Ark, however, is entirely wrong. The Ark had members of *all* animal species on board and two of each (so the myth goes); Australia really only had marsupials. It was no Ark but it certainly was a life raft and one without any really nasty carnivores. Whatever the answer to the early placental story, their short history here is a mystery, as are the evolution, composition and species radiation of Australia's Palaeogene marsupials. Two of those early marsupials have also been identified at Murgon: *Thylacotinga bartholomaii*, which was about the size of a native cat or quoll, and *Djarthia murgonensis*, a mouse-sized insectivore, both well adapted to the rainforest conditions that prevailed in Australia at that time. Unremarkably, both these animals seem related to South American marsupials although only dental characteristics are known; such a relationship is unsurprising considering Australia was still attached to Antarctica and through that to South America at the time of the Murgon deposits. Because of this geographical association it has been suggested that perhaps animals ancestral to or descendent of those like *Djarthia* may have returned to South America, and that is where the confusion comes in. Murgon has also claimed the oldest evidence for songbirds, Australia's oldest bat (*Australonycteris clarkae*) and a madtsoiid snake (*Alamitophis* sp.) that was related to those living in Patagonia, another faunal link to South America. Because of Australia's plate trajectory tracking from Polar to Equatorial regions, the environmental changes experienced on the journey were somewhat different from those experienced on other continents, most of which were moving laterally, east–west. That proved decisive in the pattern and form of speciation and development among Australia's marsupial population as the continent drifted even north, a course that would turn its temperature from cool to cold and its very wet environment to a much warmer and drier one. Nobody can say that marsupials are conservative creatures because the environmental change they experienced and survived throughout Tertiary is certain testimony to their strength of adaption and design.

Australia's Faunal Dark Ages: 55–25 Ma

The faunal Dark Age that followed may turn up another placental mammal again one day but one thing we do know is that marsupials did all right. Incredibly, the next 30 million years (55–25 Ma) mark a Black Hole in Australia's fossil fauna record that extends from the Eocene deposits of Murgon to the Oligo-Miocene boundary of Riversleigh. In effect, we skip two of the longest epochs in the Tertiary. However, while our understanding of the faunal record is almost non-existent, we do know a little more about what Australia's environment was like and how it was changing during that time. Tectonically and volcanically Australia was comparatively quiet during the Tertiary compared to some continents; nevertheless, it had its moments. As a large moving plate, it is almost impossible to slide around the world and not feel a few bumps, experience upheavals and pass over a few hot spots. Today, Australia is the only continent in the world without an active volcano, although it was not always this quiet, with one volcanic episode between 50 and 80 Ma that was active north to south across northeastern Queensland and another between 14 and 23 Ma that stretched from Tasmania to central Queensland.

Australia's position in high latitudes during the Eocene Optimum meant that it experienced very high rainfall, perhaps as much as 20 m annually in some places. This may have played an important role in our missing fossiliferous deposits of the period. One of the factors in the poor preservation of faunal and macro- and micro-fossil plant assemblages dating to the Palaeogene and Eocene is the extensive weathering and erosion of land surfaces that occurred during that time. High amounts of rainfall experienced before, during and after the Eocene Optimum were ideal for encouraging rainforest growth and marsupial diversity but it scoured the landscape, probably confounding the forces of preservation and probably sweeping away much of our palaeontological heritage from that time. As testimony to the wet

conditions, a network of very large and very long river channels, termed palaeovalleys today, extend out from and stretch across central Australia in a large, often inter-
connecting, network of channels, many of which fan out to the south and west of the
continent. Some channels were over 1000 km long and they obviously carried broad
expanses of water that occasionally ponded into large lakes before moving on. This
broad network of continental drainage largely took water to the west and south of the
continent as well as inland through the Lake Eyre drainage basin. The sinuous course
of some still remains in the form of comparatively recent channels that have re-
excavated into the old channel valleys, often following original palaeovalley topog-
raphy. Some of the modern channels include the De Grey, Gascoyne, Fortescue and
Ashburton Rivers in Western Australia, all forming fingered end segments of ancient
valleys. Those palaeovalleys that drained south remain as fossil drainage systems
with names such as the Throssel, Noonina, Meramangye and Kadgo Palaeovalleys,
all of which originally emptied into the Eucla and Eromanga Basins on the Nullarbor
Plain. Since then, continental uplift in the south of the continent has prevented modern systems flowing south and cutting into the ancient valleys, leaving the latter as
the only reminder of the vast watershed that once was Central Australia. Australia's
inland was probably warmer and more humid than coastal regions and it was due to
this that between 20 and 30 Ma, a high water table and extensive swamps gave rise to
the deposition of thick peats that formed brown coal deposits such as those in the La
Trobe valley of Victoria.

   The fossil plant deposits at Nelly Creek and other localities in the southern Lake
Eyre basin clearly testify to the variety of plants, particularly tree species, that were
growing in the southern part of the continent *ca.* 45 Ma; they show the dominance
of *Nothofagus* pines and temperate rainforest species in response to the Eocene
Optimum and high rainfall (White, 1994). But it was all about to change as world
temperatures began to fall, chilling, high-latitude environments and ushering in
a more steeply graded ecological differential from the Poles to the Equator. In the
southern ocean this trend was compounded by the opening of the Drake Passage
between Antarctica and South America *ca.* 35 Ma. The Circum-Polar Currents
that formed added to the cooling Antarctic Ocean, and shark teeth and varieties of
warm-water shells found along the Antarctic coast show how warm that ocean had
been 15 million years earlier.

   The sharp decrease in ocean temperatures brought the first signs of ice formation on the Polar continent. World cooling reduced rainfall but the southern cooling may also have pushed rain systems further north. Collectively these changes
steadily changed Australia from a continent of temperate rainforest to one of more
diverse environments, including the first hints of semi-aridity and the beginnings of
embryonic deserts. Australia's fossil pollen record from the Oligocene is scanty and
what there is largely represents what was happening in the southeast of the continent.
However, that tends to suggest a decline in plant variety and the rising importance of
eucalypts, with the first occurrence of *Acacia* pollen probably in response to decreasing rainfall and the concomitant increase in natural fire frequency. An increase in
natural fires must have been a strong adaptive force faced by marsupials in terms of
changing floral patterns and distributions on a continent that had up to now been a

wet forested place. Now landscapes changed into more open and drier forests. With these environmental alterations came niche differentiation with increasing variety, and from that we can infer the kind of adaptive radiation that must have been going on among Australia's fauna. Adaption to a drier continent was going to be the biggest challenge that Australia's marsupials had ever faced and it would play a significant role in testing their ability to persist and adapt. One thing was certain: the first series of extinctions were about to take place as those that were unsuited to the new emerging conditions were sorted from those that were not. Another challenge was that those conditions were constantly changing and this would favour only animals that could keep up: specialist animals would be the first ones to suffer.

So we see environmental diversity taking place through climate change for the first time on the independent continent of Australia. The original palaeoaustral rainforest now began to slowly shrink and later fragment. It was now somewhat warmer inland in contrast to the cool to cold of the south, and this meant the beginning of semi-arid parts in the north. Areas of the old rainforests remained in the south and east but they were contracting to patches as well as gallery forest along the large palaeochannel drainage systems that were, in turn, contracting to smaller channels. New ecosystems formed based on eucalyptus species or on variable regional rainfall, particularly along the uplifted eastern Australian Great Divide, which also offered a varied range of niches. Together this amounted to opportunities for speciation, adaptation, movement and re-distribution of animals and the emergence of variety and specialisation among marsupials living in different parts of the continent. Regional differences among these animals would now begin to slowly emerge.

The mid–late Oligocene saw the process of new soil development in Australia, the first since mountain building over 300 million years earlier. Central to this process was the second wave of volcanism that ran along the Great Divide. Regular eruptions did not occur till mid–late Oligocene but continued through the Neogene, stretching from Queensland to Tasmania. As the Australian Plate pushed north, it passed over a series of hot spots, a process that began around the Nebo, Peak and Springsure Ranges (27–30 Ma) of Central Queensland before continuing south as the continent moved north-northeast, leaving what we today call the Glasshouse Mountains in southeast Queensland (24 Ma), the Main Range (23 Ma) and Border Ranges of northern New South Wales (22 Ma), the Nandewar Range (18 Ma) and, further south, the Warumbungles (17 Ma) as volcanic remnants in its wake. The pattern continues through southern New South Wales into Victoria and on to Tasmania. The curved pattern of the volcano line shows Australia's interesting plate trajectory wiggle as it continued north during that time. With new soils came a variety of vegetation. That meant more environmental choice and niche formation and it marks the time of the first occurrence of *Acacia* pollens. By the early Miocene, world temperatures began to rise once again and Australia underwent a more humid 'greenhouse' phase because of its north nosing into the subtropics, which brought vegetation changes boosting plant variety. But by this time we have, once again, evidence of the fossil history of the marsupials themselves, with fossils sites emerging in different parts of the continent.

One rather underwhelming claim Australia can make is that over a large part of the continent its soils are mineral and nutrient poor and this also goes for the seas

that surround us. In those terms we are probably the least productive of all continents. With the exception of the eastern edge of Australia where late Cretaceous–
Tertiary volcanics and formation of the Great Divide rejuvenated soils, the lack of
widespread volcanism and mountain building has left us with poor soil replenishment, turnover and development. To make things worse, we have also experienced
widespread erosion and leaching of soils that included the Palaeogene washing of the
continent, reducing even further their nutritional quality. Even in the face of this history, however, our marsupials have thrived and they certainly did in the Palaeogene,
thanks mainly to the temperate rainforests and marsupials' propensity to adapt so
well within them; we can see this in the range of genera and species that have been
found in the late Oligocene and early Miocene fossil record. The rich forest environments made a wonderful cradle for the evolution of so many species that were naturally adapted to such conditions and were on average rather small creatures, able to
take advantage of the multi-storey niches available to them.

They Are Still There!

Australia's earliest marsupial fossils after the 30-million-year 'Dark Age' span the
late Oligocene and early Miocene. They emerge from locations around Lake Eyre,
the Northern territory and North Queensland. Other sites are scattered in New South
Wales and Victoria, but none of this age have yet been found in Western Australia.
The few finds at Quanbun in the Kimberley are probably the earliest, although they
likely date to the Pliocene. The earliest and most prominent assemblages come from
sites in the Tirari Desert, east of Lake Eyre and the expansive Riversleigh site in
Queensland's Gulf country (Tables 5.1 and 5.2, Figure 5.1). Palaeontologists working at Lakes Palankarinna and Ngapakaldi in the Tirari Desert needed to use their
imaginations because recovering a forest fauna from a desert can be disconcerting
and can play with your mind. Many of the earliest sites lack an accurate chronology
and are located only to early, mid- or late stages of a particular epoch, although sites
in the later Neogene do have a more accurate range of dates. This, of course, follows
the rule of accuracy stating that sites are more accurate the younger they are.

Over 40 genera of marsupials are known from the late Oligocene and early
Miocene, although species allocation within these is often difficult because of the fragmentary nature of the fossils. Nevertheless, they suggest a diverse set of fauna probably adapted to an equally wide variety of environments and niches (Table 5.3). Site
location in all arid areas of the continent and some in the most arid part (Lake Eyre)
attest to the very different environmental circumstances in which these animals lived.
But whatever they were adapted to, it was very different again from the ecological surroundings that *their* ancestral stocks were used to – which is in itself a measure of the
transition Australian faunal undertook in the Tertiary. Some Oligo-Miocene specimens
have posed real problems for interpreting what role they actually played and the lifestyles they led. It seems there must have been some adaptive forcing in specialised
directions and it is likely that these were determined by the climatic roller coaster that
was operating at the time. The various types of marsupials that emerged also shows

Table 5.1 Australian Cretaceous and Tertiary Fossil Sites and Their Ages

Fossil (Local Fauna) Site and Age	Epoch
Dinosaur Cove – 110–115 Ma	*Cretaceous*
Lightning Ridge – 100–111 Ma	
Winton – 90–100 Ma	
Tingamarra (Murgon) – 55.6 Ma	*Eocene*
Pinpa (Lake Frome) – 25.5 Ma	*Late Oligocene*
Ericmass (Lake Frome region) – 25.3 Ma	
Tarkarlooloo (Lake Frome region) – ~24 Ma	
Ditjimanka/Ngapankaldi (Lake Eyre region) – ~25 Ma	
Riversleigh System A – ~24 Ma	
Kangaroo Well (NT) – 22–23.5 Ma	
Geilston Bay (Tasmania) – 22.4 Ma	*Early Miocene*
Riversleigh System B – 20–22.5 Ma	
Kutjamarpu (Lake Eyre region) – 20–22 Ma	
Wynyard (Tasmania) – 21 Ma	
Canadian Lead (NSW) – 17 Ma	
Riversleigh System C – 14–17 Ma	*Middle Miocene*
Bullock Creek (NT) – 12 Ma	
Encore Site (QLD) – 10 Ma	*Late Miocene*
Alcoota (NT) – 7–8 Ma	
Beaumaris (Vic) – 6–7 Ma	
Ongeva (NT) – 5–6 Ma	
Sunlands (SA) – 3–5 Ma	*Pliocene*
Riversleigh (Rackham's Roost) – 3–5 Ma	
Big Sink & Bow (NSW) – 3–5 Ma	
Hamilton (Vic) – 4.46 Ma	
Quanbun (Kimberley) – 2–4 Ma	
Parwan Local Fauna (Victoria) – 4.1–4.2 Ma	
Bow (Vic) – 4.0–4.3 Ma	
Bluff Downs (NT) – 4.8 Ma	
Palankarina (Lake Eyre region) – 3.9 Ma	
Coimadai (Vic) – 3.6 Ma	
Chinchilla (QLD) – 3.5 Ma	
Allingham Formation (QLD) – 3.6 Ma	
Kanunka (Lake Eyre region) – 3.4 Ma	
Fisherman's Cliff (NSW) – 3.2 Ma	
Smeaton (Vic) – 3.2 Ma	

Source: After Rich et al., 1985; Long et al., 2002 and Prideaux, 2004.

their adaptability when left to their own devices. Marsupial variability must also have been a product of the wide swings in temperature, humidity and rainfall that caused a variety of rapid environmental changes in some parts of the continent but not so in others. It is now obvious that such changes drive evolutionary processes much faster than the steady state of environmental and climatic conservatism. As an example of these changes, it was around this time that the earliest radiation of the eucalypts occurred.

Table 5.2 Australian Fossil Sites with Epoch Allocation Only

Other Faunas and Fossil Sites	
Etadunna Formation Faunal Zones A, B, C, D and E (SA)	*Late Oligocene to*
Pinpa Formation (SA)	*middle Miocene*
Tarkarooloo Local Fauna (SA)	
Yanda Local Fauna (SA)	
Wadikali Local Fauna (SA)	
Namba Formation (SA)	
Kutjamarpu Local Fauna (SA)	
Ngapakaldi Local Fauna (SA)	
Bunga Creek (Vic)	*Early Pliocene*
Lake Tyers (Vic)	
Tara Creek (North QLD)	
Curramulka Local Fauna (SA)	

Source: After Rich et al., 1996.

Figure 5.1 Principle fossil sites of pre-Tertiary and Tertiary Australian fauna. 1. *Lake Eyre Sites* – Lakes Ngapakaldi, Palankarinna and Etadunna Formation (Zones A–E); 2. *Lake Frome Sites* – Lakes Namba, Pinpa, Tinko, Tarkarooloo and Yanda; 3. Alcoota; 4. Bullock Creek; 5. Riversleigh; 6. Bluff Downs; 7. Tingamarra Formation (Murgon); 8. Lightening Ridge; 9. Sunlands Local Fauna; 10. Hamilton; 11. Beaumaris; 12. Wynyard and 13. Geilston Bay.

Table 5.3 Australian Marsupial Species of the Late Oligocene to Late Miocene

Late Oligocene (n = 41)	Early Miocene (n = 28)	Mid-Miocene (n = 23)	Late Miocene (n = 17)
Ankotarinja tirarensis	Balbaroo fangaroo	Balbaroo camfieldensis	Alkwertatherium webborum
Apoktesis cuspis	Cercartetusb sp.	Barinya wangala	Dorcopsoides fossilis
Badjcinus turnbulli	Durudawiri anfractus	Bettongia moyesi	Ekaltadeta jamiemulvaneyi
Balbaroo gregoriensis	Durudawiri inusitatus	Burramys brutyi	Ganbulanyi djadjinguli
Balungamaya delicata	Ekaltadeta sp.	Dasylurinja kokuminola	Hadronomus puckridgi
Brachipposideros sp.	Ektopodon litolophus	Djilgaringa gillespiae	Kolopsis torus
Burramys wakefieldi	Ektopodon serratus	Dorcopsis sp.	Mayigriphus orbus
Chunia illuminata	Ganawamaya acris	Ekaltadeta ima	Palorchestes annulus
Chunia omega	Ganawamaya ornata	Hypsiprymnodon bartholomaii	Palorchestes painei
Ektopodon stirtoni	Ganguroo bilamina	Litokoala kutjamarpensis	Plaisiodon centralis
Ganawamaya aediculis	Litokoala kanunkaensis	Macroderma sp.	Pyramios alcootense
Galanarla tessellata	Marlu kutjamarpensis	Maximucinus muirheadae	Pseudokoala curramulkensis
Gummardee pascuali	Neohelos tirarensis	Muribacinus gadiyuli	Thylacinus potens
Ilaria illumidens	Ngamalacinus timmulvaneyi	Mutpuracinus archibaldi	Thylacinus megiriani
Keeuna woodburnei	Nimbacinus dicksoni	Neohelos tirarensis	Tyarrpecinus rothi
Kuterinjta ngama	Nimiokoala greystanesi	Nimbadon lavarackorum	Wanburoo hilarus
Madakoala devisi	Nowidgee matrix	Nimiokoala greystanesi	Zygomaturus gilli
Madakoala wellsi	Paljara tirarensis	Petramops sp.	
Marlu praecursor	Pildra tertius	Propalorchestes novaculacephalus	
Miralina doylei	Priscileo roskellyae	Propalorchestes ponticulus	
Miralina minor	Pseudocheirops sp.	Strigocuscus reidi	
Muramura williamsi	Rhizophascolonus crowcrofti	Trichosurus dicksoni	
Nambaroo couperi	Thylamcinus macknessi	Wakaleo vanderleueri	
Nambaroo novus	Wabulacinus ridei		
Nambaroo saltavus	Wakiewakie lawsoni		
Nambaroo tarrinyeri	Wynyardia bassiana		
Namilamadeta snideri	Yalkaparidon coheni		

(Continued)

Table 5.3 (Continued)

Late Oligocene ($n = 41$)	Early Miocene ($n = 28$)	Mid-Miocene ($n = 23$)	Late Miocene ($n = 17$)
Ngapakaldia tedfordi	*Yarala burchfieldi*		
Ngapakaldia bonythoni			
Palaeopotorous priscus			
Perikoala palankarinnica			
Perikoala robustsus			
Pildra antiquus			
Pildra secundus			
Pilkipildra handae			
Pitikantia dailyi			
Priscileo pitikantensis			
Purtia mosaicus			
Raemeotherium yatkolai			
Silvabestius johnnilandi			
Wabularoo naughtoni			

Source: After Rich et al., 1996 and Long et al., 2002.

One example of the 'weird' creatures that existed is the ektopodontids (*Ektopodon* sp. and *Chunia* sp.) that lived through the Tertiary and disappeared in the early Quaternary. These strange seed-eating possums with their meat-pounder dental surfaces were first discovered in the Tirari Desert but were later found in Victoria and then Riversleigh. They had wide dispersal across central and northern Australia but it is likely that their range contracted with the increasing aridity gradually spreading out from the continental centre during the Neogene. Another creature, named 'Thingadonta' by the long-time senior investigator of the Riversleigh research, Mike Archer, probably represents a very old lineage of marsupials that lived in Australia before the final split with Antarctica 45 Ma. Named originally precisely because of its strange appearance and the puzzlement that ensued among those who found it, 'Thingadonta' is now named *Yalkaparidon*. It has now been added to a range of genera such as *Yingabalanara*, *Elkatadeta* and *Wabularoo* that, as a group, were endemic to northwest Queensland, perhaps separated from central parts of the continent by some sort of environmental or ecological barrier (Archer et al., 1991). However, other species found at Riversleigh also lived in central Australia which, given the large time frame in which these creatures have been placed, may indicate the opening and shutting of migratory corridors across Australia as ecological changes followed the rise and fall of world temperatures that occurred when the earliest Riversleigh deposits were being formed. The principle of ecological change and corridor closure implied here is taken up again in Chapter 7 when I discuss the mid-late Quaternary. In discussing a continent renowned for its kangaroos, it is worth mentioning the late Oligocene–early Miocene genera *Nambaroo* and *Balbaroo*, both early kangaroos belonging to the Balbarid ('strange') family of kangaroos that are not related to anything around now. They sported very large canines for such small animals and were probably browsers, a behaviour that became popular among much larger varieties of kangaroos.

Besides odd-looking marsupials from Australia's earliest available fossil deposits, there are some familiar faces also, although they changed and speciated as they crossed the Tertiary. Two of these families are bandicoots and koalas, both of which broadened their species variety and have survived to the present with only comparatively small changes in appearance. However, the modern koala is somewhat more specialised than its ancient ancestors, of which there were a number of different varieties. It is also larger and that was another of the changes that were taking place among our marsupials – size increase. The marsupial lion was just one animal that grew larger as it passed through the Tertiary. It is seen first as the cat-sized *Priscileo pitikantensis* in the late Oligocene before developing through various enlarging stages through the *Wakaleo* genera and then to the fierce, lioness-sized *Thylacoleo carnifex* of the Quaternary. Another example was the Thylacine, a marsupial dog often referred to as a wolf. The early Miocene *Thylacinus macknessi* was probably less than half the size of its Quaternary descendent *Thylacinus cynocephalus* and the late Oligocene–early Miocene *Litokoala* was similarly a much smaller animal than its modern counterpart *Phascolarctos*. By the early Miocene all the early genera had gone extinct, giving rise to further diversification among Australia's marsupials and the emergence of new species. Many went extinct throughout the Miocene only to be replaced by others who were no doubt more adapted to Australia's changing climatic and environmental conditions. It was at that time that Australia really began its vast changeover from a continent of temperate rainforests to one of arid deserts. The long Ice Age that began at this time, together with the augmenting effects of the new Circum-Polar Current, the continuing elevation of New Guinea forming an increasing rain shadow for Australia's north and the fact that Australia had begun to enter the world's driest latitude belt between 20° and 30°, all accumulated to push the continent towards aridity. It is arguable that no other continent of its size underwent such radical one-way change to its environment and climate as did Australia during the Neogene.

Australia's Miocene, a Window to the Future

The Miocene extinctions saw the loss of many genera as the filtering process weeded out those unable to adapt to drying environmental conditions and those that clung to receding rainforest and denser woodlands. Moreover, although a slow process at first, these dwindling habitats gradually reduced the overall size of the marsupial population. The process of range shrinkage reduced genera, species in each genera and then numbers of animals in each species. The decreasing frequency of southern Beech (*Nothofagus* sp.) pollen in the mid-late Miocene marks the reduction of the temperate rainforests that had flourished so widely in the Palaeogene. That acted not only to eliminate species but also to reduce the populations of those that survived. Nevertheless, the process left open niches into which those that were quick to adapt could move and from which new species emerged. The key, however, was to be a little more arid-adapted than those that had gone before and this was a trend that increasingly became the norm. Fortunately fossil-bearing sites in the southern half of the Northern Territory such as Kangaroo Well, Bullock Creek and Alcoota span the Miocene. Their assemblages show a range of woodland-/forest-adapted animals but also some that might be interpreted as being in a transitory stage as the central parts of the continent began to unevenly dry out.

Miocene sites also contain remains of giant flightless dromornithids which, as a group, go back to the Eocene. These were Australia's answer to the 'Terror Birds' of North and South America but probably not as blood-thirsty, lacking as they did the vicious hook on the end of their beaks typical of the American birds. They were, however, large, strong, tall, fast runners with enormously strong legs, and they possessed very large, deep beaks, something not developed for sport alone or munching on fruit bushes! In line with the marsupial fossil record, little is really known of these creatures till the Miocene after which several species have been identified. They include *Barawertornis tedfordi* of the early–mid-Miocene, *Bullockornis planei* of the mid-Miocene and *Ilbandornis lawsoni*, *I. woodburnei* and *Dromornis stirtoni* of the late Miocene (Murray and Vickers-Rich, 2004). The latter (*D. stirtoni*) originates from Alcoota and was the largest of the dromornithids (Figures 5.2 and 5.3). While these giant birds are a little away from our story of marsupials, their sparse to non-existent fossil history in the Palaeogene reflects that of all Australia's fauna and suggests that bone preservation may have been limited due to the rainstorms of the Palaeogene. The paucity of bone must have been at least partially caused by destruction and erosion of the sediments which might have preserved them. On the other hand, the fossils may still be out there somewhere – we just have not found them. One of the descendents of these big birds does play a part in later chapters, and that is *Genyornis newtoni*, an animal counted in the megafauna species list and the last of the giant flightless birds in Australia – no offence to large specimens of emu, of course.

The Alcoota fossil locality, which spans 12–7 Ma, is situated 200 km northeast of Alice Springs. The pioneer field work carried out there over many years by Peter Murray, the late Dirk Megirian and many dedicated volunteers has produced examples of ancestors to some of the most well-known megafauna of the Quaternary. They include genera such as the diprotodontids; an example of this is the small browser *Pyramios*, a smaller version of and probable ancestor to the much later and much larger *Diprotodon*. Another, *Alkwertatherium*, also from Alcoota, was the size of a small horse, smaller than *Pyramios*, and possibly ancestral to another later diprotodontid, *Zygomaturus*. These animals were probably accustomed more to forest browsing than savannah grazing, as was *Kolopsis torus*, another 'Alcootan' diprotodontid. Yet another ancestral group of diprotodontids, the Palorchestids, have also been found in Alcoota assemblages, with *Palorchestes painei* being the late Miocene representative of this group. However, another species, *P. annulus*, from Riversleigh

Figure 5.2 The skeleton of *Dromornis stirtoni* the large flightless dromornithid of late Miocene, Australia's answer to the 'terror birds' of South America. This example is displayed in the Central Australia Museum, Alice Springs close to where it was discovered by Peter Murray and his team at the fossil site of Alcoota in the Northern Territory.
Source: Photograph by Steve Webb.

Figure 5.3 Close up of the head and deep beak of *Dromornis stirtoni*.
Source: Photograph by Steve Webb.

1000 km to the northeast in northwestern Queensland, may predate *P. painei* as a mid-Miocene representative of the genera. Others such as *P. parvus* have been recognised from the early Pliocene of the Darling Downs in southeastern Queensland and Bluff Downs in northeastern Queensland.

What we can see is one very important development among the marsupial fauna of the time: increasing body size. That was the special adaptation they were undergoing and a dead giveaway of the onset of arid conditions. A larger body was not only capable of ingesting more of the reduced-nutrient fodder typical of arid-adapted vegetation, it also usually came with a bigger gut for metabolising the larger input of material required to extract the necessary nutrition. The large forest-dwelling quadrupeds of Alcoota were not alone: they had the company of *Dorcopsoides fossilis*. This was a forest-dwelling kangaroo now confined to New Guinea but, interestingly, a genus of sthenurine kangaroo (*Hadronomus puckridgi*) which, in turn, was probably a flag species of another macropod also in transition. *Hadromus* was more adapted to open forests and was probably a forerunner of the later sthenurines that were to become more successful as aridity increased through the Pliocene, although most remained browsers (rather than the modern grazing macropods), probably of the arid-adapted tree and shrub species.

Experimentation of species production as a product of rainforest dominance was over. The trend now was in the opposite direction towards the development of animals that could make their way in a continent of much-reduced forest and growing aridity and which had the ability to live in encroaching desert landscapes. If they could adapt to these conditions then they would have a distinct advantage. That in itself signalled that there would be a net reduction in faunal variety in line with normal biogeographic principles recognising that fewer mammal species occupy arid environments than forested ones. Tables 5.3 and 5.4 show the reduction in Australian genera in the late Miocene, a similar pattern to elsewhere around the world. By the Pliocene, far fewer genera were disappearing, which may reflect the loss of variety and the end of the clearing out of species that could not make the necessary adaptations. Although our picture is probably far from complete, there seems to be little doubt that what occurs in the Neogene, particularly in the first half, is an overall reduction in the numbers of Australia's marsupial species. The accuracy of the tables provided here has to be tempered with the fickle nature of the fossil record and the understanding that discoveries are made all the time. There are probably genera or species that I have missed but this will not change the basic trends shown. At any one time the picture is only temporary, till the next site is discovered and another group of fossils emerge. The aim here, however, is to provide a basic view of broad trends that took place in the continent's faunal turnover during the lead-up to the megafauna extinctions of the later Quaternary. The trend does, however, seem to reflect the result of adaptive filtering that by this time had eliminated most species that were not going to make it through the climatic changes that had been cooling the world from the mid-Miocene and promoting aridity, particularly in Australia; both of these changes would continue.

Australia's Tertiary extinctions are charted in Appendix 1, which shows the loss of 213 species in that time as well as the trends of extinction timing. The obvious reduction of genera and species that took place in the Miocene has little explanation beyond the widespread climatic and environmental changes that were taking place not only in Australia but on a world scale. It was, however, only a continuum of what had occurred in the Oligocene and continued into the Pliocene and Quaternary. However, the environmental changes, which were comparatively slow and benign, had a strong enough effect on the adaptive capabilities of many animals to put them beyond the turnaround in adaptive requirements that was now required. Australia

Table 5.4 Australian Marsupial Species of the Pliocene and Early–Mid-Quaternary

Early Pliocene (n = 34)	Mid-Pliocene (n = 14)	Late Pliocene (n = 9)	Early–Mid-Quaternary (n = 39)
Archerium chinchillaensis	Antechinus sp.	Bettongia sp.	Baringa nelsonensis
Burramys triradiatus	Glaucodon ballaratensis	Dasyroides sp.	Bohra paulae
Darcius duggani	Isoodon sp.	Diprotodon sp.	Burungaboodie hatcheri
Dasycercus cristicauda	Kolopsoides cultridens	Kolopsis rotundus	Congruus congruus
Dasyurus maculatus	Nototherium watutense	Macropus mundjabus	Congruus kitcheneri
Dendrolagus sp.	Palorchestes parvus	Pseudocheirus sp.	Dendrolagus noibano
Dorcopsis winterecookorum	Phascolarctos maris	Sarcophilus sp.	Euowenia grata
Ischnodon australis	Protemnodon bandharr	Simosthenurus brachyselenis	Euryzygoma dunense
Kolopsis yperus	Ramsayia sp.	Troposodon kenti	Lagorchestes sp.
Koobor notabilis	Simosthenurus cegsai		Lasiorhinus augustidens
Kurrabi mahoneyi	Sthenurus sp.		Macropus thor
Kurrabi merriwaensis	Thylacoleo sp.		Metasthenurus newtonae
Kurrabi pelchenorum	Wallabia sp.		Osphranter sp.
Macropus pan	Zygomaturus trilobus		Perameles sp.
Macropus pavana			Petaurus sp.
Meniscolophus mawsoni			Petrogale sp.
Milliyowi bunganditj			Phalanger sp.
Palorchestes selestiae			Phascogale sp.
Petauroides stirtoni			Phascolonus sp.
Petauroides marshalli			Prionotemnus sp.
Phascolarctos yorkensis			'Procoptodon' gilli MIS6
Prionotemnus palankarinnicus			Propleopus chillagoensis
Protemnodon chinchillaensis			Propleopus oscillans
Protemnodon devisi			Propleopus wellingtonensis
Pseudokoala erlita			Protemnodon sp.
Simosthenurus antiquus			Pseudokoala catheysantamaria
Simosthenurus cegsai			Ramsayia magna

(Continued)

Table 5.4 (Continued)

Early Pliocene (n = 34)	Mid-Pliocene (n = 14)	Late Pliocene (n = 9)	Early–Mid- Quaternary (n = 39)
Sthenurus notabilis			Sarcophilus moornaensis
Thylogale sp.			'Simosthenurus' baileyi MIS6
Trichosurus hamiltonensis			Simosthenurus euryskaphus
Troposodon bluffensis			Simosthenurus gilli
Troposodon bowensis			Simosthenurus maddocki
Troposodon gurar			Simosthenurus newtoni MIS6
Wallabia indra			'Simosthenurus' pales MIS6
			Sminthopsis floravillensis
			Tropsodon minor
			Warendja wakefieldi

Source: After Rich et al., 1996 and Long et al., 2002.

was starting to dry out and many of those late Palaeogene and early Neogene genera disappeared. It was a trend that seems to have brought the faunal population almost to its knees by the late Miocene–early Pliocene, but the adaptive strength of marsupials, however, is no better shown than at this time. They seem to pick themselves up by adopting new body styles, becoming larger, learning how to graze effectively, becoming more mobile and speeding up their design. These changes are the cornerstones of adapting to receding forests, spread grasslands, spreading aridity and the need for moving long distances in response to a more changeable and fickle climate and a more contrasting geography and environment. In particular, this is also when we see the real emergence of the macropods that now dominate our modern marsupial population with possibly the best biological and behavioural design for a continent changing into a desert.

The present fossil record suggests that an extinction event took place during the first half of the Neogene with the passing of smaller and narrowly adapted rainforest species, leaving those that could make the transition (Table 5.5). Figure 5.4 shows that with the exception of possums, gliders and cuscuses (PGC), there is a general stepwise reduction in the numbers of species of macropods, diprotodontids, wombats and koalas, a trend which resulted in a comparatively limited variety of these groups by the late Quaternary (Long et al., 2002). What we also see is the general sinusoidal extinction trend reflecting the climatic and environmental roller coaster of change at the time. The Miocene stands out as a time of adaptive change as one group of animals moves over for another; the sieving out of those that cannot make the right

Table 5.3 Last Known Appearance of Some Australian Tertiary Marsupial Genera

Late Oligocene (n=26)	Early Miocene (n=21)	Mid-Miocene (n=18)	Late Miocene (n=10)	Early Pliocene (n=12)	Mid-Pliocene (n=2)	Late Pliocene (n=3)	Early–Mid-Quaternary (n=14)	Late Quaternary (n=14)
Ankotarinja	Cercartetus	Balbaroo	Alkwertatherium	Archerium	Glaucodon	Dasyroides	Baringa	Diprotodon
Apoktesis	Durudawiri	Barinya	Dorcopsoides	Darcius	Kolopsoides	Euowenia	Bohra	Metasthenurus
Badjicinus	Ektopodon	Dasylurinja	Ganbulanyi	Dorcopsis		Kolopsis	Burungaboodie	Palorchestes
Balungamaya	Ganawamaya	Djaludjangi	Hadronomus	Ischnodon			Congruus	Phascolomys
Burramys	Ganguroo	Djilgaringa	Kolopsis	Jackmahoneya			Euryzygoma	Procoptodon
Chunia	Marlu	Dorcopsis	Mayigriphus	Koobor			Lasiorhinus	'Procoptodon'
Pilkipildra	Namilamadeta	Ekaltadeta	Plaisiodon	Kurrabi			Macropus	Protemnodon
Galanarla	Ngamalacinus	Hypsiprymnodon	Pyramios	Meniscolophus			Metasthenurus	Ramsayia
Gumardee	Nimbacinus	Litokoala	Tyarrpecinus	Milliyowi			Nototherium	Simosthenurus
Ilaria	Nimiokoala	Maximucinus	Wabulacinus	Petauroides			Prionotemnus	'Simosthenurus'
Keeuna	Nowidgee	Muribacinus	Wanburoo	Prionotemnus			Propleopus	Sthenurus
Kuterinja	Paljara	Mutpuracinus		Pseudokoala			Pseudokoala	Thylacoleo
Madakoala	Pildra tertis	Neohelos					Tropsodon	Vombatus
Miralina	Pseudocheirops	Nimbadon					Warendja	Zygomaturus
Muramura	Rhizophascolonus	Nimiokoala						
Nambaroo	Thylacinus	Propalorchestes						
Priscileo	Wabulacinus	Strigocuscus						
Ngapakaldia	Wakiewakie	Thylamcinus						
Palaeopotorous	Wynyardia	Wakaleo						
Perikoala	Yalkaparidon							
Pildra	Yarala							
Pitikantia								
Puria								
Raemeotherium								
Silvabestius								
Wabularoo								

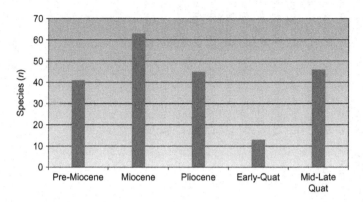

Figure 5.4 Australian Tertiary–Quaternary marsupial extinctions.

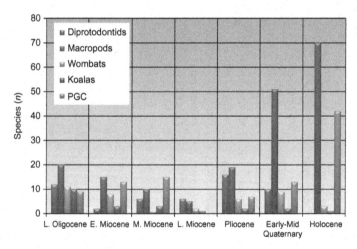

Figure 5.5 Tertiary–Quaternary extinctions among five groups of marsupials. PGC, possums, gliders and cuscuses.

adaptive moves and the ongoing populations of those who can. When the animal groups are broken down we can see a reduction in the number of species among macropods and koalas to the late Miocene (Figure 5.5). Diprotodontids fluctuate at this time but revive in the Pliocene, while the sudden collapse of the tree-dwelling PGC group in the late Miocene is probably a result of rainforest losses of that time. In the Pliocene and Pleistocene, kangaroos spectacularly revive with over 60 species in existence by the Holocene, while diprotodontids gradually become extinct and wombats falter. One interpretation of these trends is that while kangaroos succumb to the drying of the Palaeogene and early Neogene they make the best adaptive moves for the changing conditions from the Pliocene onwards, especially among the long-faced varieties as we shall see. The rise of long-faced Kangaroos in the

Holocene may also have been boosted by the loss of the many short-faced varieties during the late Quaternary. Diprotodontids make a brave effort to combat increasing aridity by body size increase (which probably included metabolic and digestive changes) but that was probably not enough to combat the stress from continued environmental change towards the arid. Koalas also succumb in the early Neogene, probably because of reducing rainforest coverage, but ultimately hang on although losing species as they pass through the increasing aridity of the continent as the Miocene Ice Age strengthened. The PGC group also flourishes till the mid-Miocene, perhaps in response to the loss of other competitive animals that shared their favourite niches which resulted in them being able to broaden their choice. Even so, they also suffer a downturn in the late Miocene, in the same way as all other marsupials, before picking up again and thriving into the Holocene.

There seems little doubt that Australia's early–mid-Neogene extinctions occurred as a response to enviro-climatic changes taking place to different degrees on different parts of the continent. Many of the animals involved had had a long history of rainforest adaptation on a cool, stable continent with plentiful rainfall. Some had become so specialised within their particular rainforest and dense woodland habitats that when even minor environmental changes occurred, as the first faint whispers of continental drying took place, they succumbed, starting a cascade of faunal changeover that continued to the late Miocene. We saw some examples of those very specialised types earlier that had been given odd names to match their odd appearance. As usual, species reductions are not nicely uniform but vary within groups and across space. Smaller varieties are eliminated at this time while the late Miocene holds the key to the new order – bigger and bigger animals. By that time, all groups succumbed to conditions, slumping to their lowest number of varieties, and that was when macropods and diprotodontids in particular began to flourish. These two chose two ways of coping with increasing aridity. Diprotodontids took on a steady growth in size; the more you eat the better chance you have of surviving. Kangaroos also chose size increase to a certain extent and this helped them in their main choice, speed, but that required a balance between the two. The underlying factor here seems to be, as usual, extinction of species that could not adapt to the new enviro-climatic circumstances. The few that remained in the late Miocene were the ancestors of those that would survive the increasing aridity of the continent over the next 5 million years. In the end it was always climate and its environmental consequences that drove the changes in marsupial evolution in the Neogene as it did in other animal populations around the world.

But What About the Others?

That reminds me: before leaving this discussion, we must ask how Australia's Miocene marsupial turnover compares with those occurring elsewhere around the world. It is not the aim of this book to try and explain all faunal extinctions and subsequent trends on every continent during the Tertiary, and, indeed, I will now focus mainly on Australia. But in general terms, Australia's Miocene marsupial transition seems to parallel those happening elsewhere although it is by no means regionally uniform unless seen in the long term (see Figure 3.11). It has long been recognised that an extinction event took place throughout the world in the Middle Miocene as it did in Australia, although that has not been broadly recognised as happening here. Termed the *Middle Miocene Disruption*, it has been suggested that up to 30% of mammal species may have been lost at that time, which coincides with the equally termed *Middle Miocene Climatic Transition*. The link between the two is not disputed but many details associated with the extinctions that took place are still not clear, including the overall impact on animal populations. The climatic events were underpinned by sea level changes and the joining of landmasses as tectonic journeys brought continents together. These actions formed corridors through which migrations occurred, switching animal groups from one place to another and bringing about *apparent extinctions* of genera in some areas only because animals moved away, following familiar habitats as they grew and shrank in the face of the broader environmental changes taking place. We'll see that same process occurring in Southeast Asia in the late Quaternary in Chapter 10. Continental docking also brought competing genera together as well as introducing new predators, which had their own consequences for extinctions and animal dispersal. Volcanic events and bolide impacts also saw local changes to animal populations but these were by no means felt throughout the world. The selective process that resulted from this major climatic shift to a cooler world brought a turnover of genera that had more effect on some continents than others, and there followed a filtering out of those animals unable to adapt to the environmental shifts that were taking place as well as a movement of animals to occupy landscapes and take their niches. Some had become so specialised that they were probably the first to succumb to this process; others disappeared as their particular habitats shrank beyond a size that could support them and/or their individual populations reached an unsustainable level in shrinking ranges.

It was noted above that in Australia there was not only a substantive turnover of genera in the Miocene but also a trend towards large body size that took place towards the end; similar adaptive stories occur elsewhere in the world. Although many mammals had large bodies in the Palaeogene, these seem to have been derived more from nature's experiments on the recently dominant mammals. By the mid-Neogene the trend to grow a bigger body was different; it seems to have come from the necessity to keep up with climatic and environmental changes that became increasingly challenging as the world entered the Pliocene and continued to plunge into an Ice Age. Increasing body size as an adaptation in Australia already had its spectacular equivalent elsewhere, with the development of the horse among other examples. Its adaptation included growing a larger body, jaws and teeth with thicker enamel in order to cope with grazing and the accompanying dental attrition associated with the coarser grasses that were replacing leafy woodlands as the world became colder and drier. Like the diprotodontids, a larger gut was required for food processing and to cope with a greater fodder intake. Unlike diprotodontids, however, longer running legs were necessary for out-manoeuvring and escaping predators on spreading, open savannahs.

What we see in the Miocene extinctions is not just the inability of animals to keep up with changing environments, a lack of adaptive qualities if you like, but the forced reduction and subsequent extinction of genera and species as they literally ran out of their favourite habitats – places that, in some cases, had probably changed too rapidly for them to be able to adaptively keep up. Therefore, the speed of local and regional environmental change was probably a central factor in some extinctions, while others could not be avoided whatever happened. In particular I think of Antarctica *ca.* 12 Ma, beginning to freeze over for the last time. There must have been a day when the last furry animal was still alive, the last mammal (marsupial or otherwise) on the continent .... Did it spend its last days sitting on a rock, huddled in bare branches, hiding under a short, scrubby bush or peering out from under a rock across an increasingly cooling ocean towards lands where other animals, extinct or not, were continuing to thrive – at least they had a chance! All things are comparative. I wonder where that fossil is and what the animal looked like ....

The Origin of the Megafauna

The origin of Australia's fauna that gave rise to the Quaternary megafauna is outlined above but I now want to take a brief look at the transition of animals from the Miocene to the Quaternary. To do that, however, we have to return to the late Oligocene and early Miocene. The Quaternary Ice Ages described in the next chapter were superimposed on a much larger Ice Age that began in the late Miocene and continues today. For the last 12 million years the average world temperature has been falling as we slide deeper and deeper into this mega Ice Age (Figure 3.2). Whatever climate change takes place, hot or cold, natural or man-made, the mega Ice Age in the background continues at an exceptionally slow pace, not measurable in human time scales but rather in hundreds of millennia. World cooling actually began in the later Palaeogene and that trend slowly continued into the Neogene with environments also slowly changing across the world in tempo with the cooling. But 12 million years ago with the beginning of an Ice Age, rainforest retreated, savannas expanded (only to contract later), deserts appeared and grew, seas lowered and ice caps formed at the Poles over several million years. It was a process that began slowly, speeding up somewhat as the Pliocene closed and we entered the Quaternary. Cooling brought ice formation and reduced precipitation rates, grasslands spread and nutrition levels fell as a consequence. Browsing animals began to be disadvantaged by these changes which manifested themselves mainly in the loss of vegetation and the assortment of plant communities. The big drawback was that spreading grasslands contained reduced nutrition, as I have already mentioned, and during the next 8–10 million years our animals underwent the transformation from smaller and medium-sized varieties of the late Miocene to those of the Quaternary that we now call megafauna.

Diprotodontids

The Australian fossil record shows clearly the development of a number of marsupial groups growing into bigger animals from late Miocene to early Pliocene after the

Figure 5.6 *Kolopsis torus*, the sheep-sized diprotodontid browser of the late Miocene, from the Alcoota fossil beds. It is placed against a perspex outline of its original shape.
Source: Reconstruction by P. Murray and photograph by Steve Webb.

onset of the Ice Age. The Miocene initiated the widespread and fundamental change of continual continental drying in Australia. As the world cooled, Australia's leaders in response to these changes were the diprotodontid, quadruped browsers. As they moved from the late Oligocene through the Miocene, they are represented by a number of species of increasing size. The fossil record has the group first appearing in the late Oligocene as the sheep-sized *Silvabestius johnnilandi* and *Bematherium angulum* found at Riversleigh in northwestern Queensland, both of which represent a step in the evolution of the Diprotodontiae that eventually gave rise to the largest marsupial ever, *Diprotodon*. Two other founding members of this group are the similar-sized *Pitikantia dailyi* and *Ngapakaldia* sp., both of which lived in what is now the Tirari Desert east of Lake Eyre and 2000 km southwest of Riversleigh. They show the wide geographic dispersal of these forest-dwelling animals and obviously the presence of a forested environment in these areas which probably consisted of large patches. Another diprotodontid of the Tirari Desert found at Lake Palankarinna is the early Pliocene *Meniscolophus mawsoni*, an animal possibly related to two other diprotodontid ancestors, *Pyramios* and *Euowenia*. Further diprotodontid genera emerged during the Miocene, such as *Nimbadon, Neohelos, Kolopsis* and *Plaisiodon*, who were slightly larger than their sheep-sized earlier ancestors but, like them, were hard or coarse vegetation browsers (Figure 5.6). Michael Archer has informed me, however, that *Nimbadon* was probably a tree climber in the vein of sloth, using its long claws to scale the rainforests for at least some of its food. They were joined by three other browsers, *Pyramios, Alkwertatherium* and *Meniscolophus*, all of which lived in the centre of the continent east of Alice Springs (Alcoota) at a time when much lusher vegetation prevailed than they would find today. This was another example of the lush vegetation that grew in the middle of the continent at that time. These environments were also favoured by the earliest zygomaturine diprotodontids such as the late Oligocene *Sylvabestius* through to the late Miocene *Kolopsis torus*. These were rainforest-dwelling browsers but their ancestors, like the Quaternary cow-sized descendent *Zygomaturus trilobus*, favoured wet sclerophyll forests or

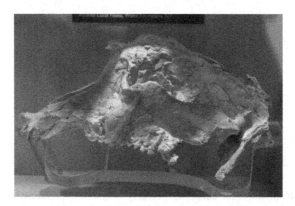

Figure 5.7 Cranium of *Palaorchestes painei* from Alcoota, NT. Note the slim nasal area that contrasts with the same region on other diprotodontid crania. There are places for attachment of muscles that probably supported a small tapir-like trunk.
Source: Photograph by Steve Webb.

riverine gallery forests with no evidence they were ever happy grazers living in open savannahs or had adapted in any way to the continental drying. *Euowenia grata* was another derived zygomaturine that lived in the Pliocene and possibly early Pleistocene; it was cow-sized, however, and much larger than earlier diprotodontids. As far as we know it lived only in southeast Queensland and was another animal favouring wet sclerophyll habitats. It is obvious that body size increase took place among diprotodontids during the Neogene and that they were adapting to drier landscapes and open forests conditions, but this was not enough because species were also slowly disappearing. By the mid-Quaternary the diprotodontids had also produced several palorchestid diprotodontids, mentioned earlier in the chapter (Figure 5.7). They were essentially large marsupial termite eaters with short trunks and large claws probably for ripping bark, similar to the South American *Megatherium*. They had restricted movement at the elbow possibly to strengthen or brace their arms during removal of bark and general fossicking in concrete-like termite mounds which can be quite difficult to break open. The largest member of this group was *Palaorchestes azeal*, weighing around 500 kg, that seems to have lasted till the end of the mid-Quaternary before going extinct, the last of its genera.

Macropods

Kangaroos evolved from possums during the Early Miocene and moved out of forests later as the continent dried. They also made important skeletal changes not just in their familiar overall shape but in small things like ankle joints. These transformed their ankles from one the was flexible (vital moving along branches) to one that was locked by a specially shaped stepped cuboid bone that kept their ankles rigid during that famous bouncing macropods do. Among the macropods, the late Miocene *Hadronomus puckridgi* of central Australia was a browsing animal, slightly larger than any of the modern kangaroos and the forerunner of the more arid-adapted

Figure 5.8 A comparison of the mandibles of a grey kangaroo (*M. robustus*) (front), one of the biggest modern kangaroos and the short-faced giant *Procoptodon goliah*. Note the large bicuspid and crenulated dental crowns of *Procoptodon* as well as the massive bony jaw structure highlighted by the thickness of the mandibular symphysis (arrow). Both these were required to process and supply large amounts of leafy fodder for this very large browsing kangaroo.
Source: Photograph by Steve Webb.

sthenurine kangaroos, a genus that contributed more than its share to the number of species included in the later Australian megafauna species list (see Table 7.1). The sthenurines were a group of short-faced kangaroos very distinct from the longer faced *Macropus* groups everyone is familiar with today. They represent a group that seemed to keep pace with a changing climate and environment through a series of adaptive morphological stages. While *Diprotodon* developed to the size of a rhino with a maximum male weight of 3.5 tonnes, the macropods tried to do the same, producing a kangaroo over 3 m tall, namely *Procoptodon goliah* (Figure 5.8). This 250-kg monster was half the weight of the full-grown *Zygomaturus* and it displayed bodily proportions to back it up, but it seems that kangaroos had a size limitation. As one of the sthenurines, *Procoptodon* had a short face that differed from modern kangaroos but its body design was almost exactly like that of the modern red kangaroo (*Macropus rufus*). The exception was its robust cranial form and jaws which were both deep and had a much great lateral thickness. They supported a row of very large teeth with crenulated surfaces most suitable for crushing leaves, berries and possibly nutty fruits. It seems doubtful, however, that the anatomical similarity between the *Procoptodon* post-cranial skeleton and that of the modern red kangaroo, particularly in the shape of their hind limbs, would have allowed the giant to hop or move at the same speed for as long as the red kangaroo, given the weight it had to shift. For macropods the balance between size and speed must have been reached in *Procoptodon*, which had surely the largest body that could be lifted with a hop. Well, we presume it hopped because its leg bones are almost exact replicas of those of today's red kangaroo, only much larger and more robustly built. We have little idea of how fast *Procoptodon* might have travelled, though; perhaps it took large but slower and fewer hops. Hopping such a large body along is a far different proposition than that faced by the biggest red kangaroo, one-quarter of the size. But even if it could be lifted in a hop, landing is another problem entirely, with a vast amount of weight put on the legs and weight-bearing joints. Such behaviour would have imposed enormous stresses on legs and hips but muscle mass and tendon strength and size may have alleviated this problem. But, again, there is a trade-off between muscle mass and the weight that comes with it and the ability of that mass to move the creature up and forward. Whatever their behaviour in this regard they would not easily be missed crossing the landscape and would have been very audible as a mob of them approached. I have always wondered how big a *Procoptodon* 'joey' was before its mother tipped it out of the pouch – and, of course, what size was the pouch? It is not unreasonable to assume that the young animal must have reached the size of a wallaby before leaving the shelter of its mother's 'handbag'.

Almost all sthenurine genera produced a range of different-sized species that were large to very large animals, ranging from 50 to 150 kg. Such a range is not particularly impressive on a world megafauna scale but for Australia with its generally small fauna, they are giants, although both the largest modern grey kangaroos (*Macropus giganteus* and *M. fuliginosus*) and red kangaroo (*M. rufus*) can reach weights of up to 85 kg – well within the minimum megafauna size of 50 kg used in Chapter 7. Put another way, there were six species of 'megafauna' kangaroos that were smaller than the large modern kangaroos. So, this was part of the scene in which Australia's marsupials were making their biological way towards surviving in a drying world. Morphological change did not take place overnight, but its main aim was achieving a size suitable to combat changing climatic and environmental conditions over millions of years, although the rate and extent of size increase varied among different genera and species. But was this enough in the face of what there was to come, namely increasing severity and depth of rapidly cycling and erratic Quaternary Ice Ages? How far could such adaptation take them, or would it lead them up a blind alley? How quick could they adapt to an increasingly disruptive and cooling climatic system that was becoming stronger over a shortening time frame? And even if adaptation could take place quickly enough, was it the right adaptation for long-term climate fluctuation?

Setting the Stage for the Quaternary

Throughout the Neogene the Australian environment totally changed, so much so that it probably had no equal in the Tertiary world, with the exception of the complete glaciation of Antarctica. Initially, the environmental changeover towards the arid was a slow process that, while there had been some false starts, began properly in the late Miocene. The Pliocene was a very short epoch compared to those preceding it, but it was during this time that there was acceleration in world cooling. For all its diminutive size in comparison with other epochs, however, it was long enough to produce humans! On top of that splendid achievement, its 3.2-million-year span also saw the most rapid part of Australia's environmental shift yet towards becoming the second most arid continent on Earth after Antarctica. This amounted to a vast shift in environmental construct. It changed forests to desert, drastically reducing the amount, variety and types of available niches for animals, and began to push most animal varieties out of the centre of the continent. It reduced humidity and increased temperatures, greatly reducing the availability of surface water and bringing about net water loss in some areas; desiccation now became part of the Australian ecosystem. All these placed enormous selective pressures on animals. It was a time of sink or swim, adapt or perish, but on a much smaller time scale than ever before: things were happening fast. The Pliocene opened with a virtual halt to the genera losses that had begun in the Miocene: now there was the emergence of new genera.

Aridity, however, did not necessarily spread uniformly across the continent and weather patterns around the coastlines remained varied. Large amounts of rain still continued to fall in many areas, particularly along the north, east and southeast coasts of Australia. One stark reminder of this rainfall was the forming of the 55,000-km^2 Lake Bungunnia, megalake that lay across the South Australia–Victorian border, *ca.* 2.4 million years ago. Its water was derived from rivers entering from the east whose basins lay in the Snowy Mountains, suggesting very high rainfall/snow melt patterns across the far southeast at that time. McLaren and Wallace (2010), who have studied this lake, have determined that it dried completely by 1.2 Ma, marking the onset of serious aridity in its catchment around that time with an increasing amplitude of arid climatic cycles. So even one of the wettest parts of southern Australia had become much drier by that time. That doubled the 700,000 years previously assumed to have been the time when aridity first began in the region. But what did this do to animal populations moving out of a drying centre?

Australia's general increasing aridity marched hand in hand with the drop in world temperatures, translated here into the gradual formation of a semi-arid and, later, arid continental centre that would eventually take up between 75% and 90% of the landmass during the Quaternary glacial cycling. With the reduction of rainforests and the spread of grasslands, several families of marsupials began to disappear while wombats and macropods spread. It is not clear where wombats originated in terms of their preferred environment, but their ancestors must have begun evolving in forested or even rainforest areas. They were possibly one of the first marsupials to feel the winds of environmental change, although they also developed a larger body form along with the koala. Australia's macropod hoppers now represent our greatest variety of marsupials but their origins are comparatively new. The earliest evidence for them appears in the mid-Pliocene *ca.* 4.5 Ma, while many, like the Agile (*Macropus agilis*) and Swamp (*Wallabia bicolor*) wallabies as well as the Common Wallaroo or Euro (*M. robustus*), did not appear till the early Quaternary (Archer et al., 1991). This was now a time for grazers; browsers retreated with their woodlands, shrub lands and other forested areas and their numbers were reduced accordingly. That is not to forget those macropods like forest wallabies (*Thylogale* sp.) in Papua New Guinea and the tree-climbing kangaroos (*Dendrolagus* sp. and *Dorcopsis* sp.) which still preferred a rainforest or heavily wooded habitat and had withdrawn to those areas that still had rainforest and continued to survive in them.

Was it possible that growing larger bodies would eventually put those animals that did so in a position that when the enviro-climatic squeeze was increased, they could not adapt further. Perhaps those same bodily proportions that had worked before would now actually mitigate against their long-term survival? The effort to grow larger in order to survive worked for many creatures, although a number of species disappeared attempting this. The strategy was now climaxing so that by the end of the Pliocene the stage was set for the emergence of the largest marsupials ever to live on the continent, although they comprised comparatively few genera. There was, however, a limit to growth as an adaptive strategy among quadruped diprotodontids and this culminated with the rhino-sized *Diprotodon optatum* as well as other smaller varieties such as *Zygomaturus trilobus* and *Euryzygoma dunense* that rarely reached even half the size of *Diprotodon*. Perhaps *Procoptodon* was an animal that reached its maximum size and could not adapt further. Reaching the limit of bodily proportions may stand as an example of the problems facing other Australian megafauna in the Quaternary. An increased body size had its limitation for most animals, however, and this form of adaptation to increasing aridity may not have been enough to withstand what was to come. Without termination of the headlong plunge of the world into cooler stages, the adaptation of body size increase may have taken some species up a blind alley. In terms of climate and environmental change, worse was approaching and size increase may have had deleterious outcomes for some animals against the new extremes. While Australia may have experienced radical environmental change during the Tertiary, those changes had only laid the foundations for what was to come next in the Quaternary: the Ice Ages. The environmental changes that were to come would render a large part of the continent generally uninhabitable for larger varieties of animal. There began a slow change in the general distribution and demographic profile of those species as they slowly drifted coastward, effectively depopulating large parts of the continent. In effect, aridification effectively reduced the number of animals living on the continent. This would also select for the fittest of the fittest but not before another extinction took place.

6 Australia and the Quaternary Ice Ages

Some years ago I wrote a description of possible late Quaternary environmental succession and biological consequences in Central Australia (Webb, 2006). That work was based on data assessed several years prior to publication as a book which was a lengthy process. More data was accumulating even as I wrote, and that is now well over a decade old. Work at Lake Eyre and other lakes since has changed the neat interpretations I made back then. It is not the first time research has leapt before it took a longer look, but it happens in the imperative to publish, with universities eager to accumulate financial points gained from research output and publications. There are dangers in this when field work is reduced to a few weeks a year and data is taking even longer to accumulate than you suspected. Sometimes the data seems to make sense even though there are gaps in it, and that can lead to academic premature ejaculation. However, there are occasions when, without building such models and putting them through the stark reality of publication, the next step forward is difficult to see or find. Publishing limited data is, however, a useful tool for laying foundations, showing possibilities for ways forward and building models so that ideas can take shape even if some corners of it are somewhat prefabricated or standing at odd angles. Standing back and looking at the structure can then show much more clearly that the roof does not fit properly or that the data from the following years makes a wall taller than it was meant to be and all the other walls now need raising or altering in some manner. The house construction analogy is not a brilliant one but I am convinced that one should not be condemned for building such 'houses'. Indeed, they not only need construction but also need to be put out there so that directions and outcomes can be more easily understood. In some cases the 'houses' will merely need to be altered or refined; in others, they will need to be rebuilt completely perhaps using the same material, or by adding new and different data. This also gives others a chance to view the structure and see how to make improvements. The dialectic process of pushing forward then continues, where adding and subtracting knowledge gradually reinforces, extends and modifies our ideas. It seems to me that building a rather flimsy house is not the problem; rather, the problem is not reinforcing or modifying it at a later date if new and better building materials arrive.

My research since the beginning of the twenty-first century has shown me how much more complex the whole picture of Quaternary glacial and climatic cycling was and what that meant for Australia's environment and the biogeography of its fauna, particularly the megafauna. One of the weaknesses of the previous interpretation was not seeing that fact as clearly as it can be viewed now and appreciating more fully the consequences of widespread environmental change associated with glacial cycling. There is nothing as clear as hindsight, so they say. Speaking of

Corridors to Extinction and the Australian Megafauna. DOI: http://dx.doi.org/10.1016/B978-0-12-407790-4.00006-9

biogeography, another problem with those earlier interpretations, was that biogeographic principles had not been applied to the megafauna extinction issue in the face of the extreme and rapidly changing circumstances fundamental to glacial cycles. But I am the only one to have missed this opportunity. They are principles that apply to all living things, not least large, complex animals. Such principles had not been applied in this case because the dominant arguments surrounding Australia's megafauna extinctions emphasised an anthropogenic cause for them. One argument that seemed to support that was the assumption that these animals had survived previous Ice Ages, so why did they disappear during the last one? That argument, however, avoided asking the question, 'What did previous Ice Ages do to them?' Obviously nothing, according to the argument.

Megafauna might be regarded as special only because they have gone, but that is where 'special' stops. They behaved, moved and responded to their surroundings just like any other animals. They should never be separated from the exacting pressures of the natural world or the principles that apply to other animals just because they tend to be regarded as a group of *special* biological entities, which of course they were not. Australia's megafauna have not been studied in terms of how they may have responded, not to the last Ice Age or the one before, but to each Quaternary Ice Age and the cumulative effects of that climatic see-saw. How did this series of over 100 extreme climatic cycles change the biogeographic parameters within which the Australian megafauna lived? How did they cope with a climatic switchback of reversals – some abrupt, others slow – from warm to deep cold and back again? Thresholds were reached that threw environmental and, thus, biogeographic switches radically changing environments to the detriment of this species or that. There were also major climate shifts embedded within major cycles. They could be just as large but took place over a shorter time period. Additionally, there were the smaller reversals similar to those we are in fear of today with our modern climate change predictions of 1°C or 2°C. In light of these events it is particularly important to view the actual conditions megafauna went through during the Quaternary. Therefore, I want to track and examine glacial climatic change in some detail to present a qualitative and quantitative view of the Quaternary glacial cycles in order to assess how radically they changed Australia and, in turn, what the consequences of those changes might have meant for megafauna.

In this book we have looked at the *very ancient* but we are about to look at the comparatively *very recent*, and that is the book's central aim. We have looked, albeit briefly, at the extinction process in the very distant past. That included the passing of animal species in the light of climatic and environmental change. Those events have benchmarked the natural movement of species turnover through a pattern of continuous extinctions that took place even without humanity's help. We have seen how these changes caused species modification (such as the horse) in response to environmental change brought about by climate change as the Earth became drier. So it is on this type of process that I now want to focus. Climate change and the subsequent environmental changes it brings have promoted certain species over others many times over because of inability of some to adapt, or to adapt fast enough or in the right direction to changing conditions. But it is not easy to investigate the past and the farther back in time we go the harder it becomes. Methods used to carry out such investigations vary and change. As methods become obsolete so others are improved or discovered and there is always something new being tried. Sometimes it works and sometimes it has only limited value or applies only to one particular landscape or situation. Geochronological dating is peppered with different methods that apply to different sediments and are useful only for specific time frames of the past.

There is, however, one method of investigating the past that will never change: the researcher tramping over kilometres of country in heat and cold, snow and ice, rain or dust, using his or her eyes and accumulated experience, drawing on an innate enthusiasm and instinct to spot the particular data, clues or fossils he or she is searching for. Sometimes it is something far removed from any of the desired objects that will provide a clue to his or her individual goals. The first thing that you have to do in this sort of field work is to 'get your eye in', a term that refers to the focusing of the field worker's eyes to the landscape and sediments. That may take several days after beginning field work, and then clue spotting can really begin. Searching for fossil bone, particularly small fragments and odd teeth, is very much dependent on getting your eye in: several days are required before you are really able to sort pieces of bone from other natural junk bits and pieces such as rock fragments, sticks, lumps of sediment and other debris strewn across your particular landscape. It's like finding the corners and edges of a jigsaw puzzle, after which the long process of understanding the picture begins to unfold, slowly. Instead of being able to monitor an oncoming car, subliminally recognise a traffic light that has gone red, avoid the person hurtling towards you head down submerged in their mobile phone, or spot speed cameras and red tail lights, you are now in a remote field work place constantly looking at the ground, at the edges of ancient, dry river beds, along the edges of rock formations etc. It is quiet, almost silent except for a breeze rustling vegetation or a bird peeping somewhere far or near. You feel slightly lost, alien almost, but you know you will settle in, you always do, and so will your eyes; you have done it before. More quickly than we might rightly expect, those eyes become programmed for other things than traffic and people, computer screens and the dross of everyday living in cities and working in universities. The quiet brings a peace and very gradually your hearing becomes more attuned to the few sounds around you and anything odd is immediately flagged: that bird becomes several with different calls, and is it worth a look? Definitely, because it could mean another tick or new sighting in your bird book!

I am often asked by students as well as lay people, Where do I get my information? How do I know such things occurred and when they occurred? How do we know these things are true? Without gathering empirical data and gradually piecing it together we would not know, but it does not come from one source alone. The formula is that a lot of disciplines overlap or are locked together at various points on their boundaries just like chemistry diagrams, in order to feed off the particular expertise that each offers. This is often the only way to achieve results, particularly in the study of the physical Earth. The vast complexity of our planet, whether it involves the lithosphere, biosphere, atmosphere, hydrosphere or all of them (which it normally does), requires patience and time. Those that have anything to do with investigating those Earth systems will almost certainly at one time or another have felt dwarfed by the task they have set themselves; indeed, that is healthy. I remember vividly the times I have stared into a field work camp fire and wistfully said, *'The more we learn the less we understand'*, and I know others standing around a laboratory bench or sitting having a coffee in the laboratory tea room will have said the same. Frustratingly, as I close in on the end of my career, I know I won't see the answers to many questions that still haunt me. When I was young there was always hope, usually irrepressible hope, that next year, the year after or in 5 years or a decade they could be revealed.

The problem with studying the Earth, however, is things are usually buried. Yes, clues do stick out of the ground, normally in a very small way. That is why you have to keep your head down looking, always looking. The odds of finding those clues can border on the astronomical, although you always hope they are not going to be that high. The trick is to optimize the odds, something usually done by accumulating field work experience and avoiding all the things that can pop up that look like the things you are looking for but are not – if that make sense. Also, there is often frustration in field work when, just as you think you have done enough to say what you want to, examination of the results back at the university plunges you into the realisation that you have to go back to the field to check or try something else or look at something that you now know you should have looked at before you left the field but because you so needed a shower and that non-rationed beer you bolted .... The real frustration is that normally you cannot just do that, particularly if you work in remote areas. You have to wait for next year's field work season (or more grant money). That return may then be frustrated further by bad weather, other commitments or duties or the inability to gather together the equipment or, even worse, the personnel that you need to carry out a new field season. That can go on over several years; hence the length of study that is normally required to understand our planet and its workings or even gather some basic data to write up a half-decent paper. Generally the best and most complete information lies buried in the most ancient of geological repositories for the very reason that it has been locked away and often undisturbed since the time it was buried. One central investigative method that informs us about both very old and recent times is, therefore, worth mentioning at this point: *drilling*.

Drilling for the Foundations

For decades scientists from many disciplines and many nations have worked together on *drilling* projects of various kinds in different parts of the world and this work has turned out to be something very special and exciting. *'Drilling* is *not* exciting', I hear you say. Perhaps you have done some? But this drilling *is* exciting because it has reached into those hidden repositories of data that have been locked away deep below the Earth's surface that humans would otherwise never see, and exciting because it shows us more and more about how the world works. Drilling penetrates old environments, actually sampling landscapes long gone, tens of thousands and millions of years in the past. Those caches of information hold the key to understanding the history of the Earth because they place samples of the past before our eyes. The challenges that drilling and sample extraction have faced over the last 40 years have helped hone its methods; increased sophistication of methods and techniques, and of drill design itself. All that has allowed greater penetration deep into the Earth's crust and the Kola borehole in northwestern Russia is an extreme example of possibilities. It has reached almost 13 km down into the 2500 Ma Archaean Baltic crust, but even this is just a pinprick in the Earth's crust, which can reach a thickness of 200 km in some places. However, heat becomes a problem when drilling to great depths: temperatures in the bottom of the Kola Hole reach 300°C, preventing further drilling because the drill does not function in such conditions.

Other drilling work includes a number of projects centred on the world's ocean beds. Oceanic drilling has been carried out by the Deep Sea Drilling Program (DSDP) and the Ocean Drilling Program (ODP), the latter using specially fitted-out vessels like the *JOIDES Resolution*, built to carry a very large cargo of equipment and personnel across the world's oceans. During nearly 25 years of the ODP, the project has travelled almost 600,000 km taking drill cores from 669 sites scattered from the Arctic Ocean to the Antarctic's Weddell Sea in waters up to 6 km deep. Essentially the work examined stratified sea bed deposits in order to provide a better understanding of Earth's history. But a variety of results came from the project, including an improved understanding of plate tectonics and continental drift over the past 120 Ma. It has also shown that the world's oldest ocean crust is younger than the Jurassic, older crust having been long recycled down into the Earth's natural fiery furnaces by the forces of subduction. A clearer picture of climatic variability back to the Cretaceous is now available, including solid evidence of the 'hot house' Eocene Optimum. The timing of the opening between Australia and Antarctica and the advent of the Circumpolar Currents that ended the Optimum has also come from the work, as has a greater understanding of the role of the Himalayan uplift on planetary cooling.

Ocean drilling samples contain the skeletons or shells of tiny bottom-dwelling foraminifera which come in a variety of exotic and elegant designs, but what is important is that those shells contain two oxygen isotopes: $\delta^{18}O$ and $\delta^{16}O$. Atmospheric oxygen contains both isotopes in the proportion 99.76% for $\delta^{16}O$ and 0.24% for $\delta^{18}O$. Atmospheric water vapour is enriched in $\delta^{16}O$ relative to sea water so when ice accumulates on land, the $\delta^{18}O$ component in sea water increases at the minute rate of 0.7 parts per thousand per degree of mean ocean temperature. The amount contained in foramina shells also changes as their uptake rates vary with ocean temperature change. Thus the presence of foraminifera in well-dated deep sea cores can be used to reconstruct climatic change over hundreds of thousands or even millions of years, particularly the timing of glacial cycling. All these findings and more show the diversity of data retrieved during the program that impinges on much of the biological workings of the planet – all from studying the ocean floor. But even more importantly, they have been crucial for tracing the story of this book. Perhaps even closer to our hearts are results from the ODP that showed extreme drought in Africa between one and two million years ago that may have given impetus for our early ancestor, *Homo erectus*, to leave the continent and begin its migrations across the world. It is a shame that such work almost never hits the headlines or emerges in everyday news bulletins. But it does highlight the dogged nature of long-term science in gradually accumulating the data with which we can cement together the scientific sensations that do occasionally manage to appear in the media and that teach us about the planet we live on.

The ODP project, as well as many other similar scientific endeavours, is the broadest of empirical data gathering. Such work provides not only the cement but the very foundations for that house I began building earlier. More often than not, it takes a long time to lay those foundations let alone the many bricks that will be piled on top, something else that the wider public often does not appreciate. One or two bricks do not make a house but media reporting often tries to satisfy the public's ravenous appetite to know *the* answer to every scientific puzzle *now*. Thus random 'sexy' discoveries are heralded as just that, and voila! The house is built! Almost always, however, this is not the case. The scientific process of accumulating data has no real end; rather, it is an ongoing process that is infinite and this is as much a frustration for the scientific community as it is for the public. However, that's the way of science and so is the fun that most scientists get from pursuing their particular goals. Many of those in charge of grants and other funding bodies would question whether an investigation should go on for 5 years, let alone the 25 years of the ODP work. It is inconceivable that a proposal for such a long project should 'get up' in the granting world, at least in Australia. Granting bodies want results – *yesterday* – and generally require that much should already have been achieved before the project is given the go-ahead. It almost seems as though the answers are required in order to obtain a grant for getting them! But that is what happens if bureaucrats and administrators, rather than scientists, are in charge of science.

Bygone Bubbles

Ocean drilling has largely told us about the ancient past, but what about the recent past? That brings us to a second sort of *drilling* and the one most relevant to the rest of this book: *ice drilling* or *coring*. This type of drilling is different again but also very exciting because it has provided an invaluable stream of data that has shown us a lot about the Quaternary Ice Ages which placed many signposts along the road of human evolution. It has also provided a template on which to place arguments associated with animal extinctions as well as allowing us to follow the growth and spread of humanity from Africa across the world. So ice drilling sets the scene for the next segment of this book, which is now focused on the Age of Ice Ages, the Quaternary. The key is that ice cores have shown us climate change over hundreds of thousands of years. Central to the findings is that the Quaternary world rode a climatic rollercoaster that brought with it massive environmental and ecological shifts that, in turn, drove the biogeography of animals during that time. Those that lived in Australia were no exception.

Like the ODP work, ice-core research has also been carried out over the last 25 years and more, although in shorter bursts and involving a number of different projects. It has also been more confined, to Greenland and Antarctica. What it has done, in particular, is greatly enhanced our understanding of the world's climate during the last half of the Quaternary. Ice coring cannot delve deep into the past because of the limitations imposed by ice formation. At best it can only go back *ca.* 12 Ma when the world began its present ice-up. Buried ice in the Beacon Valleys of the Transantarctic Mountains dates back to *ca.* 8 Ma but that is extremely rare because of the constant recycling of ice to water and back again. Also, basal melting under ice sheets (those suitable for producing a continuous record) are age restricted to between one and two million years maximum. But even this comparatively slim time depth can provide a set of detailed data from which we can assemble excellent profiles of changes in greenhouse gases over time and, thus, a look at past world climate. Indeed, it is the detail which is important here. The depth of such records is rarely available from any other source, although occasionally long speleothem records (secondary mineral deposits in stalagmites and stalactites) from caves are very useful as climatic indicators. One of these has been found in the tectonically formed Devil's Hole cave in Nevada which spans 568–50 ka, adding valuable confirmatory or primary data to the assembling picture of palaeoclimatology, particularly that of North America. The Devil's Hole data has also confirmed the findings of the Vostok ice-core data from Antarctica, although its data consisted of $\delta^{18}O$ records.

Measuring 2,850,000 km^2, the Greenland ice sheet is second only to Antarctica as the world's largest ice mass. If it melted completely, sea levels would rise over 7 m. Forty years ago it was decided that this would be an ideal place to begin to gather data about past world climate. That story would be gathered from ice cores drilled into Greenland's Ice Sheet and the minute bubbles of past atmosphere that were trapped inside the ice cores. Also, resources were concentrated in the Northern Hemisphere, so where better to begin than there? Sites were established in northwestern and central regions of Greenland high on the summit of its ice sheet. Prominent among them are *Camp Century*, the *North Greenland Ice Project* (NGRIP), the *Greenland Ice-core Project* (GRIP) and the *Greenland Ice Sheet Project* (GISP), among others. These investigations consisted of teams of international researchers and between them they drilled to depths of 2–3000 m at which point they reached bedrock or ice folding, which distorts the core sequence. These were great achievements but ice sheet thickness limited the time depth that could be achieved. The work focused on world climate history and the relationship between the Icelandic Low and Azores High, two weather systems which together make up the North Atlantic Oscillation (NAO), which funnels large anti-cyclonic fronts from the Atlantic into Europe, and its movement over time. But, perhaps more importantly, the Greenland work produced a good climate record for the Arctic that reached back to the penultimate interglacial almost 130,000 years ago. From this, details emerged of substantial climate switches that occurred within the last Ice Age, particularly identification of the smaller *Dansgaard–Oeschger* (D–O) cycles or small- to medium-sized interstadials of warmer climate that occur during glacials.

Named after their discoverers, the D–O cycles are of particular interest because these events show how abrupt climate change can be, moving from very cold to warm conditions and vice versa within decades or even a few years. I will come back to them later.

But how did this work reconstruct past climate? Examination of the atmospheric gases contained in air bubbles within ice cores provides a key to palaeoclimatic reconstruction. Besides their significant time depth, ice cores produce a climate record for all that time with results showing atmospheric trace–gas composition, including a record of changes in carbon dioxide (CO_2), methane (CH_4), dust, atmospheric composition and terrestrial and oceanic temperature through glacial and interglacial cycles. From those, the severity and extent of past climate changes emerges and that data provides information for reconstructing terrestrial environmental responses to glacial cycling and, in turn, changes to animal habitats. The limited length of the Greenland ice sheet record, however, has not prevented us going back much further with a palaeoclimate record.

Antarctica has been another focus of ice coring and of course its placement at the opposite end of the world from Greenland provides a nice comparative juxtaposition for data from both hemispheres. Nine sites were chosen for ice coring there. Prominent among them have been the Japanese Dome Fuji, the Russian Vostok Base and the *European Project for Ice Coring in Antarctica* (EPICA) at Dome C base in Eastern Antarctica (Figure 6.1). The EPICA project is carried out by a consortium of 10 European countries (Belgium, Denmark, Germany, Great Britain, France, Italy, the Netherlands, Norway, Sweden and Switzerland) coordinated by the European Science Foundation (ESF). EPICA's goal was to obtain two ice cores, extending to Antarctica's bedrock, and drilling was completed in December 2004. At present, only the upper 3000 m of the 3260 m ice core have been analysed. Glaciologists estimate the climate history preserved, in the even older ice, to span 900,000 years.

In 1998 the Vostok ice core reached a physical depth of just over 3600 km and a time depth of 420,000 years. As in Greenland, the key to the Vostok work was palaeoatmospheric reconstruction assembled through examination of the gaseous contents tiny air bubbles trapped in the ice. This time the bubbles contained air not from one Ice Age but almost five. Vostok drilling was stopped 120 km above the deep glacial Lake Vostok for fear of contaminating the lake. A method has now been worked out to sample the lake without contaminating it. The brilliant and no doubt arduous Vostok work marked an enormous step forward in our understanding of atmospheric and climate change spanning almost half a million years. It was able to show details of world temperature changes taking place during glacial and interglacial periods, of the way in which temperature fluctuated regularly and sometimes erratically and how Ice Ages began and how quickly they ended. Results enabled those interested in Quaternary studies to make more sense of the world during such vast climatic change. New interpretations began to flow regarding the effects of Ice Ages and the nature and length of both glacial and interglacial periods that made up the wild world temperature cycling of the Quaternary. One specific finding was the sudden temperature drops to Ice Age levels (below $-4.0°C$) that took place during warmer interglacial periods only to just as suddenly and quickly reverse back to normal interglacial

Figure 6.1 Principal ice-core drilling sites in Antarctica.
Source: After Fischer in EPICA (2006) Supplement.

temperatures. The work shows that there had been many such reversals over the last five glacial cycles.

Results obtained by the later EPICA drilling program, however, exceeded the depth reached during the Vostok work. EPICA drilled a deeper core in a different part of Antarctica further to the east and, therefore, almost doubled the Vostok record by pushing back the time to almost 800,000 years. As remarkable as that sounds, it is even more spectacular when we recognise that that time bite spans 10 Ice Ages. The EPICA results did not really overshadow those of Vostok, however; rather, they complemented one another because their results could be compared for at least the first 420,000 years, a very long record. That was important because their individual findings were shown to basically support each other in the timing and length of past glacial–interglacial cycles as well as the amplitude or strength of temperature change within each of five cycles. What that amounted to was a good record of climatic shifts associated with five complete cycles – in other words, a basic history of world climate for half a million years, and that was a fundamental reconstruction for those interested in the Quaternary. Further, the similarity between the two data sets strongly suggests that the EPICA results that continue back another 350,000 years to *ca.* 780,000 years could probably be accepted as a true indication of climatic variation back to that time because of the verification achieved for the first half of both cores.

With such a long EPICA record it was also possible to see details of the long-term picture of glacial events during the Quaternary: how each cycle changed some-what from the previous one and how individual glacials and interglacials changed from cycle to cycle in terms of length and amplitude and, indeed, severity. Individual cycle details were also available for the first time, showing how deep fluctuations and swings in temperature had taken place within both glacials and interglacials. Interglacials were interrupted by sudden cooling (a stadial) and glacials could have an equally abrupt rise in world temperature terminating a glaciation extremely quickly and returning the world to an interstadial or even an interglacial point, albeit some-times very briefly. It is worth remembering that stadials are usually contained within an interglacial. These results have been strengthened by those from the ODP work, from speleothem studies and from loess records derived from studying the buildup of loams and other dust sediments in places like China. I do not want to detail that here but those wishing to follow up on this data, particularly the ice-core drilling research, are referred to a number of authors including Jouzel et al. (1987, 1993, 1996), Winograd et al. (1992), Dansgaard et al. (1993), Petit et al. (1999), Watanabe et al. (2003), Anderson et al. (2008), EPICA Community Members (2004), Lisiecki and Reymo (2005), Lambert et al. (2008), Loulergue et al. (2008) and the Ocean Drilling Program, *Final Technical Report* (2008). The collective results of these studies have taken enormous amounts of time, energy and money to accumulate but they show broad agreement in terms of mid-late Quaternary glacial cycle structure.

I now want to apply these findings to Australia, particularly because of the prox-imity of our last Gondwana buddy, Antarctic, to Australia. The extended EPICA record as well as the Vostock work now play an important part in the rest of this book. I will use the data to give some structure to Australia's glacial cycles and how they changed the environment. The book now changes somewhat to one in which a fairly detailed picture of those changes and their consequences for our fauna is built up. To do that requires more referencing for the data cited and the conclusions drawn. This is essential because the object now is to build a case for the climatic vagaries of the Quaternary as a stressor on Australia's megafauna. Rather than seem-ingly pulling data or supporting argument out of thin air, it is now necessary to build the case using a wide range of research by many authors, although I use only the most salient of these in whose own references the deeper background data can be pursued by those wishing to do so.

Ice-Core Data, Glacial Cycle Structure and Climate Switches

The best place to start is the outstanding feature on graphs of world temperature over the last five million years and the tight saw-tooth pattern of rapid fluctuation that becomes more exaggerated as we enter and proceed through the Quaternary (Figure 4.4). A feature of Figure 4.4 is the dotted line which emphasises the strength of the downward trend from *ca.* 3 Ma. Before that time the trend was a gradual cooling of world temperatures of around 2.0°C, but after that the decline steepens. From three to one million years world temperatures drop even more steeply by just under 4.0°C. Clearly there was a continuing downward trend in global cooling during the Pliocene and Quaternary which began to accelerate three million years ago. Another feature is that the amplitude of global temperature fluctuations becomes more prominent over time, with deeper emphasis on cool cycles, before briefly returning to previous warmer temperatures. After one million years ago, cycle amplitude becomes larger with greater swings between warmer and colder periods. This is where the Ice Ages have set in and the world begins to experience a strengthening of those cyclic episodes.

Ice Ages are complicated things, and while some see them simply as cycles of warm and very cold, they are also punctuated by erratic peaks and troughs of varying amplitude temperature swings interrupting the main cycles. The EPICA cycles show up to a 10°C fluctuation up or down. So, if we stretch out the box featured in Figure 4.4 we can see the smaller fluctuations that occurred within the major glacial and interglacial cycles more clearly. Using the raw EPICA data I have plotted 10 glacial–interglacial cycles and those are presented in Appendix 2. The importance of these cycles in the context of this book is that the glacial component has been getting stronger and lasting longer, particularly over the last four cycles spanning 400,000 years. The world endured glacial conditions for 90% of that time, with interglacials marking only short bursts of warmer climate between each, and some did not reach even close to the temperature of the present interglacial. But the predominance of glacials is only part of the story because there has been a pattern of almost constant climatic fluctuation embedded within glacials and, on some occasions, in interglacials.

Many of these represent strong climate switching events which must have inflicted significant environmental change on Australia's environment as well as that of other continents. Smaller cycles can measure anything from 1°C to 4°C temperature variation and today we are well aware of what a 2°C rise in temperature can do; we are reminded constantly. Even though they are comparatively short events, interglacials are complex because most contain one or more stadials, which are major shifts to a cold period, as well as other smaller (substadial) shifts similar to those occurring within glacials. Conversely, glacials often contain interstadials when the climate suddenly warms appreciably although they may not reach even close to the level of modern temperatures. Stadials and interstadials are major events in themselves and they can confuse the picture of how long, for example, an interglacial actually lasts. These features are sub-cycles of one sort and another that are radical climatic changes in their own right but they are often ignored or not even recognised because they are buried within the greater glacial cycles. The last major interglacial was 19,000 years long but only 13,000 of those years were at temperatures similar to or greater than our present interglacial. Before that, you have to go back 420,000 years ago before you encounter an interglacial similar to the last, or the present for that matter, that lasted 19,000 years at modern temperatures. These are the longest interglacials in the last half a million years. Our present interglacial has lasted 17,000 years, so one could argue we are close to the record for interglacial cycle length and that, perhaps, we are rapidly closing in on the beginning of the next Ice Age.

In order to sort out these glacial cycles, keep track of them and pinpoint a particular glacial or interglacial cycle during the Quaternary, I use the Marine Isotope Stage (MIS) designation followed by its number, but MIS is used for short. For those unfamiliar with these, the rule of thumb is that even numbers indicate major glacials and odd numbers refer to the warmer and much shorter interglacials between. There have been between 30 and 50 glacial cycles during the last 2.7 Ma comprising over 100 identified MISs, with 34 occurring during the Quaternary. Some of the main interglacials are divided into sub-stages, usually designated by the first seven letters of the alphabet. For example MIS 5, an interglacial, is often divided into five sub-stages labelled *a* to *e* which indicate major interstadials and stadials that divide the period. Glacials are usually not divided this way but they do fluctuate from deep cold to warm times in discernible sub-stages which are usually not recognised as being divided from the major cold phase. Besides these divisions, glacial cycles include two other events named after their discoverers. They include the *Heinrich* and *Dansgaard–Oeschger Events.*. Both stadials and interstadials signal rapid and unexpected climatic switching within the larger glacial or interglacial cycle. But almost 25 years ago the palaeoclimatologist Hartmut Heinrich described ice-rafted debris (IRD) found in cores drilled on the North American continental shelf (Heinrich, 1988). It is now widely accepted that the significance of IRDs, now known as *Heinrich Events*, is that they signify the melting of glaciers into the North Atlantic causing armadas of icebergs to be set loose. Beneath the bergs was adherent debris (drop stones) frozen like fingers to a freezer wall until they encountered ocean waters. The rocky debris was released as Heinrich sediments offshore from whence the calved bergs originated. The rapid glacier base melt was precipitated by geothermal heat but as the ice moved, frictional heating was added, accelerating melting of bottom ice which additionally eased glacier movement. It happened very fast but the worst part was that the millions of cubic kilometres of ice water that poured into the North Atlantic stopped the Ocean Conveyor that takes warmer water to high latitudes, such as the Gulf Stream, and the world plunged deeper into the Ice Age. Six Heinrich Events have now been recorded during the last glaciation that go back to *ca.* 60 ka. They indicated interstadial-like or warming events that showed glaciations were not necessarily uniformly cold all the time and that they contained rapid warming episodes. The D–O Events that show up in the Greenland ice coring work also confirmed from ice-core data that the Earth had warmed at the time indicated by the Heinrich Events. This was supporting evidence for a rapidly fluctuating climate of significant proportions. The data was a very neat pair of arrows that pointed to small, rapid climatic changes and which came from two lines of very different investigation, although both employed *drilling*.

The D–O events have been designated numbers from 1 to 25 during the MIS 2–5 time frame (see Appendix 2, graphs H and I). However, they have largely been ignored for earlier glaciations because the evidence required to name them as such is missing or has yet to be found. There seems little doubt that they are D–O events and they are important for the aims of the work here because these rapid climatic switching episodes must have played a part in associated environmental switching in Australia. Perhaps more importantly, the cumulative effect of many switches compounded problems that animal populations experienced during glacials. So, while looking at the shape and size of major glacial cycles I want to examine the frequency of these intracyclic events which would have altered megafauna demographics and become medium or strong repetitive and cumulative stressors on their population structure and size. Because I focus on megafauna extinctions, I want to consider a time frame from MIS 4, *ca.* 50,000 years ago and the proposed time of their extinction, back to MIS 20, the earliest MIS we know quite well and the one marking the limit of EPICA ice-core data obtained so far. This time frame also encompasses the most extreme and abrupt glacial–interglacial cycling and climate transitions of the Quaternary Ice Ages. Because both Vostok and EPICA records mutually support the timing and structure of glacial–interglacial cycling throughout the mid-late Quaternary, I employ them combined and separately on occasion to define multiple temperature fluctuations that would have impacted Australia's environment. Two marine isotope stage groupings are looked at: MIS 11–20, which I label the mid-Quaternary, and MIS 4–10, representing the late Quaternary, using a chronological division separating the two of around 420 ka. The reasons for this are that earlier glacials were generally less regularly structured than those of the late Quaternary although both share the basic pattern of abrupt climate swings. Another difference is the apparent change in cycle structure length for both glacials and interglacials that took place at that time.

Abrupt glacial termination is a common feature of the mid-late Quaternary, in almost all cases going from glacial maximum to an interglacial within 1–4000 years. In contrast, glacial onset is almost always a slow process. All temperatures from now on are Antarctic temperatures unless otherwise stated. Each glaciation featured a series of temperature fluctuations, particularly during the first half of the cycle, that swing up to 3°C in short frequencies. I have termed these the *fluctuating glacial phase* (FGP), featuring broad temperature swings that precede a *deep glacial phase* (DGP). The latter is a quieter and more uniform time in which temperature fluctuations are missing or minimal. Vostok data show this two-phase structure clearly in each of the last four glacials. During MIS 6 the FGP (159–188 ka) ranges between −5.0°C and −8.5°C, averaging −6.7°C. The DGP (138–158 ka) has constantly lower temperatures between −7.5°C and −9.0°C (average −8.07°C) with little fluctuation. Average FGP and DGP temperatures during the last four glacial cycles are −5.9°C and −7.7°C, respectively. Vostok data show these phases as an overall cooling trend over the last four glacials but this is not as prominent in EPICA data. Previous glacials were shorter and less structured and lacked the DGPs that occur in later glacials. Early cycles also display smaller amplitude temperature swings but greater fluctuation than later events. As usual with climate there is always one exception to the rule and the exception to the smaller temperature swings was MIS 16, which was equally as cold as late Quaternary events. Also, interglacials were shorter and cooler at that time, usually remaining below −1.0°C. Late Quaternary cycles are also noted for their marked increases in depth and length over those of the mid-Quaternary (Jouzel et al., 1987, 1993, 1996; McManus, 2004; Petit et al., 1999).

The graphs presented in Appendix 2 show the structure of the above glacial–interglacial cycling. They were compiled using EPICA data graphed at 1000-year temperature increments positioned at or close to the mid-millennial point from 722 to 50 ka. While Vostok data terminates at 422 ka, both data sets have been compared as far as they go and, indeed, show almost exactly the same events and cycle shape. The main difference between them is cycle phase timing and that is not of particular consequence for this work. It is the presence of the major climatic shifts and their strength that are important, not knowing *exactly* when they took place, and those that emerge during MIS 5–19 are noted in Table 6.1. Here glacial cycles begin when Antarctic temperatures go below −4°C consistently over several thousand years. This follows Loulergue et al. (2005), as shown in their figure 3. Sub-glacial periods lie between −2.0°C and −4.0°C; an interglacial begins at temperatures above −2.0°C and those of −1.0°C and above are considered modern.

A Devil in the Detail: Elements of Glacial–Interglacial Cycling

Those with a low threshold of boredom might want to miss the next section, which can be done without damage to the story. I include it to record glacial–interglacial cycle detail and structure during the last 800,000 years and, in doing so, highlight the strength and number of climatic changes abrupt and otherwise that were involved in what is simply called the Quaternary Ice Ages. The Devil is indeed in the detail and that is important to the discussion in later chapters. It is useful to show the many substantial climatic disconformities embedded within major glacial cycles, each with obvious repercussions for continental environments. So, apart from the effects of the main glacial cycling, what did the less recognised rapid reversals have on Australia's environment and, in turn, on its fauna, particularly the rapid climatic switching those changes involved? To assist that assessment, intra-cyclic changes from MIS 19 to MIS 5e are viewed. EPICA data from MIS 19 to MIS 11 is used, at which time the combined Vostok and EPICA data are used. The following will mean more if used with Appendix 2.

MIS 19, Interglacial (772–789 ka): Appendix 2, Graph A

The MIS 19 interglacial emerges from a deep glacial (MIS 20) with temperatures around −9.0°C rising to the interglacial boundary (−4.0°C, dashed line on Appendix 2 graphs) in 7000 years. The rise is very steep, a characteristic of all glacial terminations. MIS 19 was a short interglacial and not particularly warm, reaching a maximum of −0.5°C for *ca.* 1000 years. It remained above the modern temperature boundary of −1.0°C for only 2000 years and then stayed above −2.0°C for 6000 years before dropping into the next glacial (MIS 18). Temperatures were maintained above the full glacial boundary of −4.0°C for only 17,000 years but, in short, it was a cold interglacial.

Table 6.1 Chronology of the Last 19 Glacial–Interglacial Cycles Compiled from the Vostok and EPICA Ice-Core Data

Glacials–Interglacials (Glacial boundary −4°C)	Major Stadials/ Interstadials	Date (ka)	Cycle Length (ka)	MIS Stage	Antarctic Temperature (Notes)
Last glacial		14–80	66	*2–4*	Min. −10.0°C
Interglacial 1	Interstadial	80–86	6	*5a*	Max. −2.5°C
	Stadial	86–98	12	*5b*	Min. −7.0°C
	Interstadial	98–102	4	*5c*	Max. −3.0°C
	Stadial	102–114	12	*5d*	Min. −7.0°C
	Interstadial	114–133	19	*5e*	Modern Temp. for 13 ka, Max. 5.5°C at 125.7 ka
Glacial 1		133–196	63	*6*	Min. −9.0°C
Interglacial 2	Interstadial	196–201	5	*7a*	Max. −2.0°C
	Stadial	201–204	3	*7b*	Min. −4.7°C
	Interstadial	204–218	14	*7c*	Max. −1.0°C for 1 ka
	Stadial	218–235	17	*7d*	Min. −8.8°C
	Interstadial	235–245	10	*7e*	Modern Temp. for 4 ka, Max. 2.7°C at 242.1 ka
Glacial 2	Stadial	245–312	67	*8*	Min. −9.5°C
Interglacial 3	Interstadial	312–316	4	*9a*	Max. −2.5°C
	Stadial	316–320	4	*9b*	Min. −5.3°C
	Interstadial	320–336	16	*9c*	Modern Temp. for 12 ka, Max. 3.8°C at 333.3 ka
Glacial 3		336–393	57	*10*	Min. −9.8°C
Interglacial 4		393–429	32	*11*	Modern Temp. for 19 ka, Max. 3.2°C at 406.9 ka
Glacial 4		429–483	54	*12*	Min. −9.7°C
Interglacial 5		483–500	17	*13*	Max. −1.5°C
Glacial 5		500–560	60	*14*	Min. −8.0
Interglacial 6	Interstadial	560–580	20	*15a*	Max. −1.0°C
	Stadial	580–603	23	*15b*	Min. −7.0°C
	Interstadial	603–625	22	*15c*	Max. −1.0°C
Glacial 6		625–689	64	*16*	Min. −9.0°C
Interglacial 7		689–713	24	*17*	Max. −1.5°C
Glacial 7		713–772	59	*18*	Min. −9.0°C
Interglacial 8		772–789	17	*19*	Max. −0.5°C for 1 ka

Source: After Jouzel et al. (1987, 1993, 1996), Petit et al. (1999) and EPICA CM (2004).

MIS 18, Glacial (713–772 ka): Appendix 2, Graph A

A prominent feature of this 59,000-year glacial includes the saw-tooth descent (FGP) to a glacial maximum of −9.0°C taking up the first quarter of the glacial. The two prominent temperature reversals look like D–O events consisting of a warm phase followed by a sharp reversal downward followed by a similar cycle, each time corresponding to a 2°C temperature swing. The second quarter of the glaciation marks its maximum but in the third quarter, a very strong interstadial occurs that takes temperatures back to the glacial boundary of −4°C for about 3000 years. After that, temperatures drop again by 2.0°C to a small 1°C D–O event that is followed by a drop to almost −9.0°C in 4000 years, which seems to be another glacial maximum. That almost immediately turns to bring MIS 18 to an end, with a rise of 5.0°C to the −4°C boundary in 8000 years; temperatures lingered here for 5000 years, marking the onset of MIS 17.

MIS 17, Interglacial (689–713 ka): Appendix 2, Graphs A and B

MIS 17 was another weak interglacial that barely lasted 8000 years and, with the exception of a few hundred year maximum temperature spike at −1.5°C, it hovered around −2°C before it descended once again into the next glacial (MIS 16). It took 7000 years to reach the glacial −4°C boundary then continued on a rapid descent to glacial depths.

MIS 16, Glacial (625–689 ka): Appendix 2, Graph B

This glacial was 5000 longer than the previous one and if it is divided into three, the familiar saw-tooth pattern is seen in the FGP section of the cycle. This consisted of three D–O events that cycle just over 2°C, 1°C and around 0.5°C, respectively, over a period of roughly 15,000 years. Each typically shows a step-wise downward descent to a glacial maximum of −9.0°C. The next third of the glacial consists of a minor interstadial similar to that in MIS 18 but weaker. It actually supports three small reversals over 5–7000 years. At their peak, they take temperatures back to −6.7°C before descending again to −9°C during the final third of the glacial phase. Imbedded in this phase are another two reversals, swinging temperatures over 0.5°C, followed by a rapid ascent from −8.5°C to −2°C in 4000 years. Technically this glacial terminated in only 3000 years, which is the time it took to reach the −4.0°C glacial–interglacial boundary; this is less than half the time taken by the previous glacial termination and yet again demonstrates the rapid nature of glacial terminations.

MIS 15, Interglacial (560–625 ka): Appendix 2, Graphs B and C

MIS 15 is the first glacial divided into alphabetic sub-stages termed 15a, b and c. Looking at them in reverse order, MIS 15c was in itself a longer interglacial (22,000 years) than MIS 17, but not much stronger. It spent 10,000 years at temperatures below −2.0°C, after which it warmed somewhat to between −2.0 and −1.0°C for

8000 years. MIS 15b was a stadial event of 25,000 years at temperatures below −4.0°C. Usually interstadials last less than 10,000 years, so the MIS 15b stadial could be regarded as a full glacial. Regardless, MIS 15b contains two prominent D–O events, each swinging through a 2.0–2.5°C range: the first was a 1.5°C drop in 2000 years, the second a 2.5°C rise in 1000 years and a descent of 3.0°C in 3000 years. That is a boring list but it is nonetheless fascinating as it shows wide and abrupt climatic changes over a comparatively short period of time. The last upward swing marked the end of MIS 15b, when temperatures rose from −7.0°C to −1.0°C in 4000 years, once again the typical fast glacial termination. MIS 15a lasted 20,000 years, similar to MIS 15c, and did not reach modern temperatures above −1.0°C.

As we move on with this description it is worth making some assessment, even at this stage, of the effects these changes, which were taking place over a quarter of a million years, had on the environment as it moved from one state to another. Then we may be able to appreciate what the faunal responses might have been to these changes. Let us continue.

MIS 14, Glacial (500–560 ka): Appendix 2, Graph C

The end of MIS 15a is marked by a steep descent into the 60,000-year MIS 14 glacial as temperatures drop from −2.0°C to −6.5°C in 5000 years, rapid for a glacial descent. That was followed by a substantial climate reversal that could be either an interstadial or significant D–O event as temperatures rose 3.5°C to −3.0°C over 4000 years, then dropped 5.0°C to below −8.0°C in 7000 years. Two smaller swings of 1.0°C occurred before another interstadial emerged with temperature rises of 3°C in 4000 years. Something similar happened in MIS 16 and 18, but this time it levelled out around −4.0°C and was then followed by another drop to −6.0°C before rising again at the end of the glacial. MIS 14 finished differently than previous glacials by fluctuating slightly above and below −4.0°C. Although lasting *ca.* 60,000 years, MIS 14 was overall a warmer glacial than previous glacials, although the temperature remained somewhere below −4.0°C for almost all of that time.

MIS 13, Interglacial (483–500 ka): Appendix 2, Graph D

Although it stayed on or above the −4.0°C mark for 17,000 years, MIS 13 was another cold interglacial and rather pathetic, basically consisting of a single spike that barely reached −1.5°C and lasted for less than 500 years. The rest of the time temperatures remained between −3.0°C and −4.0°C; indeed the last 6000 years of MIS 13 were marked by a temperature drop from −1.5°C to −4.0°C.

MIS 12, Glacial (429–483 ka): Appendix 2, Graph D

Descent into MIS 12 took 6000 years to the −4.0°C level and then, after another 4000 years, they reached −6.0°C. At that point the usual saw-tooth temperature pattern arose in the FGP where temperatures rose 2°C, dropped three, rose again 2.5°C then dropped 4°C to −8.5°C; this took place within a span of 11,000 years.

The glacial maximum consisted of a series of temperature fluctuations of 2°C, similar to the previous three glacials, but this time temperatures remained at between −6.0°C and −7.5°C. Its appearance is also not as uniform as that of previous midglacial interstadials and it displays a series of at least three D–O events. The end of the interstadial was marked by a drop back to −8.5°C, followed by a 1.5°C rise in 1000 years and then the final glacial termination that took 6000 years to reach modern temperatures close to zero for the first time in over 400,000 years.

MIS 11, Interglacial (393–429 ka): Appendix 2, Graphs D and E

The beginning of this interglacial is marked by a sharp temperature drop, lowering temperatures from modern back to −2.0°C in around 1000 years. They then rose again over 6000 years to exceed modern temperatures as the interglacial took hold. MIS 11 was the warmest since before MIS 21 and was the last long, warm interglacial before MIS 5e. It signals a change-over point between previous, less severe and variable glaciations and those with no large mid-glacial interstadial events and the deepening of glacial maxima that followed. Previous glacials were warmer and shorter, with MIS 13–15 having almost equal glacial and interglacial phase lengths. MIS 11 was also remarkable by its 32,000-year span as well as its maintaining temperatures above those of today for 80% of that time, with 3.0°C above those of today recorded *ca.* 408 ka. Temperatures then dropped almost 9°C over the next 16,000 years as the next glacial began.

MIS 10, Glacial (336–393 ka): Appendix 2, Graph E

MIS 10 is the first of the deep, cold glacials in this series which was 57,000 years long. It featured abrupt cooling between 393 and 396 ka and two warmer spikes in the usual FGP saw-tooth pattern which includes −2.5°C at 376–377 ka and 380–384 ka. The FGP spanned 363–394 ka, with abrupt oscillating temperatures of −7.0°C to −3.0°C, averaging −5.1°C. The DGP was indeed deep and cold, reaching −9.5°C, from which a very rapid terminal rise to the next interglacial took place in 7000 years.

MIS 9, Interglacial (312–335 ka): Appendix 2, Graph E

MIS 9 consisted of three sub-stages, a, b and c, consisting of two interstadials comprising 20,000 years divided by a stadial phase of 4000 years. MIS 9c was the main part of the interstadial at 16,000 years and featured temperatures above −4.0°C for that time with 12,000 years at modern temperatures, including almost 4000 years where they reached 3.5°C above those of today. Glacial temperatures existed between 316 and 320 ka (MIS 9b), followed by a brief interstadial that reached −2.5°C maximum (MIS 9a) that then entered the FGP of MIS 8. Entry into MIS 8 was abrupt, similar to the exit from MIS 10 with a temperature drop from −1.0°C to −5.0°C in 5000 years.

How are those environments changing with all this and, more importantly, how are the animals coping?

MIS 8, Glacial (245–312 ka): Appendix 2, Graphs E and F

At 67,000 long, MIS 8 was 10,000 years longer than MIS 10 and the longest in the Quaternary up to that time. It sank to a low of −8.0°C before entering a 10,000-year interstadial or D–O event *ca.* 292 ka, but temperatures did not rise above −4.0°C. The glacial maximum peaked *ca.* 275 ka. The FGP occurred between 277 and 300 ka and averaged −5.3°C, while the following DGP (245–277 ka) averaged −7.1°C. The termination of MIS 8 was particularly abrupt as it displays a steep rise of 8°C in 3000 years into MIS 7.

MIS 7, Interglacial (196–245 ka): Appendix 2, Graphs F and G

In effect, this was the shortest interglacial of the mid-late Quaternary and was more glacial than interglacial. It has been divided into five sub-stages with MIS 7e (235,245 ka) as the only real part of this interglacial because it was the only time temperatures reached modern equivalents – but then only in a 4000-year-long spike. That was followed by a deep stadial at 219–232 ka (MIS 7d) which dropped to −7.5°C, equal to full glacial conditions; it lasted at least 17,000 years. There were two interstadials: a very short one at 196–201 ka (MIS 7a) and another at 204–218 ka (MIS 7c). Neither reached modern temperatures and both were separated by a very brief stadial MIS 7b between 201 and 204 ka that just reached glacial temperatures.

MIS 6, Glacial (133–196 ka): Appendix 2, Graph G

The 63,000-year long MIS 6 glacial sits with MIS 8 and MIS 10 as the three largest and deepest Quaternary glacials. All three included a prominent FGP–DGP structure, with −6.7°C and −8.1°C average temperatures, respectively. Nevertheless, this glaciation was infested with temperature reversals which ranged from 1°C to 3°C. They mark a phase of constant climate change which must have seen rapid reversal of weather conditions around the world. The glacial maximum occurred at *ca.* 135–148 ka when temperatures were comparatively stable, ranging between −9.0°C and −8.0°C, although two smaller excursions below −9.0°C occurred before this time. Again, the termination of this glaciation occurred rapidly, going from almost −9.0°C to fully modern temperatures in 7000 years and marking the beginning of MIS 5e.

MIS 5, Interglacial (80–133 ka): Appendix 2, Graphs G and H

MIS 5 was a warmer and wetter interglacial than present and probably the most normal interglacial in the mid-late Quaternary – whatever normal might mean when describing these cycles. It began with a very abrupt deglaciation from MIS 6 and continued with a series of short stadial–interstadial cycles. MIS 5e was the warmest time, reaching a maximum of 4.5°C above modern temperatures. It basically comprised the MIS 5 interglacial, lasting 19,000 years. This is the time of meg-alake formation in Australia. After this time the following four sub-stages are little more than stadials, at least in temperature. The first, a real stadial (102–114 ka, MIS 5d), dipped to around −7.0°C, reaching glacial depth, and was followed by a simi-lar length interstadial (98–102 ka, MIS 5c) that reached only just above the −4.0°C glacial boundary temperature. After less than 500 years at that temperature, a short stadial (86–98 ka, MIS 5b) followed which took temperatures back down to the pre-vious stadial level. Finally, there was a short interstadial between 80 and 86 ka (MIS 5a) which had maximum temperature of −2.5°C for about 2000 years before tem-peratures once again declined into another glacial, MIS 4.

The MIS 4 glacial was very cold, with temperatures reaching almost −10.0°C below those of today *ca.* 65,000 years ago. But in reality, MIS 4 continued through till the end of MIS 2 *ca.* 14,000 years ago because the MIS 3 interglacial was a non-event, according to EPICA data, with temperatures never reaching above −4.0°C. It is interesting to speculate that if the EPICA records of MIS 2, 3 and 4 occurred, say, *ca.* 400,000 years ago, there would not be an interglacial plonked in the middle marking stage 3! So in effect, the last glacial event (MIS 2) almost replicated the pre-vious three, lasting 65,000 years (14–79 ka) divided into FGP (34–79 ka) and DGP (14–33 ka) tiers, averaging −6.4°C and −8.2°C, respectively.

It's over! I have to apologise for the repetitive mantra of the last five and a half pages but that was partially the purpose; the repetition says it all. Looking at the climatic roller-coaster and the many mid-range stadial-interstadials shows that the mid-late Quaternary was constantly peppered with vast climatic changes and those mentioned above were only the salient ones. Many were smaller than those and have not been mentioned. So many of the changes were extreme in terms of tem-perature change and often abrupt, which would have far-reaching consequences for the environment and biota of the planet. It was not just a broad, basic structure of warm–cold (sometimes not very warm at that) but also a structure of constant stag-gered and seemingly random interruptions that comprised a repetitive hammering on the adaptive abilities of the biological world. Antarctic temperatures recorded in the Vostok–EPICA data cannot be directly extrapolated to all regions of the world but the changing temperature profile recorded from that data is a rigorous indication of the strong, frequent and sometimes erratic Quaternary glacial–interglacial climate structure; it also highlights the extremely long and cold time frames of glacials. It is that stark profile that can be applied in its broad form to all regions of the planet including Australia, the continent next door to where the data was obtained. It is also most important to point out the time spent during glacials at temperatures above and below −1.0°C during the 799,000 years of the mid-late Quaternary and shown in Table 6.2.

Multi-proxi cross-correlation with and bi-hemispheric concordance for glacial enviro-climatic changes have been well documented. From that we can assume the resulting record underpins Antarctic temperatures as a basic indicator of mid-late Quaternary world Ice Age sequence and timing as well as the environmental con-sequences that came from them. Nevertheless, we have a long way to go before we know *exactly* how those temperatures can be interpreted for any given region with

Table 6.2 Length of Time in Glacial and Interglacial Conditions Between 50,000 and 799,000 Years

Time at or Above 0°C	Time Above −1.0°C AT	Time Below −1.0°C AT
39,000 years (5.2%)	58,000 years (7.7%)	652,000 years (87.1%)

AT – Antarctic temperature.

regard to details of the vast biome changes that resulted from them. What we do know is that they flag glacial events that formed ice sheets several kilometres thick in some places and covered ~40 million square kilometres of Earth's land surface. They turned savannahs into deserts; severely reduced and shifted massive areas of rainforests across Southeast Asia and elsewhere; shifted millions of square kilometres of Boreal forest south; created widespread tundra in its place; and lowered the world's oceans by over 120 m below modern levels, changing their shape, joining landmasses and altering ecosystems worldwide. The marked increase in atmospheric methane accompanying interglacial resumption recorded in Antarctic ice cores can also be translated into broad environmental changes occurring around the world at that time. Methane production during global warming at the end of a glaciation points to the exposure of land following glacial retreat, permafrost melt, wetlands forming, and the changing position of the inter-tropical convergence zone (Loulergue et al., 2008). The correlation of the methane signal with glacial cycling, therefore, is a direct proxy of enviro-climatic changes that accompanied it. It also largely originates from environmental shifts taking place in the northern hemisphere and on the other side of the globe from Antarctica.

It may also have been the onset of aridity across Africa that squeezed anatomically modern humans into the world outside rather than a matter of them choosing to go under their own volition – a bit like boardroom directors. The expansion of aridity in two great belts north and south of the African Equator during the latter half of MIS 6 may have been one reason for the separation of the modern human mitochondrial genome which appears so obvious today, with a major difference between South African bushman and all other human groups around the world. Moreover, it might also explain why the earliest modern human archaeological sites begin after 135 ka at the termination of MIS 6, allowing humans to move from the Equatorial region south. Glacials, therefore, played havoc with various aspects of the world's biosphere and hydrosphere as well as the cryosphere. Then, with interglacials it was all put into reverse again, only to do it over again in the next glacial event; that happened many, many times. Those are no little environmental changes that resulted from the temperature swings recorded in the Antarctic data used here. The implications of such widespread and severe changes around the planet need little further explanation but their effects on the Australian continent were no less dramatic.

To a large extent the exactness recorded temperatures, their exact timing or when cycles began or ended, is not important. Rather, what is important is the turmoil of the interminable succession of climate change and what that meant in terms of environmental consequences, particularly the disruption imposed on Australia's megafauna. These changes also quite naturally prompt one to ask how they challenged the adaptability of animals which had been adapting to a drier world, particularly in Australia, for a long time before these events. For example, did they have enough genetic wherewithal and time to adapt further to an increasing see-saw of changeable conditions and cold? Moreover, could changes that were drying out most of the continent be accommodated by all the fauna? We know that Australia underwent drastic environmental change during glacial events. So, logically, severe repercussions for animal groups implied by these events cannot be underestimated but probably always has been. Glacial onset repositioned animals from higher to lower latitudes and onto landscapes released with oceanic regression. On other continents they had room to move but in Australia they did not. Nevertheless, they must have moved as deserts expanded and in response to sea level fluctuation during glacial cycling where they were pushed offshore then onshore again as continental shelves were exposed and then became inundated, reflecting Polar ice growth and melt. That was accompanied by the shifting of biomes which also moved and dispersed animal groups as the centre of Australia became uninhabitable. Such changes certainly look like the sort of ingredients needed for extinctions to take place. It was precisely these sorts of processes that are now recognised as taking place in Southeast Asia and which affected the distribution, assortment and extinction of genera there during the Quaternary, and I take this up again in Chapter 10.

The proximity of Australia to Antarctica, therefore, probably makes the Antarctic data equally or more relevant to environmental interpretations and the biogeographic alterations taking place here. The methane signal mentioned above also implies biogeographic consequences among animal populations linked to rangelands and habitat alteration with specific implications for species assortment, dispersal and distribution. It also points to population displacement and movement at glacial–interglacial end-points, particularly for those animal guilds favouring newly exposed or released environments; the drowning and emerging of continental shelves; and the emerging niches and habitats that accompanied these changes, as well as those that disappeared during glacial times. Moreover, landmass reduction at oceanic high stands eliminated large areas where animal populations had flourished for a long time during the much longer glacial phases. Such a biogeographic melee would continue as environments stabilised and plant communities once again established themselves, but only for a while. With these factors in mind, what were Australia's likely environmental circumstances during the mid-late Quaternary? To try and answer that, the Vostok and EPICA temperature proxies that have been detailed here are used in the next chapter to make inferences regarding the frequency and depth of environmental change in Australia, its biogeographic consequences and, most importantly, its impact on Australia's megafauna.

7 Who and Where: Australian Megafauna and Their Distribution

I mentioned before that the approach to Australia's megafauna has largely been to treat them as something special, separated from the vagaries of enviro-climatic change and not subject to the vast alterations to their preferred habitats and niches that took place during glacial cycles. Indeed, they are often seen as a bunch of animals that lived in the Quaternary and which did not undergo change in any way till humans arrived on the continent. In reality they were like any other animal group with the same requirements and the same weaknesses, and they faced extinction and its causes in the same way as all those animals that went extinct before or after them. Laws governing the lives of megafauna were basically the same as those governing our present fauna and *Lystrosaurus*. The difficulty of solving the megafauna extinction issue is that we have little understanding of the extinction timing of individual megafauna species, although that is becoming clearer. Also helpful would be a knowledge of their continental distribution as a group or as individual species, of their biogeography and of the number that still existed at the proposed time of their extinction. Moreover, we cannot be sure that the most recent date for any given species represents the last animal. Another problem is that a large part of the continent is yet to be properly surveyed to establish where megafauna lived, and many of the large museum collections and field assemblages remain undated. Understanding the spatial and temporal distribution of megafauna is, however, vital if we are to establish reasons for the processes and timing of their extinction and the proportion of that extinction. It all sounds as though there is not much sense in carrying on from here, but you would be wrong to think that. There is much we can do.

This chapter begins the job of trying to understand what animal species made up Australia's megafauna, their distribution and the impact that glacial cycling had on them. These aims are placed against a background of their biogeographic status and their biological capabilities for adaptation and survival using biogeographic principles derived from the study of animals worldwide. But first we look at the megafauna themselves and how many species were involved in the extinction.

Australian Megafauna: How Many Species?

The first question asked here is what animals comprise *'the Australian megafauna'*? That seems a simple task but there are different views concerning what species should or should not be included in the group. It can be argued that *megafauna* may also be the wrong term to use for some species normally included in the group.

Corridors to Extinction and the Australian Megafauna. DOI: http://dx.doi.org/10.1016/B978-0-12-407790-4.00007-0

For example, more than half the species usually included overlap in weight and size with extant Australian species, but I will return to that later. There is also a sizable inconsistency when trying to list these animals. For example, we now know there are some species that in the past were included in the list that supposedly went extinct when humans arrived, but that actually went extinct long before that. So, that means we are not trying to list all the megafauna species that ever lived, but just those that seem to have ended up on the list of usual suspects. That list varies because opinions differ, with some authors including species rejected or not included by others. Again, depending on which author one reads, it seems the megafauna range can span between 30 and 52 species. It is worth pausing to see that already there is great uncertainty in the subject and we have not really started. The longest list often includes any species that lived in the Quaternary, or even before in some cases. It is now quite apparent that some animals included in megafauna lists even 20 years ago went extinct a long, long time before MIS 4, the supposed threshold time, while others that have been included in the past should not have been. The Tasmanian Tiger (*Thylacinus cynocephalus*) and the Tasmanian Devil (*Sarcophilus harrisii*) are two often included, although both survived on the mainland well into the Holocene.

Assembling a full list, then, is often not so straightforward, but here goes. Table 7.1 presents a cohort of 52 large animals that lived in the Quaternary, all designated at one time or another, by various authors, as megafauna. It excludes animals such as *Euryzygoma dunense*, which was included in previous lists but is now known to have gone extinct before the 40,000- to 60,000-year extinction threshold. There is still a question mark over 12 other species in this regard, but they have been included nevertheless. There are also 16 species that are rare in the fossil record but that will be taken up in the following chapter. The list also excludes animals weighing <50 kg, a little heavier than Paul Martin's 44 kg weight standard (1984) delimiter for qualification as a megafauna species. It is worth noting that just over 50% (*n* = 28) of the 52 weigh between 50 and 80 kg by present estimates, which well overlaps with our three largest modern kangaroos *Macropus rufus, M. giganteus* and *M. fuliginosus*. The list also excludes animals of doubtful or controversial taxonomic status and those that are believed to have 'dwarfed' over the Pleistocene–Holocene boundary and continue to exist in a modern form. There are five of those, including the kangaroos *Wallabia indra-bicolor, Macropus titan-giganteus, M. siva-agilis, M. cooperibicolor*, and the giant koala *Phascolarctos stirtoni-cinereus*. A group of other species have also previously been rejected as megafauna (Murray, 1991). The hyphenated second species name indicates the dwarfed type. The trouble with this subject is there are always updates; for example, a paper published recently has withdrawn the koala *Phascolarctos stirtoni-cinereus* as a 'dwarfer', confirming that the large *P. stirtoni* was a legitimate giant koala that did not undergo size reduction, morphed it into the modern koala *P. cinerus* (Price, 2008b). So it has been included in the list. *S, harrisii* may also have been one of the 'dwarfers' over the Pleistocene/Holocene boundary from an earlier large species identified in the past as *S. laniaris*.

My list comprises four large diprotodontids, 37 kangaroos (12 genera), five other marsupial genera, four reptiles, and a giant flightless bird (*Genyornis newtoni*). The reptiles are the giant varanid (*Megalania prisca*), a giant freshwater crocodile

Table 7.1 Continental Distribution of 53 Australian Megafauna Species

Megafauna Species n = 52	Weight kg ave.	QLD	NSW	VIC	LEB	SA	WA	TAS
Diprotodontids (quadrupeds)								
Eowenia grata 1, 2	<3500	—	—	—	—	—	—	—
Palorchestes azeal 1 JMF	750	S	—	—	—	—	—	—
Zygomaturus trilobus JMF	350	—	—	—	—	—	—	—
Kangaroos								
'Procoptodon' browneorum JMF	50	—	—	—	—	—	—	
'Procoptodon' gilli 1 JMF	54		—	—	—	—	*ln*	—
'Procoptodon' mccoyi 2	50	—	—	S	—	—		
'Procoptodon' oreas JMF	100	—	—	—	—	—	*ln*	
'Procoptodon' williamsi 1 J	150	—	—	—	—	—	—	
'Simosthenurus' baileyi 1 J	55					—		
'Simosthenurus' brachyselenis 2	>50		S					
'Simosthenurus' cf. antiquus 2	>50					S		
'Simosthenurus' pales 1 JMF	150	—	—	—	—	—	—	
Macropus (agilis) thor 2	>50	S						
Macropus agilis siva 2	>50	S						
Macropus ferragus JMF	150	—	—			—	*ln*	
Macropus greyi 2	>50					S		
Macropus pearsoni 1, 2 JF	150	S				S		
Macropus rufogresius 2	>50					S		
Metasthenurus newtonae 1 J	55	—	—	—	—	—	—	—
Procoptodon goliah JMF	230	—	—	—	—	—	—	
Procoptodon pusio JF	75	—	—	—	—	—	—	
Procoptodon rapha JMF	150	—	—	—	—	—	*ln*	—

(Continued)

Table 7.1 (Continued)

Megafauna Species n = 52	Weight kg ave.	QLD	NSW	VIC	LEB	SA	WA	TAS
Propleopus ocillans	50	/	/	/	/	/	/	/
Protemnodon anak JMF	130	/	/	/	/	/	/	/
Protemnodon brehus JMF	110	/	/	/	/	/	/	
Protemnodon roechus JMF	170	/	/	/	/	/	/n	
Simosthenurus maddocki 1 JMF	80	/	/	/	/	/	/n	
Simosthenurus occidentalis JMF	120	/	/	/	/	/	/	/
Simosthenurus orientalis MF	>50	/	S	/	/	/		/
Simosthenurus euryskaphus 2	>50	/	S	/	/	/		
Sthenurus agilis siva	>50	/	/	/	/	/		
Sthenurus andersoni JMF	72	/	/	/	/	/	/n	/
Sthenurus atlas JMF	150	/	/	/	/	/	/	
Sthenurus murrayi 2	>50	/	S	/	/	/		
Sthenurus stirlingi J	>50	/		/	/	/		
Sthenurus tindalei JMF	130	/	/	/	/	/	/n	
Troposodon minor	>50	/	/	/	/	/		
Baringa nelsonensis 1,	>50	/	/	/	/	/	/n	
Bohra paulae 1,	>50	/	/		/	/	/n	
Congruus congruus 1,	>50	/			/	/	/n	
Other marsupials (quadrupeds)								
Thylacoleo carnifex JMF	140	/	/	/	/	/	/	/

Table 7.1 (Continued)

Megafauna Species n = 52	Weight kg ave.	QLD	NSW	VIC	LEB	SA	WA	TAS
Phascolarctos stirtoni 2	>50	S						/
Phascolonus gigas JMF	150	/	/	/	/	/	/	/
Phascolomys medius 2 JF	50	/				/		
Lasiorhinus augustidens 2	>50	S						
Ramsaya magna 1, 2	100	/	/					
Giant bird and reptiles								
Genyornis newtoni	250	/	/	/	/	/	/	
Megalania prisca M	500	/	/		/	/		
Pallimnarchos pollens	1000	/	/		/	/		
Meiolania platyceps 1, 2	300	/			/			
Wonambi naracoortensis M	>50					/	/	
Species in each area		36(6)	34(3)	21(1)	22	36(3)	25	10

Key

QLD – Queensland, NSW – New South Wales, VIC – Victoria, SA – South Australia, LEB – Lake Eyre basin, WA – Western Australia and TAS – Tasmania.

1 – Species thought to have gone extinct prior to MIS 3 (assumed time of human entry).

2 – Species with limited distribution, only known from one site or region.

n = Species found on or to the east of the Nullarbor Plain but not in the southwest of Western Australia.

#() = Number of species recorded only in that state.

JMF: J – Johnson (2006), M – Murray (1996), F – Flannery (1990), author including that species in their megafauna extinctions list.

S – Single example.

Figure 7.1 Reconstruction of the horned turtle *Meiolania platyceps*.
Source: Photograph by Steve Webb.

(*Pallimnarchos pollens*), a boa constrictor (*Wonambia naracoortensis*), and a coffee table-sized horned turtle (*Meiolania platyceps*) (Figure 7.1). In Table 7.1 the surname initials of three major previous researchers and list builders, Chris Johnson (J), Peter Murray (M) and Tim Flannery (M), have been placed next to species they included in their respective listings. Also noted with a 1 are those species assumed to have gone extinct prior to MIS 4, the assumed time of human entry to Australia, although with future discoveries that may prove a false notion. Those with a 2 had a limited distribution or are known only from one site or as a single individual. Table 7.2 shows the vanishingly small number of designated megafauna species found in sites around the continent. Note that one-third of species have been found only once.

It is worth looking further at these animals, so in Tables 7.3 and 7.4 I have eliminated these groups in different combinations as well as excluding those animals weighing <100 kg. Table 7.3 shows that by eliminating the latter, only 23 remain; that number is reduced to 16 if rare species and those going extinct prior to MIS 4 are eliminated. Table 7.4 lists 26 species after rare and prior extinct animals are left out. It is also worth noting that 50–70% of species comprise large kangaroos, depending on which table is viewed. So, by looking at the list in various ways it can be argued that the number of Australian megafauna species that supposedly went extinct at the 40,000- to 50,000-year threshold was not very large.

The Megafauna

Reptiles

The two principle reptiles in the group are the giant lizard, *Megalania prisca*, and the equally giant freshwater crocodile, *Pallimnarchos pollens* (Figure 7.2). Comparatively few *Megalania* post-cranial remains have been found so that a complete skeleton assembled from individual component parts is yet to be assembled. However, this has not stopped estimates of its length. Its size varies from 5

Table 7.2 Site Occurrence of 48 Australia Megafauna Species

Species	QLD	NSW	VIC	SA	WA	Tas	Total
Diprotodon optatum	21	10	10	23	6	2	72
Thylacoleo carnifex	7	8	11	9	7	3	45
Simosthenurus occidentalis	0	5	10	8	3	5	31
Procoptodon goliah	2	10	1	12	1	0	26
Zygomaturus trilobus	1	5	7	6	5	4	28
Protemnodon brehus	3	5	5	4	4	0	21
Protemnodon anak	3	8	4	3	0	3	18
Metasthenurus newtonae	1	3	1	7	4	1	17
Megalania prisca	6	2	0	6	0	0	14
Sthenurus tindalei	1	4	1	7	2	0	15
'Procoptodon' browneorum	1	0	2	5	6	0	14
Phascolonus gigas	3	5	2	7	3	1	21
Palorchestes azeal	4	3	4	2	1	3	17
Procoptodon rapha	2	5	3	5	0	0	15
'Procoptodon' gilli	0	2	6	4	1	0	13
Simosthenurus maddocki	1	2	3	6	3	0	15
'Simosthenurus' pales	2	7	0	3	1	0	13
Sthenurus andersoni	1	6	0	7	1	1	15
Sthenurus atlas	1	5	1	5	1	0	13
Propleopus ascillans	1	3	2	4	0	0	10
Protemnodon roechus	4	0	0	2	1	0	7
'Procoptodon' oreas	1	3	0	3	0	0	7
Procoptodon pusio	2	4	0	0	0	0	6
Sthenurus stirlingi	1	0	0	6	0	0	7
Simosthenurus orientalis	1	1	0	1	1	0	4
'Procoptodon' williamsi	0	1	0	1	1	0	3

(Continued)

Table 7.2 (Continued)

Species	QLD	NSW	VIC	SA	WA	Tas	Total
Macropus ferragus	0	1	0	1	1	0	3
Wonambi naracoortensis	0	0	0	1	0	0	1
'Simosthenurus' baileyi	0	0	0	1	1	0	2
Phascolomys medius	0	0	0	3	0	0	3
Macropus pearsoni	1	0	0	0	0	0	1
Pallimnarchos pollens	4	1	0	3	0	0	8
'Simosthenurus' cf. antiquus	0	0	0	1	0	0	1
Sthenurus murrayi	0	1	0	0	0	0	1
Sthenurus agilis siva	3	2	0	2	0	0	7
'Simosthenurus' brachyselenis	0	1	0	0	0	0	1
Simosthenurus euryskaphus	0	1	0	0	0	0	1
'Procoptodon' mccoyi	0	0	1	0	0	0	1
Troposodon minor	2	2	1	0	0	0	5
Macropus pearsoni	1	0	0	0	0	0	1
Macropus agilis siva	1	0	0	0	0	0	1
Lasiorhinus augustidens	1	0	0	0	0	0	1
Phascolarctos stirtoni	2	0	0	0	0	0	2
Congruus congruus	0	0	0	1	0	0	1
Congruus kitchenerie	0	0	0	0	1	0	1
Macropus greyi	0	0	0	1	0	0	1
Macropus rufogresius	0	0	0	1	0	0	1
Baringa spp.	0	0	0	0	1	0	1

Only 20 species have been found in more than 10 sites and almost one-third have been found only once. *Genyornis newtoni* is missing from this list because of the numerous sites where its egg shell have been found this is explained in the text, see also Table 7.4.

Table 7.3 Megafauna Species Remaining after those Weighing <100 kg are Removed from Table 7.1 Listing

Megafauna Species n = 23	Weight kg ave.	QLD	NSW	VIC	LEB	SA	WA	TAS
Diprotodontids (quadrupeds)								
Diprotodon optatum JMF	<3500	✓	✓	✓	✓	✓	✓	✓
Eowenia grata 1, 2	750	S						
Palorchestes azeal 1 JMF	350	✓	✓	✓	✓	✓	✓	✓
Zygomaturus trilobus JMF	500	✓	✓	✓	✓	✓	✓	✓
Kangaroos								
'Procoptodon' oreas JMF	100	✓	✓	✓	✓	✓		
'Procoptodon' williamsi 1J	150		✓			✓	/n	
'Simosthenurus' pales 1 JMF	150	✓	✓		✓	✓	✓	
Macropus ferragus JMF	150		✓	✓	✓	✓	/n	
Macropus pearsoni 1, 2 JF	150	S						
Procoptodon goliah JMF	230		✓	✓	✓	✓	/n	
Procoptodon rapha JMF	150	✓	✓	✓	✓	✓		
Protemnodon anak JMF	130	✓	✓	✓	✓	✓		
Protemnodon brehus JMF	110	✓	✓	✓	✓	✓	✓	
Protemnodon roechus JMF	170	✓	✓		✓	✓	/n	
Simosthenurus occidentalis JMF	120		✓	✓	✓	✓	✓	
Sthenurus atlas JMF	150	✓	✓		✓	✓	✓	
Sthenurus tindalei JMF	130	✓	✓		✓	✓	/n	
Other marsupials (quadrupeds)								
Thylacoleo carnifex JMF	140	✓	✓	✓	✓	✓	✓	✓
Ramsaya magna 1, 2	100	✓	✓		✓	✓		
Giant bird and reptiles								
Genyornis newtoni	250	✓	✓	✓	✓	✓	✓	
Megalania prisca M	500	✓	✓		✓	✓		✓
Pallimnarchos pollens	1000	✓	✓		✓	✓		
Meiolania platyceps 1, 2	300	✓			✓			✓

Only 16 remain if rare species and those suspected of going extinct prior to MIS 3 are excluded.

Table 7.4 Megafauna Species Remaining After Categories 1 and 2 Are Removed from Table 7.1 Listing

Megafauna Species n = 26	Weight kg ave.	QLD	NSW	VIC	LEB	SA	WA	TAS
Diprotodontids (quadrupeds)								
Diprotodon optatum JMF	<3500	✓	✓	✓	✓	✓	✓	✓
Zygomaturus trilobus JMF	500	✓	✓	✓	✓	✓	✓	✓
Kangaroos								
'Procoptodon' browneorum JMF	50		✓	✓	✓	✓	✓	
'Procoptodon' oreas JMF	100					✓		
Macropus ferragus JMF	150	✓	✓	✓	✓	✓	/n	
Procoptodon goliah JMF	230	✓	✓	✓	✓	✓	/n	
Procoptodon pusio JF	75					✓		
Procoptodon rapha JMF	150	✓	✓	✓	✓	✓		
Propleopus ocillans	50					✓		
Protemnodon anak JMF	130	✓	✓	✓	✓	✓		✓
Protemnodon brehus JMF	110	✓	✓	✓	✓	✓		
Protemnodon roechus JMF	170	✓	✓	✓	✓	✓	/n	
Simosthenurus occidentalis JMF	120		✓	✓	✓	✓		
Simosthenurus orientalis MF	>50		✓	✓	✓	✓		✓
Sthenurus agilis siva	>50					✓		
Sthenurus andersoni JMF	72	✓	✓	✓	✓	✓	/n	✓
Sthenurus atlas JMF	150	✓	✓	✓	✓	✓	✓	✓
Sthenurus stirlingi J	>50					✓		
Sthenurus tindalei JMF	130	✓	✓	✓	✓	✓	/n	
Troposodon minor	>50		✓	✓		✓		
Other marsupials (quadrupeds)								
Thylacoleo carnifex JMF	140	✓	✓	✓	✓	✓		✓
Phascolonus gigas JMF	150	✓	✓	✓	✓	✓		✓
Giant bird and reptiles								
Genyornis newtoni	250	✓	✓	✓	✓	✓		
Megalania prisca M	500	✓	✓	✓		✓		
Palimnarchos pollens	1000	✓	✓			✓		
Wonambi naracoortensis M	>50					✓	✓	

They are reduced to 17 species if animals weighing <100 kg are also removed.

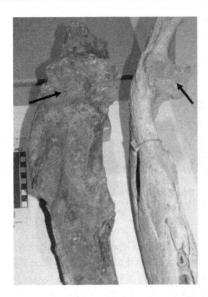

Figure 7.2 Comparison of the lower articular jaw joints (arrowed) of *Pallimnarchos pollens* (left) and a fully grown *Crocodillus porosus* (right) highlighting the large size disparity between these two crocodiles.
Source: Photograph by Steve Webb.

to 8 m although recent reconstructions from elements found in obscure museum boxes favour the lower end of this range (Hocknell, personal communication, 2009; Molnar, 2004; Wroe, 2002). Nevertheless, even with its recent downsizing *Megalania* was an impressive lizard, much larger and of course much heavier than its Komodo cousins, making it a fierce predator without enemies. It seems to have been part of the original ancestral reptile stock, individuals of which left Australia and gave rise to other very large lizards in southeast Asia, including the Komodo dragons (*Varanus komodoensis*) (Hocknell et al., 2009). We can only assume that as a larger creature, *Megalania* was as efficient a scavenger as a Komodo, a possible pack hunter and, because of this and its size, quite capable of attacking the largest marsupials. Some support for this conclusion comes from eyewitness accounts of a group of 3-m Komodos observed to attack a buffalo and horses (Auffenberg, 1972). It obviously had no terrestrial enemies, although a large *Pallimnarchos* would have offered a major challenge, certainly in the aquatic realm. Alternatively, *Megalania* may have been a solitary animal most of the time and because of the lack of skeletal elements found in fossil assemblages it has been suggested that they were never very common, especially compared to other megafauna (Figure 7.3). *Megalania* remains consist almost entirely of vertebrae and teeth, although isolated fragments of post-cranial bone are also found in long-forgotten boxes of fossil bone detritus in obscure parts of museum collections. It was widespread in the southern Lake Eyre basin (SLEB), with vertebra and teeth the most common elements found in seven sites along the Warburton, Cooper and Kallakoopah Creeks that feed into Lake Eyre, as well as on

Figure 7.3 Comparison of a *Megalania vertebrae* with that of a full grown Goanna (right). *Source:* Photograph by Steve Webb.

the edge of a small lake in the Tirari Desert east of Lake Eyre. Evidence for this animal has also been found northeast of Lake Eyre in Queensland as well as southeast of Lake Eyre in New South Wales.

Megalania's only challenger would have been *Pallimnarchus pollens*, a giant freshwater crocodile of which there may have been two species, the other being *P. gracilis* (Molnar, 2004; Willis & Molnar, 1997). It has been found in various areas of the upper Lake Eyre catchment in northeastern Queensland, at Glen Garland, Rosella Plains, Tambo and Riversleigh. Large feeder streams flowing southwest probably acted as entry/exit corridors for this creature, allowing it to move into the Lake Eyre region during the very wet times that occurred during some interglacials. Described as an aquatic ambush predator, the average length of *Pallimnarchus* was 5–6 m, although some jaw fragments and teeth suggest they grew much larger, with some specimens achieving 7–8 m. They obviously lacked enemies except other large crocs, a situation that probably helped them achieve very large proportions. The remains of these creatures mainly consist of vertebrae, pieces of cranial and jaw bone (which are usually dense and very thick) and teeth – similar skeletal elements to those of *Megalania*. The salt water or estuarine crocodile, *Crocodylus porosus*, becomes heavier when living in freshwater habitats rather than tidal rivers but it is unlikely to have penetrated 1400 km into Australia's interior where *Pallimnarchos* is found. However, today they are known to swim 200 km upstream in the Fitzroy River in the northeastern Kimberley as well as similar distances down the Daly River in the Northern Territory (Webb, personal communication, 2009; Webb & Manolis, 1989) and considerable distances up other northern Australian rivers. *Pallimnarchos*, then, is probably the only candidate to have penetrated far into central Australia. It has been included in the megafauna list but it had a limited distribution and relied on the existence of inland waterways and long-term lake fillings in the centre of the continent; although big and fierce, it would have been the first to succumb to the environmental change and dispersal of its prey.

The third reptile is *Wonambi naracoortensis*, a python that is large but not related to the present constrictor family of the genus *Morelia*. It probably reached 5–6 m and seems to have been a thick-bodied snake rather than a long one; it was restricted

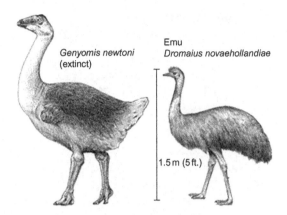

Figure 7.4 Comparison of *Genyornis newtoni* and an emu.

to South Australia and southwest Western Australia. *Wonambi* is an Aboriginal word which is associated with their Dreamtime ancestral Rainbow Snake and points to a possible link between the early Australians and this giant Madtsoid family, which is now extinct.

Genyornis

Because of an intensive continent-wide survey of the distribution and prevalence of *Genyornis* over the last 20 years, we probably know more about its geographic range, extinction threshold and what it ate than for any other species in the megafauna listing. It used to be commonly referred to as a giant emu but *Genyornis* was a Dromornithid, or Mihirung, and was one of a group of five genera, all large, with a broad weight range between 60 and 500 kg that first appeared in the mid-Tertiary. It was the last representative of the group and lived right through to the Quaternary. It weighed between 200 and 300 kg and probably stood at least 2 m high. Although often likened to a giant emu (*Dromaius novaehollandiae*) and having many skeletal similarities with the rattites (emus, ostriches and rheas), *Genyornis* was really a giant goose (Figure 7.4). It probably looked something like a large version of the Magpie Goose (*Anseranas semipalmata*), especially in the shape of its beak and head; mummified remains from Lake Callabonna helped confirm this resemblance. That head and beak shape strongly contrasted with the enormous deep beaks of its giant Tertiary relatives *Bullockornis planei* and *Dromornis stirtoni* which more closely resembled those of the Tertiary terror birds of the Americas, although without the frightening hook on the end. For a full and extremely erudite description of the Dromonorthids in Australia, Peter Murray's and Pat Vickers-Rich's exquisite book *Magnificent Mihirungs*, published in 2004, is certainly worth a read.

Genyornis' fossil presence is easily recognised by its egg shell fragments which are similar in appearance and thickness to emu eggshell but smooth, lacking the usual raised granular bumps on the external surface of emu egg shell. The shell is

Figure 7.5 A large section of *Genyornis* egg found in the Lake Eyre region of South Australia. *Source:* Photograph by Steve Webb.

very durable during interment although it usually fragments further from the time after hatching has occurred, particularly when exposed on the surface. Complete and semi-complete eggs have been found along with a number of fragments bearing holes with internal spalling around the edges showing they were made by piercing from the outside (Figure 7.5). The origins of these penetrations are a mystery and are regularly debated around field work camp fires. Animal predation has been suggested but the jaw pressure and gape required to make the hole without shattering the egg makes this rather unlikely. Moreover, the dental arrangement of the only animals that may have had a jaw gape capable of encompassing the large egg also had dental cusps that were too large to have made a neat hole 4.6 mm in diameter. A larger hole has never been found. They may have been made by humans or by the birds themselves to help free chicks, but the arguments go on. Burned fragments have also been found; their colouration indicates high-temperature burning typical of being left or thrown in a camp fire rather than a bush fire (Figure 7.6A and B). Burning flattens the natural curvature of the shell and these pieces often show a colour gradation from black through dark and light grey, similar to the colour of cremated bone. These have obviously been in high-temperature camp fires rather than caught in bush fires. So the likelihood of human contact with these birds seems difficult to dismiss, and dating of the shells shows that the most recent specimens of *Genyornis* existed between 40,000 and 50,000 years ago, overlapping with human settlement.

While fossil bone of *Genyornis* has been recorded in only 21 sites, we are fortunate that it laid eggs. Immense scatters of egg shell have been found in some places that obviously indicate this bird sometimes congregated in large nesting groups, each laying a similar number of eggs. It also suggests that driving these birds from their nests may have been quite a hazardous occupation if they decided to defend them in the manner of giant geese. Egg fragments are usually found in patches that sometimes stretch for kilometres. They are particularly common in dune fields where

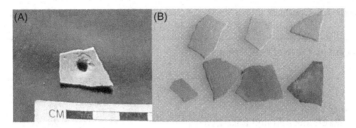

Figure 7.6 An internal view of a fragment of *Genyornis* egg shell with the hole often found in this bird's egg shell (A). It is surrounded by spalling indicating the direction of piercing from the outside. (B) shows several pieces of shell with a darker colouration indicating various degrees of burning. The colour change is only slight on the top pieces while those at bottom have dark burning typical of being in a high temperature fire.
Source: Photograph by Steve Webb.

fragment clusters emerge from dune crown blowouts that have scoured down to the palaeodune core, as well as in eroded scalds in inter-dune corridors at the base of the dunes where late Quaternary environment is exposed. During extensive field work for over 20 years the bird's egg shell has been found in hundreds of sites across the continent, with the exception of the vast desert inland. That work has been led by Gifford Miller of the University of Colorado, Boulder together with John Magee of Geoscience Australia, myself and other 'scouts' who have walked what must be hundreds of square kilometres of the continent. Table 7.5 gives the location of 206 main sites where the shell has been found, with almost every site representing multiple collection places that cover from a few square metres to several square kilometres. Egg shell is fairly common in northern South Australia, western New South Wales and along the central Western Australian coast but the most remote pieces yet found are from Lake Lewis, a medium-sized salt playa north of the West MacDonell Ranges in the Northern Territory. This location on the far eastern edge of the western deserts of central Australia is the one exception to an otherwise megafauna-free region that extends west to the coast of Western Australia. So, like the rest of the megafauna, even *Genyornis* seems to have given the western deserts a wide berth. Miller's laboratory has carried out amino acid racemisation dating and C^3/C^4 analysis on hundreds of samples which have confirmed the omnivorous nature of this bird's diet. Further, the continental distribution of egg shell suggests *Genyornis* was a broadly adapted creature that probably preferred to live in colonies. It may also be the only species at present whose extinction timing is anywhere near certain (Miller et al., 1999), although *Diprotodon* may be close to that same threshold (Roberts et al., 2001).

Diprotodontids

The Diprotodontids were the next most ubiquitous animals on the megafauna list after *Genyornis*. Primarily these were *Diprotodon optatum* and *Zygomaturus trilobus*. These giant marsupials lived in many areas across the continent from Cape York to King and Kangaroo Islands and around Lake Eyre, and single items of their

Table 7.5 Continental Distribution of 206 Sites Where *Genyornis* Egg Shell or Its Fossil Bone Have Been Found

Region or Place of Collection	Approximate Number of Sites
Lake Eyre region	58
Darling River Lakes (NSW)	31
Intradesert areas around Lake Eyre (SA)	24
Port Augusta area (SA)	23
Ningaloo, central West Australian coast	20
Lake Frome area (SA)	13
Warburton Ck (SA)	6
Willandra Lakes System (NSW)	6
Lower Cooper Creek (SA)	5
Lake Torrens area (SA)	4
Kallakoopah Creek, Southern Simpson Desert (SA)	3
Lake Lewis, Napperby, (NT)	2
Woomera area (SA)	2
Naracoorte/Mt Gambier (SA)	2
Sandringham Stn, Western Queensland	1
Cuddie Springs, (NSW)	1
Burra District (Goyder) (SA)	1
Parachilna (SA)	1
Brothers Island, Coffin Bay (SA)	1
Adelaide (SA)	1
Lancefield, Victoria	1

Source: Data from G. Miller, personal communication.

skeleton have been found in remote parts of Western Australia. *Diprotodon* is now considered a monotypic species after having been assigned various names during the last 150 years because of confusion over metrical variation caused by sexual dimorphism. It has now become a symbol of Australia's megafauna as well as being the largest marsupial ever, and if its fossil remains are anything to go by it seems to have been the most common – although, as a rhino-sized animal, it had large bones which, all things being equal, generally last longer in the fossil record and are often more obvious and recognisable in fossiliferous deposits. It occurs in sites encircling the central Australian deserts and semi-arid regions; in the Gulf country; along major river systems such as the Diamantina and Cooper that run from Queensland into South Australia in isolated spots like Winton, Barcaldine and Boulia in western Queensland; Cootanoorinna, 150 km west of Lake Eyre; and in remote Western Australian sites. Its cousin *Zygomaturus* had a similar distribution, from the Kimberley to southwest Tasmania and Lake Eyre, but its fossil remains are rarer and more often associated with riverine, previously forested and coastal environments. Even those found inland reflect the different environmental conditions that it occupied then from those found there today, a good measure of the vast environmental

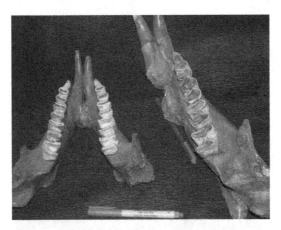

Figure 7.7 The mandible of a dwarfed *Zygomaturus* found on Kangaroo Island (left) compared to the larger mainland form.
Source: Photograph by Steve Webb and personal communication R. Wells.

changes that have taken place. *Zygomaturus* obviously moved on to islands during times of low sea level because it has been found in Tasmanian and on Kangaroo Island. The Kangaroo Island inhabitants, however, underwent some degree of dwarfing, probably as a result of isolation there during interglacials when sea levels rose and cut off the island from the mainland (Figure 7.7) (R. Wells, personal communication, 2007). The other two diprotodontids include *Eowenia grata* and *Palorchestes azeal*. *Eowenia* was smaller than its *Diprotodon* cousin and is not known beyond the Darling Downs of southeast Queensland. Its limited range would have always made it vulnerable to climatic and environmental changes and in that regard, some debate exists over its actual relationship to *Diprotodon*. *Palorchestes* has been described as a marsupial anteater but is probably better imagined as Australia's answer to the giant ground sloths of South America, although smaller. It possessed a fierce set of claws on its front paws, strongly built forelimbs, a short trunk and a long protrusible tongue similar to that of a giraffe. Its anatomical equipment suggests it fossicked for termites and ants in logs and upright tree trunks. Its hind limbs also had large claws that may have been useful for gaining purchase when its front claws were tearing bark off trees. Such a specialist is always vulnerable when ecosystems change, even of the smallest kind. It is also unlikely that such animals ever existed in large numbers; in fact, they were probably shy animals living a lonesome life or in sparsely dispersed small families. The fossil remains of this animal are not common, with only a few representative samples found. Nevertheless, it has been found in all parts of the continent except the Lake Eyre basin, and seems to have been most common in Queensland.

Diprotodon distribution could indicate that it was a more adaptive animal than Zygomaturus and that it may have had a behavioural and/or physiological capability enabling it to withstand at least semi-arid habitats. It may also have had

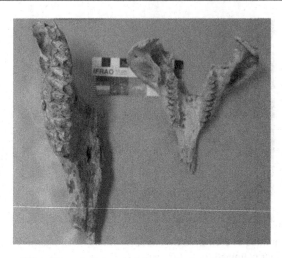

Figure 7.8 The mandible of the giant wombat *Phascolomys gigas* compared to its modern counterpart (*Vombatus*) (right).
Source: Photograph by Steve Webb.

physiological adaptations to semi-arid conditions, such as urine concentration and a low metabolic rate similar to modern wombats. This could also explain the widespread distribution of the giant wombat *Phascolonus gigas*, which may also have been adapted to semi-arid conditions like its smaller modern counterpart, the hairy-nosed wombat (*Lasiorhinus* sp.) (Figure 7.8). *Diprotodon* distribution not only exemplifies a broad adaptation to many Australian environments, but collagen isotope study of their dental enamel also indicates an opportunistic browser and grazer of C_3 and C_4 plants (Gröcke, 1997). If this is true, any climatic or environmental change switching between these plant groups would not have particularly disadvantaged it *per se* and this may be a clear difference between *Diprotodon* and *Zygomaturus*. Dietary evidence recovered from the stomach region of well-preserved fossil *Diprotodon* from Lake Callabonna indicates a mixed diet of chenopods, sclerophyll shrubs, grasses and daisies, indicating a grazing animal which was also know to be a browser (Johnson, 2006; Stirling, 1900). It may also have been semi-aquatic, as has been suggested for *Zygomaturus*, living much of the time along or very close to rivers and lakes, almost like a marsupial hippo. Its widespread presence around Lake Eyre, with its many late Quaternary palaeoriverine environments, might also be explained that way. However, while diprotodontid remains in the LEB are overwhelmingly found close to or associated with lacustrine and riverine environments, it is hard to discern whether they kept close to water for all the usual reasons; either they actually needed to wallow in it or such environments redeposit bone assemblages to make them look like water junkies. One hazard diprotodontids faced was crocodiles. While crocs and hippos usually avoid each other in Africa, deep crush pits on diprotodontid long bones found in the Lake Eyre region suggest that they

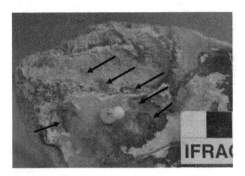

Figure 7.9 Crocodile tooth crush pits on a *Zygomaturine* humerus. The tooth row is marked by arrows.
Source: Photograph by Steve Webb.

were made by *Pallimnarchus* and that, while they both shared a riverine environment, they did not enjoy a similar arrangement (Figure 7.9).

Rare or Common?

The widespread distribution of *Genyornis* and *Diprotodon* contrasts with most species in the megafauna list. Canonical distribution predicts that within a biological community there are some very rare species, many that are fairly abundant and just a few that are very abundant (Preston, 1962). That seems to be the case with Australia's megafauna. They too were distributed in a similar manner with some being common but many more not so common or rare. None was very abundant, although *Genyornis* probably came close, together with *Diprotodon* (Appendix 3). Having said that, what does 'common' mean? Probably not much, because while fossil remains of one creature or another might look common, they are usually spread over time – a lot of time. So until those remains are individually chronologically dated, 'common' cannot be proven. We will have to settle for 'widespread' instead. *Zygomaturus* was also fairly widespread together with the giant kangaroos *Procoptodon goliah* and *Protemnodon brehus*, some of the sthenurine kangaroos, the very large wombat *Phascolonus gigas* and Australia's largest marsupial predator, *Thylacoleo carnifex*, which seems logical as it was probably the top predator of larger species. On the other end of the scale, we have a group of 13 species (25%) that are geographically restricted, being found in only one part of one state or represented only by single examples. There are five of these in Queensland, three in New South Wales, one in Victoria and three in South Australia. Such confinement also suggests that they were specialised animals tied to a particular niche or habitat. Today these would probably be designated as rare and at least *vulnerable*, perhaps even *endangered*, and susceptible to even relatively small environmental or habitat changes. Another four species are found in only two states, bringing the total of rare or very rare species to 33%. Table 7.1 also shows 14 species that at present are believed to have gone extinct before MIS 6.

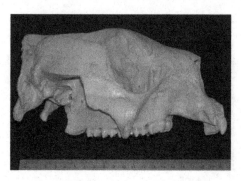

Figure 7.10 Lateral view of a *Sthenurus tindalie* cranium. This large species was one of the biggest sthenurid short-faced kangaroo browsers.
Source: Photograph by Steve Webb.

Kangaroos

Marsupials make up 90% of the extinct megafauna species in the list of 52 animals and Kangaroos comprise 71% of those, the largest single group. Of those, 23 (44%) are short-faced sthenurines with 14 species making up long-faced varieties. The sthenurines include the genera *Procoptodon, Sthenurus, Simosthenurus, Metasthenurus, 'Procoptodon'* and *'Simosthenurus'* (Figure 7.10). Genera enclosed in quotes have emerged from a recent reassessment of the subfamily Sthenurinae (Prideaux, 2004). Their adaptive radiation probably began around the time of the Miocene–Pliocene boundary, with *Sthenurus* and *Simosthenurus* becoming the most widespread genera by the early Pliocene (see also Chapter 5). The radiation of these browsing animals seems to have been driven by the gradual appearance of arid vegetation communities and open habitats resulting from the continued drying of the continent. Sthenurine distribution in the mid-late Quaternary shows these animals as an extremely successful and ubiquitous group, with the largest varieties occurring in South Australia (*n*=18), Queensland (*n*=14), the LEB (*n*=12) and New South Wales (*n*=18). In contrast, only six species are found in the southwest of Western Australia, with another six on the Nullarbor, all of which are found in South Australia although none have been found in the southwest. It must be pointed out that three sthenurids found in New South Wales (*'Si.' brachyselenis, Si. euryskaphus* and *S. murrayi*) are only single examples of their respective species, as is *'P.' mccoyi* in Victoria and *'Si.' sp. cf antiquus* in South Australia. It is difficult to believe that an animal so widely distributed across the eastern half of Australia should not have lived in the northwest or Northern Territory unless they were temperate creatures. Only four lived in Tasmania, which also indicates that sthenurines may have preferred warmer climes and were more suited to open woodland, and that some did make the journey south to Tasmania when Australia was in a glaciation and sea levels were low. Sthenurids came in a variety of sizes but just over 65% were on the small side even though widely accepted as megafauna with a weight range between 50 and 80 kg. Larger varieties weigh twice that, with *Procoptodon goliah* the largest at around 230 kg.

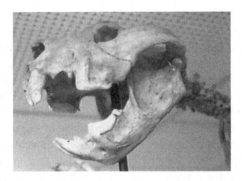

Figure 7.11 The head of *Thylacoleo carnifex*, a marsupial lion that possessed the strongest bight among modern mammals. The single paired carnassial molars are prominent together with the knife-like ripping lower incisors.
Source: Photograph by Steve Webb.

Long-faced macropods comprise the genus *Macropus* and *Protemnodon*. Extinct examples of the former are not common and those that are have confined ranges; indeed, they are almost all rare species with five of the six found only as single examples with three in Queensland and two in South Australia and one in New South Wales. Only two weighed around 150 kg; the other four weighed around 50 kg, well within the modern kangaroo weight range. The three *Protemnodon* species were much more common, spread from Queensland to South Australia with one found on the Nullarbor and a second in the southwest. All three species weighed in the 110–150 kg range. Other macropod genera were *Baringa*, *Congruus* and *Bohra*, all weighing around 50 kg or over, the latter being a giant tree kangaroo.

It is worth briefly mentioning something about the lioness-sized *Thylacoleo carnifex* the only large and totally dedicated marsupial predator on the list. This was Australia's largest mammalian carnivore and probably the most specialised (Archer, 1984: 684; Finch, 1982; Wells, 1985; Wells et al., 1982; Wroe, 2002). It was strong, although probably not swift, but it may not have needed to be. It may also have had semi-arboreal habits. Its specialised carnassial dentition, strong enough to guillotine through thick long bone, convergent incisors for piercing or stabbing and pseudo-opposable thumb with a very large ripping claw on all thumb digits made it a fearsome creature (Figure 7.11). It is unsurprising that this animal is common to most parts of the continent because wherever other large animals resided, so did *Thylacoleo*.

Australian Megafauna: Where Did They Live?

A picture of Australia's megafauna cohort begins to take shape but there are two questions to ask for each species. One, did it go extinct before the threshold? And, two, how common was it? Those questions encompass our biggest problem with

Australian megafauna, which has been trying to understand their temporal and spatial distribution. Future research needs to focus on these two issues, particularly the temporal placement of individual species, in order to establish individual extinction thresholds. Temporal and spatial distribution of an animal is a cornerstone for knowing its proper context in the environment and life history. To study modern animals we need to know where they live, what environment they prefer, how many there are, whether they are rare or common, highly specialised or generalists, whether they are tropical or temperate animals, what their range of variation is, what they eat, their reproductive parameters and so on. From those parameters we can then make an assessment regarding their vulnerability or whether they are endangered. Many of the basic questions that a zoologist or biogeographer might ask about a modern animal cannot be known for megafauna but we need to get as close to answering them as possible if we are to understand how and why they went extinct. We may not know the extinction timing for many megafauna species but we can try and find out more about their spatial distribution and in so doing answer other questions concerning their biogeography and thus their lifestyle and degree of vulnerability.

In order to begin the task of putting together a spatial picture of the megafauna, 350 fossil megafauna sites in all Australian mainland states and Tasmania have been surveyed (Webb, 2008) (Appendix 3). Papua New Guinea has been excluded from the survey although it was part of the Greater Australian continent (Sahul) during times of low sea level. However, the general lack of understanding of its species range, distribution and structure – made extremely difficult due to its dense jungle terrains and difficult access to many areas – makes it hard to provide a meaningful assessment with regard to its fossil megafauna population. What we do know is that the fossil record from there is very limited and probably does not in any way represent the total number of megafauna species that may have lived there. Or perhaps it does, but the difficulty of working there makes a broader survey of the country almost impossible; therefore, little in the way of a conclusion can be offered at this time.

Australia's site distribution shows that megafauna lived from Cape York (15° south) across to Western Australia and into the now-arid southern Lake Eyre basin (SLEB) (Appendix 1). In eastern Australia, they are found on either side of the Great Divide from the coast to Queensland's inland Darling Downs and the central west of New South Wales. During low sea level, 10 species spread to Tasmania, as far as the Florentine Valley (42° south). They moved onto King Island, off Tasmania's northwest coast, during times of low sea level and onto Kangaroo Island off the coast of South Australia. Presumably, they also moved onto other parts of the exposed continental shelf during times of low sea level. With the exception of *Meiolania*, all megafauna species can be found in southeast Australia, a distribution probably reflecting more reliable fodder and water resources in the well-watered temperate and subtropical environments. Site distribution in eastern Australia is extensive but becomes patchy further north, petering out in northwestern Queensland and the far northwest of New South Wales. At present the evidence suggests that, unlike their earlier ancestors, Quaternary megafauna seem to have avoided much of central Australia but the environment there had changed. They are also absent from a large part of western Australia, with none recorded in the Western Deserts or the whole of the Northern

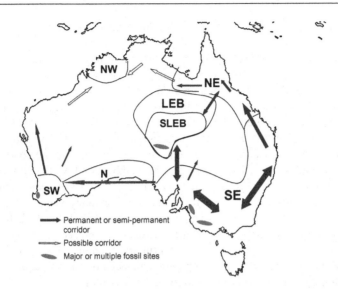

Figure 7.12 Six basic areas of megafauna distribution across Australia and listed in Table 7.1. These include the southeast (SE), northeast (NE), northwest (NW), southwest (SW) and the Lake Eyre basin (LEB). In Table 7.1, the Nullarbor section (N) is included in the SW column and the Lake Eyre column refers only to the southern section of the Lake Eyre basin (SLEB) (hatched area).

Territory (with the exception of a few pieces of *Genyornis* egg shell), which covers an area of more than 2,000,000 km², over a quarter of the continent. Altogether megafauna seem not to have occupied perhaps as much as two-thirds of the continent. It is possible that we have just not found the fossil evidence there but at this stage we can only go on what we have. At present, then, I have divided their occupation areas into southeastern Australia, northeastern Australia, the southwest of Western Australia, a small pocket on the northwest and the Lake Eyre basin (LEB) (Figure 7.12).

Southern and Southeastern Australia

The southeastern section runs from southeastern Queensland through eastern New South Wales and Victoria and includes eastern South Australia. The environments of this region varied but generally consisted of temperate mosaic vegetation made up of grasslands, wet and dry sclerophyll forest, savannah woodland and shrub lands as well as temperate and subtropical rainforest in areas along the east coast. Such variety offered a wide range of browsing and grazing habitats and niches and these were underpinned by permanent lakes, major rivers and streams and wetlands associated with many large river systems across the region. Range extension northeast and southwest would have been dictated by the permanence of food and water resources and the degree of aridity, which fluctuated enormously across inland regions and spread coastward during glacials. Geographical continuation of favourable environments along the east coast allowed animals to populate those areas as well as some far inland, depending on conditions. It is logical, therefore, that because megafauna required good water and fodder and would have favoured a stable environment, southeast Australia offered optimum conditions for habitation. As such, it would also have acted as a feeder of animals to other regions in good times.

From the standpoint of fossil visibility, southeastern Australia also contains the biggest late Quaternary sites. Prominent among those are Lancefield, Reddestone and Cuddies Springs as well as the fossil lake systems of the Darling and Willandra, Lake Victoria and Lake Callabonna. A series of cave systems in the southeast such as Russenden, Wellington, Naracoorte, Mount Gambier and McEachern have also yielded many megafauna fossils, as have a range of open sites across the southeast, so the region has been a great fossil resource for many years. Cave environments offer good preservation, particularly vertical sink holes, which have acted as natural pit traps, in many cases securing exquisitely preserved whole skeletons of megafauna; the Naracoorte caves are a primary example of such preservation. Assemblages from these areas have all greatly enhanced our understanding of megafauna morphology, variation, species variety and distribution.

Southern Lake Eyre Basin

The SLEB region's present environmental and climatic profile belies its appearance at certain times in the mid-late Quaternary, particularly MIS 5. It lies in the most arid part of the most arid continent with an average 120 mm of rain annually and over 4 m of evaporation. But during MIS 5e the region was quite different, boasting large river systems that brought enough water to form a megalake structure that expanded Lake Eyre beyond its present boundaries and joined it with smaller neighbouring lakes, including Frome and Callabonna. The lake and channel systems presented an ecological setting that was home to 22 of the 52 species (42%) in the mist, which made it the second most important area for megafauna variety on the continent after the southeast (Table 7.1, Appendix 3). Megafauna have been recorded from over 40 main sites around Lake Eyre, along main river channels and into the southern Simpson Desert. The main channels are those of Cooper Creek and the Diamantina/Warburton River, which stretches from Lake Eyre 500 km northeast to Birdsville and then up into northern Queensland. Other fossil areas are found in the Tirari Desert and along Kallakoopah Creek, an anabranch of the Diamantina River that circles in a large loop north into the southern Simpson Desert before running south to join the Diamantina again just before it enters the northern end of Lake Eyre.

The first fossil discoveries in the Lake Eyre region were made to the southeast on Lake Callabonna in 1885 and in terms of size, this was the first of the great megafauna discoveries in Australia (Stirling, 1896, 1900 1913; Stirling & Zeitz, 1896, 1899, 1900). The fossils consisted of a large collection of well-preserved *Diprotodon* skeletons in excellent condition together with footprints, lake floor body impressions and even gut contents. Other megafauna included *Genyornis newtoni*, some with patches of gastroliths or 'gizzard stones', and the giant wombat *Phascolonus gigas*. *Diprotodon* and *Genyornis* were already known from the discoveries made by Thomas Mitchell in the Wellington Caves, New South Wales in 1836 and which were subsequently sent to the famous comparative anatomist Sir Richard Owen in London who named these creatures as well as *Megalania*. Since then, extensive palaeontological work has been carried out in the region, showing it to be one of the richest megafauna and Tertiary fossil regions in Australia, with discoveries documented in a substantial set of papers and volumes (Stirton, 1967; Stirton et al., 1961, 1967; Tedford & Wells, 1990; Tedford et al., 1992; Webb, 2008, 2009, 2010; Wells & Callan, 1986; Wells & Tedford, 1995). Many prominent researchers such as Tom Rich, Paul Lawson, Mike Plane, Neville Pledge, Mike Archer, Mike Woodburn and the late Dominic Williams cut their palaeontological teeth in the southern parts of the Lake Eyre system. They made up the core of researchers who have helped tease out the marsupial story of the region. But none of these noted scientists would forgive me if I omitted the names of two great pioneers of that research who began the first serious and systematic field work in 1953: Professor R.A. Stirton and Richard 'Dick' Tedford.

Southwestern Australia

The next largest megafauna population of 14 species is found in the comparatively isolated southwest of Western Australia and it is probably no coincidence that all these animals also lived in the western section of southeastern Australia on the other side of the Nullarbor. Eleven species have been found in caves or sink holes on the Nullarbor Plain, which stretches from western South Australia west into Western Australia as a biological bridge between the east and west of the continent (Table 7.1). So, Western Australia's megafauna are concentrated in two basic regions: the far southwest corner and far to the east out on the Nullarbor Plain. The Nullarbor has a couple of special deposits where superb fossils have been found. These, oddly, include an example of the giant tree kangaroo *Bohra* – I say 'odd' because Nullarbor means 'treeless'. Almost no trees grow there today, in an interglacial, so it might logically be expected that none would have grown there during a glacial. Perhaps the presence of *Bohra* indicates a time when environmental conditions favoured the growth of trees that must have prevailed at one time, and that time was around 250 ka, the date for this particular fossil on the Nullarbor Plain. Like the Nullarbor, the southwest also has cave deposits to thank for its megafauna record; indeed, without caves such as Mammoth, Devil's Lair, Strong's, Tight Entrance, Skull, Moondyne and many others, Western Australia would look as though it was almost devoid of megafauna.

The megafauna list from the southwest has no surprises. It contains some of the most ubiquitous animals which have been found in all other areas. They include *Diprotodon*, *Phascolonus gigas*, *Thylacoleo* and six of the commonest sthenurids. It is odd, however, that none of those six species has been found on the Nullarbor, which could be assumed to have been a natural staging point for animals crossing to the southwest. Perhaps they are yet to be discovered. What does stand out is the fact that none of the animals are unique to the region. All must have migrated over at some time in the past. It may have the case that some of these species were actually moving to a sub-species level after long bouts of isolation and the effects of founder effect, genetic drift and a small gene pool.

Northwestern Australia

Only four species have been recorded in the northwest Kimberley region of Western Australia, namely *Phascolonus gigas, Diprotodon, Zygomaturus trilobus* and *'Procoptodon' browneorum*, but the latter two were found as Aboriginal artefacts (Akerman, 1973). These four species are found throughout the continent but as artefacts the *'Procoptodon'* and *Zygomaturus* fossils could have originated elsewhere, perhaps traded over hundreds of kilometres, so they must be regarded as difficult to definitely assign to the region. Quanbun in the Kimberley has yielded *Protemnodon brehus* and *anak* and *Phascolonus gigas*, but this site is believed to be Pliocene, although it still awaits formal dating, which leaves only *Diprotodon* as the single local. Nevertheless, for the purposes of this work I have accepted the first four species as representing the northwest's small cohort of megafauna.

The west coast of Western Australia between the southwest and northwest has yielded only a few fossils including a *Zygomaturus* on each of the Murchison and Greenough Rivers and *Diprotodon* at Karatha and Dampier along the northwestern Pilbara coast. Three other *Diprotodon* have been recorded from the Oakover River, also in the Pilbara but 300 km from the nearest coast. Another two have been found far to the south and again inland, one at Karonie, in the Goldfields district, the other near Lake Darlot east of Leinster. These few exceptions are the only localities where fossil bone has been found north of the southwestern corner. The fossil picture for an area into which you can fit India makes the region, with the exception of Diprotodontids, look megafauna empty with only one extinct macropod (and that may not be firm). One saving grace is *Genyornis*, which lived along the west coast (see above). It seems unbelievable that this situation is what it seems across such a broad area of the continent.

Northeastern Australia

Only nine species of megafauna have been found above the Tropic of Capricorn in Australia's northeast, in addition to *Pallimnarchos* and *Meiolania* which seem to have been endemic to the region. Both these animals, as well as eight of the northeastern mega-marsupials, are found in the SLEB including the four found in the northwest, but those are widely distributed as are *Protemnodon anak* and *brehus*, *Thylacoleo carnifex* and *Genyornis*. The trend seems to be that fewer species are found the farther north we go on both sides of the continent, although this is less so in the east. The shape of the megafauna distribution takes on the image of a pair of horns stretching out north and tapering. To emphasise this analogy, no megafauna are found in the central north of the Northern Territory. A recent survey of dunes on the eastern tip of Arnhem Land at Yirkalla produced no evidence for these animals, not even *Genyornis* egg shell. The one salient environmental factor that stands out in terms of animals reaching the northeast is the lack of desert and generally arid country for great distances that exists in the west. Thus a corridor for the movement of megafauna north along the east coast was almost certainly present and probably open most of the time (see below).

The statistics above show that the quality and quantity of Australia's megafauna population was largely dubious in terms of the continental picture. Species quantity varies wildly, and species variety is inconsistent, with some species veritably rare or almost non-existent. Before leaving this section it is worth reflecting on the not-so-surprising distribution of Australia's modern marsupial fauna. Apart from 17 species that have gone extinct in the last 150 years or more, the distribution of modern species seems to reflect that of the megafauna. Out of 112 species, 145 of them live in eastern Australia, with 69 in the west of the continent (Strahan, 2000). Most (131) are concentrated in the southeast with 67 in the northeast, 45 in the southwest, 39 in the northwest, 22 in the Lake Eyre region and another 30 in the central deserts, a distribution broadly reflecting that of the megafauna, although we now have animals that are now fully adapted to a dry continent with many quite happy living in desert conditions.

Megafauna Demography: Patches, Corridors and Feeders

Successful continuation of an animal population relies on its flexibility and movement in the face of change. That flexibility is, in turn, a reflection of the robustness of its biological makeup and adaptive qualities, which are linked also to its viability and genetic vigour, and we need to examine these variables more closely.

Animal movement depends on area and distance effects, particularly population size relative to the size of the area they occupy. Different species require different range and patch sizes to support their population and keep it viable. With a mixture of animal types, some may survive in a given patch size while for others it is too small. In many parts of the world, corridors have been created to join up neighbouring populations where forest and other environments have been encroached by humans farming and logging and through development. The purpose of these 'green belts' or 'green corridors', it has been argued, is also to keep lanes open for animals to travel from the remaining miniscule areas kindly left them in the face of civilisation's onslaught. These corridors are meant to enable animals to move – thus joining populations, in effect – so patches are expanded by joining them up. It was hoped also that maintaining a connection between populations would allow them to continue genetic contact and thrive. It is very obvious, however, that such corridors are not as functional as they were once thought to be. Corridor width and length, as well as the patch size at either end, are important to their success as ecological lifeboats. If they are too small for the designated animal it might not use them and they become useless to that particular animal group. Corridors can also be open to invasion at their edges by pathogens or predators and become places of danger. The size of the animals under concern is also important: the larger the animal the bigger the corridor needed, as well as the patch at either end. But what about the past?

There is no reason to believe things were any different in the past. Therefore, I want to use the concepts of *corridor* and *patch* in a wider context and apply them to the Australian megafauna. We can look at corridors and patches as they might have occurred in the past across the continent. In this case, they are not products of human expansion and encroachment in the living space of animals but rather products of natural episodes of geographic and biogeographic change driven by glacial cycling. The size of an animal and its environmental and fodder requirements determines how much country it needs; thus, different species require different patch size. The situation is magnified when considering megafauna rather than smaller species living on oceanic islands, for example. We can only assume that Diprotodons needed bigger grazing areas than smaller kangaroos and that different species of kangaroo had their own range sizes which were different from those of others. To look at this in more detail I want to begin with *patch size* then move on to the *corridors* that opened and shut between them.

Patches

What emerges in the above discussion is the importance of *patch size*. Patch size is the area required by a particular species to thrive and that varies across species but as a general rule larger animals need bigger patches to support a given population. Larger patches also hold larger numbers of animals but there is a distribution across the size range of the animal population inhabiting a given patch. That usually follows with more small, fewer medium and few large animals in any one patch. Doubling the patch size, moreover, does not necessarily mean the possibility of doubling the population of large animals, although smaller species may be able to accomplish this. So patch reduction eliminates large animals first while preserving smaller varieties. That of course depends on the size of the population and the patch size in the first instance. Small changes to a patch area may displace large animals (or eliminate them) while smaller species are not especially affected by the same changes.

Consider a long-term drought which has displaced a group of Diprotodons. They then find a patch with food and water to move into. It might be assumed that because they have found a refuge they can sit down, have a pant and wipe their brows – they have been rescued! But that is not necessarily the case. We have to ask what size is the animal – big or small? Then, is group size large or small? It can be a large group of small animals or a large animal with a small group size; each will need a different patch size to survive. If group size is too large, the patch becomes unsustainable so animals will be lost over time as the species–area relationship operates and equilibration of the group takes place. Refuge size and equilibration may then reduce the population to a size which is too small to survive in the medium or long term, so even though the group moved away from a degrading environment, it is still not safe. Again, this is particularly important for larger more so than smaller animals, many more of whom can survive on a given size patch than big animals, so if viable land becomes a premium big animals suffer first. Even with a few large animals surviving in the relatively small patch, there is another problem: viability.

One consequence of being a small, isolated population is viability, with different species requiring different minimum numbers to maintain their population (Shaffer, 1981, 1987, 1990; Shaffer & Samson, 1985). The arrangement and shape of patches is important because they can shrink and/or food resources can change in terms of quantity and/or quality. A drought situation in a small patch or across a number of small patches can be worse than in larger patch. Patch shrinkage at these times increases the distance between them, reducing the chances of successful movement from one to another, reducing the numbers of fertile animals and mating chances and making gene transfer harder. This process reaches its ultimate when complete isolation compounds any pre-existing viability problems. Moreover, a small group is always in danger of becoming inbred. A cluster of patches may contain more genetic variance than a single large patch of equal size but if animals cannot meet one another, a dozen isolated populations on a dozen patches cannot rescue the species. But all are subject to uncertainty through human-induced systemic pressures and/or stochastic events which include demographic as well as environmental factors, natural catastrophes and genetic morbidity, such as the accumulation of harmful alleles or 'genetic load' (Gilpin, 1987). Genetic load is an additional issue here and something that takes some time to enact. The genetic load of a large population is spread thinly and expressed rarely, but in small or reducing populations it can lead to inbreeding, population depression and infant mortality through limited breeding pairs. Over time, a range of dysfunctional developments or adverse biological or morphological traits emerges as rare or harmful recessive alleles become homozygous. With population fragmentation comes founder effect, where small populations have a reduced expression of the parent gene pool, which compounds the consequences of isolation. Isolation also causes genetic drift such that an unrepresentative gene pool moves even farther from the original stock. Generally, patches have a homogeneous structure with low genetic variety because of their limited gene pools. Long-term isolation prevents natural rates of turnover, thus compounding these problems further and increasing the predisposition of the population to extinction. Thus, small patches have less species diversity, host fewer animals, have more extinctions, are more vulnerable, even to small environmental disturbances or change and they receive animals rather than exporting them. Large patches cope better with environmental disturbance and can support large groups which do not necessarily suffer from many of the consequences outlined above, resulting in more stable populations that can survive enviro-climatic episodes that would put other populations in danger of extinction or in difficulty of coming back from the brink of extinction.

The principles that operate in patch populations described above are derived from basic and well-understood biogeographic tenets. As such it is time to apply them to megafauna and their demographic dynamics expected during the enormous enviro-climatic changes they experienced in the Quaternary. Approaching Australian megafauna biogeography through the concept of corridors and patches can be closely likened to island biogeography and its principles. This seems a useful way to envisage the situation of population distribution and the consequences of that distribution during glacial cycles. So, it is time to consider the other element of this duo: *corridors*, the connection between patches.

Corridors

The theme of population disjunction and isolation requires discussion of the making and breaking of linkages between patches: those linkages are corridors. Corridors come in various shapes and sizes but basically they include *short-broad, long-broad, short-narrow* and *long-narrow* forms with, of course, variations. These routes between patches opened and closed at different times and for different lengths of time in response to variation in local regional and continental environmental conditions. Corridors allowed animals to move freely between distant patches as well as occupy the land between, but that movement depended on the length and width of the linking corridor and the patch size at either end with movement from larger to smaller patches. Wide corridors had more habitable area available for interchange. Thus they enhanced successful passage and better withstood lateral contraction during environmental downturns with the widest ones probably holding animal populations in their own right. Narrow corridors had a more limited habitat area than wider ones, thus making movement more difficult and less secure; they closed easily by lateral contraction and were less likely to hold animal populations of any size or variety.

Corridor length is an important closure factor because the length of a corridor is inversely proportional to the success of movement along it. Successful movement along a corridor from one patch to another reduces as the distance between them grows. Long corridors also close more easily than shorter ones in the same way that military supply lines become more vulnerable the longer they are because they are more vulnerable over distance and open to something going wrong over that longer distance. For animals, the longer the traverse the more effort is required to reach the end, particularly when the corridor is narrow. There are also more opportunities to perish from thirst or starvation from closures ahead and behind. So, fewer animals achieve success and gene flow becomes tenuous or non-existent. Long, narrow corridors are also unstable because they confine lateral movement even during optimum conditions. They represent, and in fact operate like, highly confined ranges. Once again, however, the chances of gene exchange increase if a large feeder patch lies at one end, feeding animals along it. A small patch has fewer animals that can make the journey so those that take the chance through a long corridor are more open to failure which, in effect, drains a small patch with limited species variety and/or numbers. Short-broad corridors are more open and allow constant movement between patches under most environmental circumstances because they are less likely to close and they usually connect large patches. Therefore corridor width and length produces varying degrees of passage difficulty but movement success can only be assessed when coupled with patch size at either end. This principle is shown in Figure 7.13 where the abscissa or animal transfer numbers increase as patches become larger and corridors shorter and wider.

In summary, large patches have more animals and wider corridors have more habitable area for animals to move within, thus enhancing passage success. Short-broad corridors allow constant movement between patches under almost any environmental circumstances. This type of corridor connects large patches usually with large feeder populations. Conversely, small or remote patches have fewer species and

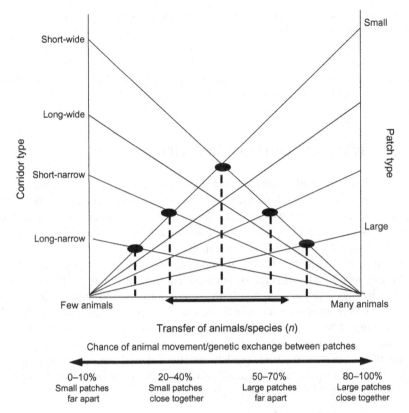

Figure 7.13 Animal transfer between patches and along corridors in a species–area relationship.
Source: After MacArthur and Wilson (1967) and Webb (2008).

smaller populations and narrow corridors close more easily and have a limited habitat passage area than wider ones thus making transfers more restricted and are less successful. Long corridors have the same effect by extending passage time required to move from one end to the other. They are less likely to support large animals and they yield fewer migrations, suffer more extinctions, do not offer long-term safety and are not good refuges. Small patches on the end of long, narrow corridors are the most disadvantaged to receive and support a wide variety of animals, particularly large ones, and thus are places of extinction. So *the probability of extinction is inversely related to the area of the patch* (Gilpin, 1987: 136).

Instead of oceanic isolation, as described by MacArthur and Wilson, the model is applied here to fragmented megafauna populations across Australia as spreading aridity during glaciations closed corridors and reduced and eliminated habitats in the same way that oceanic waters isolate islands. These factors could have caused broad demographic disjunction, separating groups for long periods and so producing a series of 'metapopulations' or small groups isolated from one another (Gilpin, 1987) and vast changes took place throughout the late Quaternary. The effect of the biogeographic mechanisms described above on Australia's megafauna during glacial cycles needs to be seriously considered in any discussion of their extinction because they were part of the normal function of the ecology of these animals. In combination, these principles are used below to interpret how glacial cycles could have affected megafauna populations across the continent.

Palaeopatches and Corridors in Action

The effect of vast climatic shifts brought on by glacial–interglacial cycling brought with them concomitant environmental change to the Australian continent; of that there is no doubt. Such changes naturally triggered the emergence and disappearance of the means by which megafauna moved around different parts of the continent as well as regularly reducing their various populations, literally and at the species level. The way this took place was through the opening and closing of movement corridors and the expansion and shrinkage of patches. The results of the megafauna survey described in Chapter 7 provide a pattern for the presence and absence of different species as well as providing an overall picture of where megafauna lived. But that pattern could not be regarded as constant with the reduction of areas of habitation, sometimes to an extraordinary limited range. Nevertheless, the pattern presented does allow us to make some basic predictions about animal movement, and the dominant continental megafauna movement seems to have been from east to west and from south to north. Major examples include:

- East to west across the Nullarbor (long-narrow corridor);
- Limited south to north movement along the Western Australian coast (very long-narrow corridor);
- From western, southeastern Australia north into the southern Lake Eyre basin (short-narrow corridor);
- Northeastern Queensland into the Lake Eyre basin and Gulf region (long-narrow corridor);
- Darling Downs (inland Queensland) linked populations between south and north Queensland (short-wide corridors).

Figure 7.11 shows those corridors together with some likely but smaller ones that possibly linked various parts of the continent. Arrow thickness indicates possible animal flow strength through the corridor and hollow arrows indicate only the possibility of a corridor. The map also shows a basic distribution of megafauna sites around Australia. The demographic shape of the Australian megafauna population centred on the southeast which held the largest and most varied population. That area may have acted as a hub from which corridors, like spokes, formed temporary or intermittent connections with other areas. In that way the southeast could have been *the* feeder patch, with gene flow largely radiating out. Rather than hosting wall-to-wall megafauna, the southeast consisted of a cluster of large to intermediate size patches, some closely associated and others less densely packed, supporting populations of varying size and species composition in a pattern that thinned towards the inland edges. A look at Appendix 3 certainly suggests this. Some southeastern patches were probably joined by short, very broad corridors extending east and west linking South Australia to southern Queensland through large river systems that flowed across the region and acted as natural corridors. Rivers like the Darling System of western New South Wales provided small corridors, moving animals north at certain times (see below), and that was probably the case for cross-border systems such as the Warrego, Paroo and Barcoo Rivers. That configuration would have provided the opportunity for animal movement supporting far inland populations at optimum times that also relied on groundwater soaks and natural spring systems which occur in that country. Connectivity across the southeast also allowed almost unimpeded population mixing across the region, forming a centre (the main centre?) of genetic vigour in Australia. While one main connective corridor led northeast into Queensland, another more intermittent corridor connected Lake Eyre to the southeast at South Australia and an extremely tenuous and very intermittent corridor linked South Australia to the southwest across the Nullarbor Plain. But this was at optimum times during short interglacials and not all those were optimum. Glaciations were never optimum times and would have brought population contraction even to the southeast.

The division between the southwest and southeast suggests each functioned as an independent entity during glacials, something more easily done in the east than in the west because the southwest was a small patch and was kept that way by its approximate position to the southwestern desert. The large patch feeder of South Australia occupied the western quarter to one-third of the southeast and supported many resident species, certainly enough to supply animals further west to the Nullarbor. The Nullarbor was, however, was part of a long, narrow corridor and not easily crossed, with a tendency to close easily and quickly with environmental change typical of an area closely stood over by a vast desert to the north. It was unreliable enough to provide a barrier to all but 14 species of the 53 in Table 7.1, exemplified by it having only six of the 18 species found in the west. Moreover, there are another 11 megafauna species found in caves on the Nullarbor that are found only in the east. That indicates megafauna only moved east to west from South Australia because no species is unique to the southwest and strongly suggests the Nullarbor corridor opened and shut at various times, depending on environmental conditions, and prevented many travellers from the east going further west. It also supports the principle mentioned earlier that large patches export animals and small patches receive them. Gene flow to the southwest was obviously rare and intermittent, taking place only when optimal corridor conditions prevailed. Although 14 species made the crossing, they may have done so when the glacial–interglacial cycles were in their early stages, possibly in the late Pliocene and early Quaternary. Interestingly, the southwest was a large enough patch to be self-sufficient once seeded with its few species and able to survive when the Nullarbor corridor closed. The southwest has a similar environment to the southeast with reliable food and water supplies and it built a population large enough to provide genetic rigour and population viability, but it did not make it big in the megafauna stakes. The species and size limitation of the southwest probably acted against it exporting species or large numbers of animals north. Difficulties further north including large stretches of arid environment would not have helped in this regard, providing a barrier to northerly distribution. It was a long and narrow corridor with only *Diprotodon* and *Zygomaturus* found to the north, living as outliers both along the coast and in some places inland.

The paucity of megafauna in the northwest possibly shows the difficulties animals faced reaching this area. While four species may have made it there, it seems that few others did. A corridor may have linked the northwest with the southwest although the few fossils found in the northwest only represent four species, suggesting very tentatively that it may have been totally isolated, a long-time refuge neither exporting animals nor receiving them. It may also have been a place where the first complete extinction of megafauna took place sometime before it occurred elsewhere. To get there from the south required crossing the edge of the Great Sandy Desert even along the coast, with few watering points now let alone during full glaciations, although the exposed continental shelf during those times probably contained soaks and other supplies of freshwater. Pollen records from cores taken on the adjacent Exmouth Plateau suggest that during the interstadials of 100, 80 and 70 ka, open grass-rich *Eucalyptus* woodland occurred in the mid-coast region, probably watered by high summer rainfall of between 300 and 450 mm (van der Kaars & De Dekker, 2002, 2003; van der Kaars et al., 2006).The dates roughly coincide, however, with MIS 5a, and two interstadials in MIS 4 (Appendix 2, Graph H). Nevertheless, the Great Sandy region further north was probably always a difficult area to traverse, experiencing frequent, long-term closures as desert moved to coastal edge. It probably remained open only during interglacials and even then only sporadically. During glacial maxima, aridity probably expanded further south, presenting an even more extensive barrier for animals moving north. One thing that must be emphasised is that some corridors might open for only very short periods of time of which only the most opportunistic animals would have taken advantage. They also would have had to be there to move through them. If the surrounding region was already sparsely populated or even unpopulated there would be few or no animals to take advantage of them and the fossil record found so far indicates this was the case. A further danger was that corridors could close just as quickly as they opened, isolating animals, and with minimal numbers they could easily perish.

Another mystery of the northwest is the possibility of a corridor leading east into the Top End of the Northern Territory. Any corridor spanning that region remains doubtful because of the complete lack of megafauna fossils from that broad region, including *Genyornis* egg shell. While some limited survey of this remote area has taken place, taphonomy could be playing a part in the apparent absence of any megafauna evidence as it often does in dictating the visibility, composition and preservation of fossil assemblages. Fossils may be present in the region but so far lie unexposed; contrasting conditions do not preserve them; they are destroyed on exposure; or there are no exposed sediments of a suitable age. Moreover, differential rates of preservation, survey bias, environmental conditions including regular monsoons, as well as a variety of other factors including destruction by domestic livestock, can all skew the picture. However, there are other areas where similar conditions prevail and fossil megafauna bone has been found, including the tropical northeast of the continent. So, our present understanding of megafauna demography requires that, until proven otherwise, the 'empty' areas represent places where megafauna did not live.

It is interesting that Lake Gregory just south of the Kimberley certainly underwent megalake status during MIS 5 but both that and the surrounding area are totally devoid of megafauna bone and *Genyornis* egg shell. A recent survey I carried out along Lake Gregory's main tributary, Sturt Creek, also found no trace of egg shell or megafauna bone. Normally such a tributary, which feeds Lake Gregory from the northeast, would act as a natural corridor for animals to reach the giant lake as they seemed to have done at Lake Eyre. Support for the absence of fossil bone and egg shell in the region as well as in central parts of the continent to the southeast also comes from results of personal survey. For example, Lake Mackay lies 250 km southeast of Lake Gregory in the western Tanami Desert at the very heart of central Australia and it experienced megalake fillings in the same way as Lakes Eyre and Gregory did during MIS 5. It has a comparatively small catchment which does not display any significant or obvious rivers or creeks that could have provided a corridor linking it with any patch feeder, although large palaeochannels do cross the desert to the north but have yet to be investigated. In 2007 I, with others, conducted a survey of Mackay's late Quaternary sediments and MIS 5 palaeobeach ridges on its western edge. Over many days of footslogging (and quad bike riding), visiting and peering at suitable exposures over an extensive area, not one fossil megafauna bone or piece of *Genyornis* egg shell was found. A similar experience occurred a few years prior to that when again I and colleagues of long standing experience surveyed exposures on 13 lakes strung across the Great Victoria, Tanami and southern Gibson Deserts further to the south of Lake Mackay. The lakes included Rason, Carnegie, Meramangye, Buchannan, Burnside and the Serpentine Lakes as well as others. Most had mid-late Quaternary age sedimentary outcrops around the lake edges but we had the same results: nothing. There was no evidence of megafauna or *Genyornis* egg shell, both of which we were familiar with, although the sediments were suitable for fossil preservation

While megafauna have not been found across the Top End of the Northern Territory, the presence of *Genyornis* at Lake Lewis in the Bottom End represents the only species found in the Territory so far and that is 1000 km south of where megafauna had to pass from northern Queensland to enter northwestern Australia. At the moment the only explanation for the presence of *Genyornis* at Lake Lewis is that it must have moved there through a narrow corridor from Lake Eyre heading to the southeast, which is where the nearest evidence for *Genyornis* has been found to date – but that is also over 1000 km away. Perhaps the answer to the puzzle of the absence of megafauna around the megalakes of Gregory (with a suitable corridor) and Mackay (without an obvious one) is that in both cases there were no megafauna in the north to find them. Moreover, all the reasons offered above to explain the absence of fossil remains in northwest, north and central Australia could also apply to the Lake Eyre basin but megafauna are found there, in abundance.

With the exception of one kangaroo (*Procoptodon pusio*) and two reptiles (*Pallimnarchos pollens* and *Meiloania platyceps*), all megafauna found in the SLEB are endemic to the area further south in South Australia rather than to the upper

Lake Eyre catchment of northeast Queensland. It is, therefore, difficult to believe that many species found in Lake Eyre did not move between these two regions. In one respect *Megalania* and *Meiolania* should not be counted because these aquatic species were endemic only to northeast Queensland and moved to Lake Eyre via the large drainage channels that begin in the northeast and were part of the Lake Eyre catchment. The megalake stage at Lake Mackay roughly corresponds to the 125 ka Lake Eyre filling which covered 332 km^3, although those that followed at 80 and 65 ka were incrementally smaller at 216 and 74 km^3, respectively (DeVogel et al., 2004; Magee, 1997). The 125 ka filling must point to an enormous water flow through the basin's 1500-km-long river system far bigger than any of which we have any experience. It was these massive systems that probably formed a series of natural corridors besides the main channels that were taken by *Pallimnarchos* to reach Lake Eyre. *Meiolania* was more of a land-based turtle that must have waddled its way down, probably using adjacent swamps and wetlands as places of rest and recuperation to help it on its journey. The presence of these two reptiles in the SLEB is also the farthest west on the continent that they have been recorded. Today, the Cooper and Warburton/Diamantina systems are the main channels and although they have probably changed in terms of their morphology and position, they played the part of a corridor as did the Willandra and Darling systems further to the southeast. The LEB corridors were, however much bigger but reliable only during regular catchment flow and local precipitation events which took place during interglacials and interstadials.

Zygomaturus lived at Lake Eyre as it did on the Murchison and Greenough Rivers in Western Australia, which suggests an animal that probably favoured water courses. If so, at Eyre they must have been largely confined to rivers and riverine environments by surrounding semi-arid country experiencing tethering at dwindling water holes when systems dried. Riverine corridors were convenient ways of travel but they also had their drawbacks. Regional drying paralysed the area around Lake Eyre for tens of thousands of years during glaciations, preventing habitation by megafauna of any kind. But subsequent relief from these conditions opened up possibilities for opportunistic forays with environmental upturn, with animals travelling up from southern refugia where better environmental conditions prevailed during glacial conditions. The only drawback with these demographic cycles was that the southern population would slowly but inevitably run out of saved species. From cycle to cycle the population as a whole was reduced by repetitive Ice Ages and the demographic pressures they brought. Megafauna slowly lost numbers as they had done during previous events, in an incremental fashion. Thus replacement and survival was becoming increasingly difficult to achieve in the long term. Losses occurred through range reduction, corridor closure and shrinking or total loss of patches during glaciations which predisposed populations to differential species loss and loss within species. It was from this impact that they had to recover; that was not easy and in most cases was not really possible. Hence, the incremental pattern of population reduction although differentially across the range of species. In that way, by MIS 5e times only small numbers of animals were returning to the Lake Eyre region after enduring many glacial cycles during the previous million years.

The Willandra–Darling river systems were extensions of what might be expected to be well-populated areas through the Murray-Riverina region of southeastern Australia. The Willandra system encompasses 13 interconnected dry lake basins, varying in size from 6 to 350 km², located along the lower reaches of Willandra Creek, a palaeo-distributary of the Lachlan River. It was only a comparatively short retreat of a couple of hundred kilometres from arid areas to the north of the Murray River, which must have made it a suitable refuge in times of environmental downturn. The only species found along the Willandra system, however, are *Procoptodon goliah*, *Zygomaturus*, *Sthenurus* sp. *Protemnodon* sp. *Macropus ferragus* and the ubiquitous *Genyornis*, a motley lot. Once again, where there is a river there is a *Zygomaturus*, but only three varieties of kangaroo, two of whom are unknown at species level, have been found there. It is, however, a depauperate group, considering that they were living along an anabranch of the Lachlan River, a major southeastern river; even as an anabranch the Willandra was at certain times large enough to fill five medium to large lakes strung along it. Megafauna remains are comparatively rare on the Willandra, with most only very fragmentary, single examples. This is especially puzzling considering that intense archaeological investigation and monitoring of the region has been carried out there since 1969, so the paucity of remains cannot be due to the lack of professional interest and intense survey over 40 years. The special archaeological focus arose from the discovery of Australia's oldest human remains along the palaeoshorelines of the ancient lake system, but while ample evidence of the menu of the people who made their campfires around the lakes and across the region has been found and analysed, the camp sites and hearths reveal no evidence that they hunted and cooked megafauna. It could mean, however, that emerging as it is from sediments spanning MIS 2–5, the spectacularly few pieces of megafauna bone found there represent a much-reduced species variety by the time humans settled the area – certainly not enough to feed them. It also suggests that the Willandra corridor as an intermittent creek system did not carry many megafauna genes to the area from either the south or the east, indicating the nearest patch population was small. That must have been located somewhere along the Murray River corridor which together with the Darling River is the biggest Australian river system.

The Darling River also has a fossil lake system which lies 100 km west of Willandra. That could have been another small patch on the end of a corridor leading north from northeastern South Australia. The Darling system also has a series of lakes but the range of megafauna found there is far greater. Like the Willandra, most fossils are found in association with distinct sedimentary layers that have built up around two of the biggest lakes, Tandou and Menindee, through alternating wet and dry periods of the lakes. The 11 species found there are *Thylacoleo*, *Diprotodon*, *Procoptodon goliah*, *Protemnodon anak* and *brehus*, '*Procoptodon' oreas*, *Sthenurus atlas*, *tindalei*, and *andersoni*, *Propleopus oscillans*, and of course *Genyornis*. At Lake Victoria, 160 km to the south, adjacent to the Murray River and closer to the patch centre in South Australia, 14 species occur and they reflect a similar range to the Darling lakes with *Thylacoleo*, *Diprotodon*, *Zygomaturus*, *Macropus ferragus*, *Protemnodon anak*, and *brehus*, *Procoptodon goliah*, '*Procoptodon' oreas*, *Sthenurus andersoni*, *murrayi*, *tindalei* and *atlas*, the giant wombat *Phascolonus gigas*, and *Genyornis*. Once again *Zygomaturus* is found where there is a river. The similarity of the fauna of the Darling lakes and Lake Victoria strongly indicates a corridor existed between them probably along the Darling River. They both contrast, however, with the minimal number of Willandra species. The Darling corridor may have been another finger-like extension out of South Australia with its broad species variety whereas the Willandra Creek exited into the Murrumbidgee 200 km east of the Darling and was associated with a region of fewer species. With full lakes along the Willandra at many times during MIS 2–4, it seems water did not necessarily mean megafauna, a reflection of the Lakes Mackay and Gregory experience; it is also possible that in both these cases there were no megafauna to visit them. There was certainly enough food and water to attract humans – they were living there from *ca.* 45,000 years ago. Perhaps the difference between the Darling and Willandra systems actually lies in the way the two were fed. The Darling received and still receives water from the Warrego, Culgoa and Barwon systems. These are extensive rivers that can run huge floods out of southern Queensland and northeastern New South Wales. In fact, the drained water from these rain events ends up in a naturally arid region far to the west that in the past may have provided niches and refuges for megafauna, much like the lakes themselves did, as well as associated swamps and smaller lagoons placed peripherally to the lakes. Willandra Creek's water, however, came from the main Lachlan system that, during glacial conditions, brought melt water run-off from a small glacier on the Snowy Mountains 600 km to the east. That glacier is only one of two that formed in Australia during glaciations; the other was on Tasmania's central highlands. Both provided the only direct icy feel Australia received during glacial events. So water flow through the Willandra was during wet or, in other words, colder times which were probably when the megafauna population was at a low ebb and restricted in the region.

Megafauna variety is more limited than might be expected in the lush tropical environments of the northeast, and far more limited than that of the southeast. But it is there, in great contrast to the north and northwest. There is, therefore, a clinal reduction in species variety along the corridor that went from south to north in Queensland and then out west from there. However, further work may reveal a tropical homeland for some megafauna, possibly because these areas were probably just a touch more stable during glacial periods. Reduced numbers of megafauna in northern Queensland could, however, reflect limited adaptation to rainforest/tropical/sub-equatorial environments – in other words, a fauna generally adapted to a drying continent, which its Tertiary background would support. The area from the Great Divide to the coast was also a place of freshwater run-off and lusher pastures; in other words, it was more stable and reliable in terms of fodder than over the range to the west. Animal movement along this region was possible because environmental circumstances made the east coast corridor a short, wide system that probably contained a number of patches acting as a chain of boosters to animal movement, a stark contrast to the west coast. One of those patch boosters was the Darling Downs region of southeastern Queensland.

It is worth reflecting that today Australia has two major types of bovine domestic stock. One is a species descended from the European *Bos taurus* that is adapted to a temperate environment, its diseases and parasites. The other is *Bos indicus*, an animal adapted to higher temperatures, humidity and the parasites that go with those – in other words, the tropics. The division of these types for commercial farming is not absolute in terms of where they are stocked, but neither is it accidental. Generally, the farming of these two domestic cow species is divided between northern (*B. indicus*) and southern (*B. taurus*) Australia, with some overlap. Perhaps we can see something like this in the megafauna. Is it possible that most megafauna were not particularly comfortable with the tropical conditions of northern Australia? That would explain their relative abundance in the southern half of the continent, particularly the temperate southeast. However, the fact that some species such as *Diprotodon* have been found in northern Queensland, and that Arnhem Land rock art depicts a few other megafauna species, suggests that at least some species may not have had too much trouble with tropical environments. It is worth suggesting that habitation of the tropics by megafauna may also have taken place during glacial times when conditions were not so tropical.

Megafauna Demography and Continental Shelves

While we are up in the north, we can take a closer look at what was going on up there regularly during the last million or so years. Each glacial phase must have effectively pushed animals from the inland to the continental edge by spreading aridity, with some of those moving onto the emerging continental shelf, and I will continue with that in Chapter 9. But at lowest sea levels, the continental area grew by one-third to 10,000,000 km^2 and 70% of that was across the northern coastline. When that took place, an area of central Australia almost twice its present size became arid, which could be regarded as an environmental trade-off whereby habitable land moved effectively outward and offshore. We have seen that glacial onset was almost always a slower process than its termination, usually taking many thousands of years to plunge to maximum cycle depth. Consequently, shelf exposure was probably also a slow process that possibly lagged behind the process of increasing inland aridity, producing a net deficit of available range lands. Moreover, most of the time sea levels were not at a maximum ~145 m but lay somewhere between there and today's levels. So added land almost never equalled that lost through arid expansion. This is difficult to quantify properly because we do not really understand the balance between environmental degradation and glacial cycle depth during the mid-levels of such cycles. Nevertheless, Ice Age strength, enough to reduce sea levels – let's say, to 50 km below what they are today – must have had considerable effect on central parts of our continent.

At maximum low stand, the continental shelf pushed out as far as 500 km in places like around the Admiralty Gulf in the northwest. On other parts of the Australian coastline the amount of shelf exposure varied, probably with the least exposed shelf situated along the central and northern coast of New South Wales. Submarine topographic features in the Admiralty Gulf show ancient rivers, gorges and large lake systems, which at times of an emergent shelf became places for

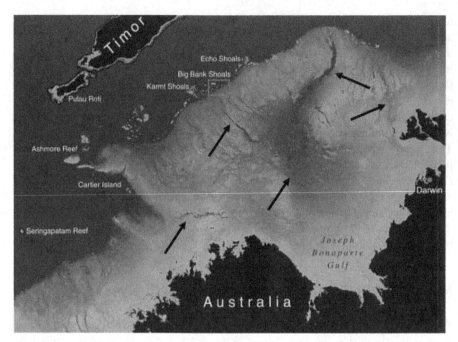

Figure 7.14 A satellite image of Australia's submerged northwest continental shelf. Long ancient river channels, large lakes (arrowed) and other topographic features are clearly visible that would have offered niche opportunities for animals during shelf emergence in glacial cycles.

the development of new environments and niches for animal habitation. These geographic features can be seen clearly on sub-surface ocean satellite images (Figure 7.14). We have little idea, however, how long these environments took to establish themselves or what their composition was like when they became established. It probably took hundreds of years to form them but exposure of the substrate upon which they would grow took thousands of years to occur in step with the slow regression of sea levels. Again, the lag between these two processes is a significant factor to consider because it would seem the emergent biosphere was always ready to house animals as it had plenty of time to develop against a slowly receding shoreline. However, it is not that easy because increased shelf slope and sudden, deep fluctuations in world temperature, which are starkly obvious in the graphs in Appendix 2, could bring equally sudden drops in sea level. There is little doubt that this was not a smooth process and we have yet to consider ocean transgression, which is a much more dangerous proposition for any animal once living out on the shelf. While animal refuges would have become vacant out on the newly exposed shelf, it is important to remember how fast desertification was spreading within the continent because, once again, animals could have been trapped between a faster moving arid fringe and slow-forming shelf environments suitable to shelter them. A biogeographic melee would follow as environments stabilised and plant communities established themselves on the exposed shelf, which to some extent complemented the loss of land within the continent; in this way, shelves may have become refuges. The bottom line is that slow development of shelf environments balanced against faster arid onset and environmental degradation may have cost many animals their chance of a future resulting in at least local extinctions.

Animals living close to the emergent shelf would have explored it without really noticing any particular transition as shorelines gradually moved out over long periods of time. The presence of *Diprotodon* on Tasmania as well as its smaller offshore islands and on Kangaroo Island off South Australia shows clearly that these animals moved across and colonised emergent shelf between the mainland and islands as did other species. Its presence with *Protemnodon anak* and *Simosthenurus occidentalis* on King Island also shows they explored Tasmania to its corners, with the northwestern corner later becoming King Island when sea levels rose again. But that is probably a different situation because, as we have seen, these animals were in far greater numbers in the southeast of the continent than in the north and so needed little encouragement to move to Tasmania once the opportunity arose. It might also suggest that it was easier for them to move into temperate than tropical environments because no fossil megafauna has yet be reported from any island off the north of the continent. So, the situation in the north is far less certain because, as we have seen, on present evidence few megafauna species lived there anyway. Nevertheless, besides *Diprotodon*, possible émigrés to the northern shelf could have been *Zygomaturus*, kangaroos like *'Procoptodon' browneorum*, *Protemnodon anak* and *brehus*, the large wombat *Phascolonus gigas*, *Megalania*, *Pallimnarchos* in the newly formed rivers, and the marsupial lion *Thylacoleo*, because all have been found somewhere in northeastern Queensland, the Gulf Country and the northwest. Many of these species also occur wherever megafauna are found and some may even have had their images recorded.

The fossil megafauna–depauperate north might not be written off just yet. Apart from a very recent find of a *Diprotodon* near Katherine (to late to include in the tables here), and more survey, there are other possibilities of a record for these beasts. While we have no real fossil evidence for them at present, some species have appeared as images on Aboriginal rock art. Such an image in Deaf Adder Gorge, Arnhem Land, situated 200 km south of the low stand palaeoshoreline, depicts a quadruped with muscular hind quarters, a long tongue, prominent claws and a coarse coat with prominent tufts of chest hair. This has been interpreted as a depiction of *Palorchestes*, perhaps another diprotodontid that could have lived on the shelf (Chaloupka, 1993). The image may not be *Palorchestes* but it is difficult to identify as any known marsupial; it is obviously big and it has a youngster with it that looks like a smaller version. It has also been cautiously suggested that other Arnhem Land images represent *Thylacoleo* and *Sthenurus* (Murray & Chaloupka, 1984). A possible artistic depiction of *Thylacoleo* has been recorded from the north Kimberley (Akerman, 1998, 2009) as well as the Pilbara but one image, again in Arnhem Land, is difficult to dismiss as anything else but the marsupial lion with its large robustly built jaws. Another depiction of megafauna from the

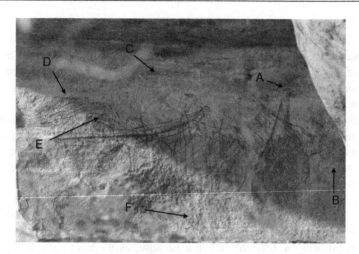

Figure 7.15 The rock art in this gallery in Arnhem Land shows a large, stocky, flightless bird (A) with a short neck, unlike an emu, and a beak similar to that of a magpie goose and like that of *Genyornis*. This bird is clearly not an emu but it is superimposed on a painting of a very large male kangaroo with a short face and hairy ears (B). Its long back (C) and thick tail are clearly shown as well as testicles (E) and possibly a very large and bulky (muscled) hind leg (F) as though in mid-stride or jump with the tail lifted for balance as occurs in modern kangaroos.

southern Arnhemland escarpment that I have photographed could be the image of a large flightless bird that has the head morphology of *Genyornis*. That also appears to overlay a picture of an extremely large and well-muscled kangaroo with a short face. Perhaps it depicts a *Procoptodon* or a large sthenurine animal (Figure 7.15). Whatever the case, the face does not reflect the long, slender, deer-like muzzle of today's kangaroos. A second image painted elsewhere in Arnhem Land also depicts a large bird with a very bulky robust body and large feet and, like the *Palorchestes* already mentioned, it seems to have a smaller individual of a similar type of bird with it, although it is more slender in its build. Another image from the Wellington Ranges in Arnhem land show that of a large-cat-like animal with a robust head an jaws, reminiscent of *Thylacoleo*. Of course, other images recorded in the region of the long-nosed echidna (*Zaglossus bruijnii*), now only resident in New Guinea, as well as the *Thylacine* and Tasmanian Devil (*Sarcophilus harrisii*), both of which were probably around in the region till 3000 years ago, are distinct and unmistakable. A large emu-like bird painted in a cave on Cape York has been interpreted as *Genyornis*. If true, this is the only evidence for it in that part of Australia where no *Genyornis* egg shell has yet been found (Tresize, 1971). We know *Genyornis* was not really shaped like an emu and these images seem to confirm that – if, indeed, *Genyornis* is what we are looking at. I do doubt these images are *Genyornis*, Thylacoleo and a very large short-faced kangaroo, however, because Aboriginal artists painted excellent and detailed likenesses of the animals they lived with which helps to identify the species.

Nevertheless, if these images do record megafauna they may all have lived out on the emergent shelf as well. If the final extinction threshold is ~45 ka, it suggests these drawings predate any other true art so far associated with humans that depicts the world around them. If *Palorchestes* became extinct before the end of MIS 6, as present dating suggests, the Deaf Adder Gorge art, on this interpretation, takes on a completely new significance for the arrival of humans in Australia as well as making the art more than five times as old as the animal cave paintings of the Dordogne! Another interpretation, however, is that *Palorchestes* did not go extinct so far back but, indeed, did quite the opposite and survived the coming of humans to Australia with some large overlap between them and the new arrivals, perhaps surviving in small pockets in the tropics and on the continental shelf. It may be that rock art will help considerably in future with determining where and when megafauna existed after the arrival of humans.

The demonstrated pattern of glacial temperature cycles is, as I have said above, slow, but its termination was fast, with temperatures rising 8°C or 9°C in a few thousand years. Sea levels would have worked in a similar way, with regression slowly adding land but ice core temperatures indicate that, at interglacial onset, the shelf would disappear comparatively quickly. Rapid inundation would have brought disastrous consequences for shelf-dwelling animals. In many places, Australia's continental shelf comprises a gentle slope, allowing comparatively quick inundation that could surround high ground-forming islands. Animals would naturally move to higher ground as water slowly enclosed their living area. Perched on a shrinking island with continuing rising seas, animals would be lost as the island disappeared. Very rapid oceanic transgression due to ice shelf collapse in Greenland and/or in the West Antarctic are now understood to have taken place during the last interglacial, suggesting a very rapid sea level rise of metres in decades (Hearty et al., 2007). Add this to a shallow continental shelf slope and the danger is multiplied for shelf-dwelling animals. The neat transferring of animals away from inland aridity onto exposed shelf, as described above, may not, therefore, have been a saviour in all cases. Moreover, it is likely that there were occasions when high sea levels occurred without direct compensatory environmental amelioration inland, thus trapping megafauna in a narrow strip of vegetation on the continental edge. Once again, the principles of island biogeography would come into play with consequential reduction in animal numbers. It is unlikely that the circumstances were ever the reverse.

When we consider the possibility of the megafauna population moving from central Australia onto the exposed northern continental shelf, it suggests migration but that is difficult to prove. Long-distance migration is not part of extant marsupial behaviour, so there is no model with which to compare such behaviour in extinct megafauna. Australia is the smallest continent and observed migration among reindeer, elk, African buffalo and possibly mammoths is not something that might readily apply to our megafauna. Moreover, migration is probably the wrong word because the movement described here is not annual, seasonal or part of a natural behaviour cycle. Instead, it is the abandonment of one area and moving to another, more or less permanently, in response to changed long-term environmental conditions. Such movement took place over long periods of time corresponding to glacial cycles and permanent environmental change. The free movement of animals between different parts of Australia was easier during interglacials, while at other times it would have been restricted. On some occasions that was very restricted or impossible, particularly when corridors closed when vegetated country was replaced by desolation and blowing dust and sand blew across Australia as the westerlies moved north during the ice ages.

In a modern drought, central Australian kangaroo populations crash, although they do move from drought-affected areas if they can. But they don't all end up on the coast, partly because local extinctions take place due to population reduction or crashes reducing pressure on food resources and because they just don't move that way. It is possible that diprotodontids were herd species, and as such moved across the landscape following food and water sources, but that is not really migration. However, the ability to move as herds could have led them eventually to coastal areas as food resources slowly dwindled behind them. All animals respond to drought or desertification and that must have been no less the case with megafauna.

What were the mechanisms that brought animals to the coasts, possibly pushing them onto the exposed shelf? Knowing so little about megafauna behaviour hinders prediction of their response and avoidance techniques to climate shifts and of how they moved within the continent. All we can do is apply the principles of biogeography and animal behaviour that have been well-researched and stand as the cornerstones by which we understand animal responses to environmental and climatic change as well as how population dynamics alter with such changes. We are familiar with these as they apply to modern animal populations, so why would they not apply to those of the past? In Chapter 8, we look at a particular megafauna population, that inhabiting the southern part of central Australia, and its biogeography.

8 Megafauna in the Southern Lake Eyre Basin: A Case Study

As a dark blue–grey curtain ascends on the eastern horizon, it draws up a soft-yellow full moon. In the west a blood-red, dark orange and yellow sunset rages, courtesy of volcanic gases and dust high in the stratosphere. They have come from an erupting Mount Pinatubo. I have seen many sunsets in the Lake Eyre basin but none have blazed as fiercely as this rare sight of immense beauty. The Philippine volcano has put on a similar show to that of the Indonesian volcano, Mount Tambora, which erupted in 1815, the effects of which influenced the English artist Joseph Turner to paint his magnificent inflamed sunsets in the early 19th century. The sight makes an extraordinary backdrop to our camp while the Earth's shadow lays a placid contrast to the east with a darkening sky of varying soft catafalque colours. Beers are opened, red wine poured; there's no white wine as it has become known as 'Yak's piss', a term invented by one of our eminent geologists and wine 'connoisseurs' who shall remain nameless (although reference to his papers occurs in the bibliography). I am cooking a curry and the excruciatingly mouth-watering smell, appreciated only by the ravenously hungry who will eat anything, drifts through the open air and across our camp site from its cooking fire. As always, the centre of the gathering is the fire, just as it was for millions of our ancestors, black, white and brindle. Chairs are drawn up around it, ready for the evening's 'conference'. The noses of the diners twitch at the curry's wafting breath; I bet a dingo can smell it on the other side of the lake! Usually, I cook in our field camps because I like it and don't like washing up. Ah, a bit of cheese before dinner….

We are settling down after much foot slogging (again) over dunes, across eroded scalds of Quaternary landscape and along salt-lined river courses that truncate ancient palaeochannels of the same age and trap unwary boots in almost bottomless black and grey, super-saline mud and ooze. We have collected, dug deep pits, taken hand-augured samples for geochronological dating, examined stratigraphy and hauled down walls of Katipiri sands last washed by rivers 100,000 or 200,000 years ago. It's all in a day's work. I cast a quick glimpse at the box of megafauna fossils I have collected lying on a nearby table. Not a great collection; none are beautiful or a collector's piece – a few teeth, plenty of long bone fragments and some pieces that are totally confusing, as usual. My collection is not endless, but it is just enough to help me say what I need to about the range of animals that lived there. But the natural quiet of the desert evening makes the world a peaceful place – we are doing what we like best: field work. A bird chirps away the last few notes of the day. The work is taxing and hard (and we moan about it on occasion) but there is nothing quite like being the first to discover an assemblage lying in the ancient mud or eroding from the

Corridors to Extinction and the Australian Megafauna. DOI: http://dx.doi.org/10.1016/B978-0-12-407790-4.00008-2

sediments and soils of an equally ancient and remote landscape now being exposed from its Quaternary tomb. In this area you may even be the first non-Aboriginal foot treading the ground. To try and unravel pictures of past climate, environments and the biological and zoological mysteries of a human-empty continent is a privilege. I have never met anyone working out here who does not like it – if you don't, you cannot work here. It will drive you mad and I have seen that happen – well almost. This is the SLEB and we are camped near the centrepiece: Lake Eyre. The sunset continues to reflect from the vast gibber desert and dunes that lie on the unseen western side of the lake, while the moon rises inexorably over the Tirari Desert. North of us is the 176,500-km^2 Simpson Desert, the fourth largest desert in Australia.

In 1902 the geologist J.W Gregory took a fossil-collecting expedition from Melbourne University into the SLEB during the summer (Gregory, 1906). Undaunted by warnings of extreme heat at that time of the year, he and his students travelled along Cooper Creek, crossed the Tirari Desert and continued their collecting west along the Warburton River. As they battled January heat and dust storms, they followed the main Warburton River collecting the bones of megafauna called *Kadimakara*, with a local Aboriginal Dieri man who was his main help. The Dieri was one of the desert tribes of the region who knew the desert intimately and were full aware of the large fossil bones found in the region. Gregory's collections were eventually sent to Glasgow University, where they lay largely unstudied and suffered partial destruction by bombing during the Second World War. It was Gregory that first published a book with the title *The Dead Heart of Australia*, a phrase that would become much echoed in the description of Australia's centre, particularly for those who cannot put up with space, peace, mystery and beauty and usually don't go out there. The Dieri knew of the large animals that once lived out there and they had a Dreaming story about them: *The Kadimakara*. In his book Gregory related the story of the Kadimakara which has an uncanny resemblance to what we are finding out about the region.

> *According to the traditions of some Australian aborigines [sic], the deserts of Central Australia were once fertile, well-watered plains... where to-day the only vegetation is a thin scrub, there were once giant gum-trees... the air, now laden with blinding, salt coated dust, was washed by soft, cooling rains, and the present deserts around Lake Eyre were one continuous garden.*
>
> *The rich soil of the country, watered by abundant rain, supported a luxuriant vegetation, which spread from the lake-shores and the river-banks far out across the plains. The trunks of lofty gum-trees rose through the dense undergrowth, and upheld a canopy of vegetation, that protected the country beneath from the direct rays of the sun. In this roof dwelt the strange monsters known as the 'Kadimakara'...*
>
> *Now and again the scent of the succulent herbage rose to the roof-land, and tempted its inhabitants to climb down... Once, while many Kadimakara were revelling in the rich foods of the lower world, their retreat was cut off by the destruction of the three gum-trees, which were the pillars of the sky. They were thus obliged to roam on earth, and wallow in the marshes of Lake Eyre, till they died, and to this day their bones lie where they fell.*
>
> *(Gregory, 1906: 3–4).*

The Aboriginal Dreaming has no dates but the *Kadimakara* story has parallels with the palaeoenvironmental and palaeontological story unravelling around Lake Eyre. But then the SLEB is uncanny; it does not take a lot of imagination to see how this gigantic arena of vast emptiness, hush and peace evoked stories about and explanations of the past among the first Australians inhabiting the region. They spent every night of their lives under the vast velvet sky with its millions of stars, stars that on some frosty nights cast 'star shine' in a way that you can see by it. Everybody who works there feels the special environments and whether they call it spiritual, magic or some other epithet, several weeks of being out here soon changes the way you think – at least till you hit the next sealed road. For me it brings a welcome, peaceful relief from the noise, stress and regimentation of things to do, and the constant electronic battering of 'civilisation' that assaults the body from every direction and urges you to do things yesterday.

The last time Lake Eyre was completely filled was in 1974. In 2011 it tried to fill again but it failed even in a year during which La Niña tried to drown parts of Australia. Water takes months to flow through the modern catchment rivers and streams as it slowly makes its way down from northern Queensland over hundreds of kilometres of extremely low gradient. Almost every year, as soon as water flows towards the lake, the media inform us that Lake Eyre is filling. It isn't, but that does not prevent thousands of tourists from coming to see if it has. The truth is it takes an extraordinary amount of water to fill it to its present shoreline, producing a lake around 9500 km². Our climate can only manage that about once every 40 years or so: the present interglacial just does not seem to have the stuff of the last interglacial. During MIS 5e, enough rain was produced that it formed great megalakes across Australia's inland and as far as we can tell at present, megaLake Eyre was the biggest, a lake of over 30,000 km² at its greatest extent.

The SLEB Megafauna

Why am I now talking about the SLEB? This is an appropriate juncture to look at what we do know about one population of megafauna, and a very important one, before we head in to look at the glacial biogeography of the animals across the continent. The SLEB provides a snapshot of that biogeography in central Australia; each region probably had a similar, story and we do not have another arid region population with which to compare it. Over 20 years of research shows us something of the lifestyle and distribution of megafauna living in an environmentally precarious part of the continent, one that would record the vast contrasts of glacial switching possibly like no other. Most importantly, it shows us something of the mechanisms operating there during glacial cycles and the violent environmental switching that arose from them. Reflecting those changes more than anything else was the megafauna; they probably experienced the greatest effects from glacial cycling compared to species anywhere else on the continent and hundreds of thousands of animals would have died as a result.

SLEB megafauna comprise a subset of 22 species of the continental population and their presence in what is now the driest part of the driest inhabited continent raises questions concerning their demographic response to Quaternary climate change. Almost all the best-known animals are represented here, which suggests they were very adaptable in terms of responding to opportunistic movement when environmental amelioration took place. Those periods took the environment from very long-term and extreme aridity during glacials to a short-term interglacial wet, savannah environment. As such, the region provides a template to examine demography, species variety and the palaeoenvironmental conditions under which these animals lived and benchmarks behaviour and demographics of a core group of megafauna. By looking at what was going on in the SLEB, we can lay a foundation for the next chapter which looks at the biogeography of megafauna and the mechanisms that could have driven them to extinction on a continental scale.

Geographic Setting

The Lake Eyre catchment stretches from northern South Australia, northeast across Queensland and northwest into the Northern Territory, spanning latitudes 32° to 19° south and longitudes 132° to 147°. The SLEB component consists of approximately 240,000 km^2 of the southern Simpson and Tirari Deserts, north and east of Lake Eyre, respectively, as well as the Lower Cooper Creek and Warburton River that run through them (Figure 8.1). Smaller lakes (Gregory, Blanche, Callabonna and Frome) extend southeast of Lake Eyre. They formed part of the giant megaLake Eyre during the last interglacial, occupying an area ~32,000 km^2 and >285 km^3 in volume, but today only rarely receive major runoff, perhaps once or twice in a century (Figure 8.2). The Simpson Desert dune fields predominate throughout the region, running in a north-northeast direction with some over 100 km long. Salt and small playa lakes are scattered across the SLEB, many with the discernible ghosts of small palaeochannels running beneath them that then disappear under adjacent dunes. Some of the eroded and truncated lake edges expose Quaternary environments a few with Tertiary outcrops at their bases.

The LEB is the only true internal or endoreic drainage system in Australia. On occasion the system gathers extraordinary amounts of monsoon rainwater in its northern catchment which then drains into Australia's dry centre. The northern catchment lies above the Tropic of Capricorn pushing into Queensland's monsoon belt. Water takes 3 months to reach Lake Eyre, carried by major rivers like the Georgina, Thompson, Diamantina and Cooper Creek. A system of dry river courses enters from the northwest consisting of the Todd-Hale, Hay, Illogwa, Plenty and Mulligan Rivers. They rarely flow and when they do, the water usually disappears into the Simpson Desert dune field, never reaching Lake Eyre far to the south. In the Quaternary they were probably much more active together with the oldest river in Australia, the Finke, and fed a network of palaeochannels that were directed southeast but now lie under the Simpson Desert. These are fascinating because they testify to a complex riverine environment that was spread throughout the region during the Quaternary at a time when few dunes existed. The presence of so many ancient

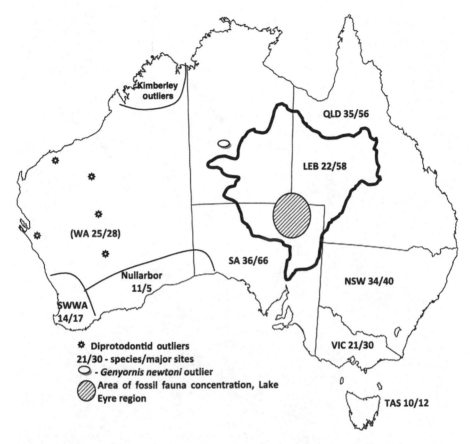

Figure 8.1 The position of the Lake Eyre basin (LEB) (bold outline) and the southern section (SLEB, hatched) together with total species and major fossil sites in individual states.

channels presents a picture of an environment that included many wetlands and billabongs and a variety of habitats that animals would have enjoyed, but they did not exist as flowing streams all the time. Each time they dried, it seems, dunes crept in, a few at a time, a process gradually building the dune field that we see today and that covers the old Quaternary environment like a carpet.

The Quaternary Katipiri Environment

The mid-late Quaternary environment and its associated sub-surface palaeochannels are found as the latest in a series of sedimentary layers spanning the last several million years. Named the Katipiri Formation, it contains a mixture of river, lake and wind-blown sands, silts and clays of various refinement and it is from these that the most recent and best preserved megafauna fossils emerge (Figure 8.3). White to orange-coloured Katipiri sediments erode out along dry modern channels and around

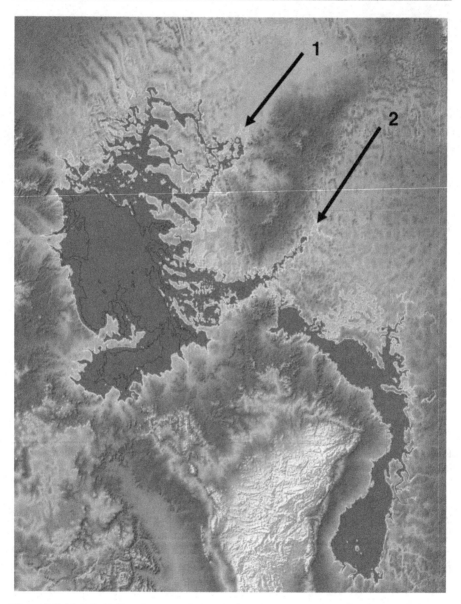

Figure 8.2 MegaLake Eyre incorporating several smaller lakes (Lakes Callabonna, Frome, Gregory and Blanche) to the southeast. Major inflows from the northeast Diamantina River (arrow 1) and Cooper Creek (arrow 2) systems are shown together with a faint outline of the present lake boundary.

playa lakes. Grey, grey–blue and green Katipiri clays mark deep-water palaeolake beds (Figure 8.4). Originally, the Katipiri was separated into Upper and Lower sections that roughly correspond to MIS 4–5e and 6–9, respectively (Callen & Nanson, 1992; Magee, 1997; Magee & Miller, 1998; Magee et al., 1995; Nanson et al., 1988,

Figure 8.3 High bank section of the Kallakoopah Creek showing well-bedded deep yellow sands typical of the late Quaternary Katipiri Formation.
Source: Photograph by Steve Webb.

1992, 1998). Recently the story has become more complicated, however, and it is now thought that these divisions each encompass a longer time frame, perhaps MIS 4–7 (60–243 ka) for the Upper and MIS 8–11 (>240–429 ka) for the Lower Katipiri (Table 8.1). While these divisions remain flexible and somewhat speculative, the Upper section spans glacial and interglacial environments alternating through a series of flood and arid cycles that changed the region from the extremes of hyper-arid desert to savannah wetlands. Katipiri sediments overlie the Kutjitara Formation, which probably marks a similar set of environmental indicators of mid-Quaternary age, possibly reaching back 800,000 years. However, little is known about the Kutjitara and it is yet to be fully studied. Below the Kutjitara lie the oldest early Quaternary and Pliocene-age Tirari and Etadunna Formations, respectively, the latter having a Miocene basal date.

Figure 8.4 Thick, greasy, blue to grey lacustrine clays typical of deep-water sediments of large lakes formed in the SLEB during the MIS 5e Lake Eyre megalake phase and associated with the Katipiri Formation.
Source: Photograph by Steve Webb.

Glacials brought extreme aridity to the SLEB with strong winds, erosion and scouring. Lakes and rivers dried and dune building took place as sands swept in from central Australia. One dry period around Lake Eyre during MIS 6 lasted for 30,000 years and was so severe that it excavated down to the Etadunna Formation (Magee, 1997). The region experienced very cold winters and hot, dry summers. Dune deposits overlying upper Katipiri Formation emphasise conditions differentiating environmental extremes. Megafauna remains are found along the main river channels entering Lake Eyre as well as north on the Diamantina River as far as Boulia in western Queensland. Forty-one major fossil sites, some incorporating smaller sub-sites, have been recorded.

Table 8.1 Proposed Time Frame of Major Lake Eyre Sedimentary Formations

MIS	Major Cycle	Sub-Phase	Range (ka)	Cycle Length (ka)	Sedimentary Formation	Sedimentary Location
5a–d	Interglacial		*80–114*	34		
5e		*Interstadial*	*114–133*	19		Punkrakadarinna 6, 8 KWB 4,6 Lake Dom1,8 WW1, 2 M95TL2 Billicoorinna Punkrakadarinna Lookout Site, Lake. Kutjitara, Lower Quana 3, 5 SDC, Sydney Harbour.
6	Glacial		*133–196*	63	**Upper Katipiri**	Lookout, Kala, SDC, Sydney Harbour, Toolapinna East
7a	Interglacial	*Interstadial*	*196–201*	5		
7b		*Stadial*	*201–204*	3		Billicoorinna, Lake. Hydra, Fly Lake. KWB17
7c		*Interstadial*	*204–218*	14		
7d		*Stadial*	*218–235*	17		
7e		*Interstadial*	*235–245*	10		
8	Glacial		*245–312*	67		Billicoorinna Creek, Lookout, Fly Lake
9a	Interglacial	*Interstadial*	*312–316*	4		
9b		*Stadial*	*316–320*	4	**Lower Katipiri**	
9c		*Interstadial*	*320–336*	16		
10	Glacial		*336–393*	57		Toolapinna, KWB2
11	Interglacial		*393–429*	35		
12	Glacial		*429–483*	58		
13	Interglacial		*483–500*	17		
14	Glacial		*500–560*	60		
15a	Interglacial	*Interstadial*	*560–580*	20		
15b		*Stadial*	*580–603*	23	**Kutjitara**	Kutjitara West Bluff (Lower Cooper Creek)
15c		*Interstadial*	*603–625*	22		
16	Glacial		*625–689*	64		
17	Interglacial		*689–713*	24		
18	Glacial		*713–772*	59		
19c	Interglacial		*772–789*	17	**Tirari**	*Basal sediments at Toolapinna Cliffs, Warburton River (Dated beyond the Brunhes–Matuyama Reversal at ~780 Ma)*

Much more work is required before the older formations are firmly positioned.

Many remains have been strewn along dry river courses and on point bars where they have been redeposited after eroding out of Katipiri deposits, truncated along high river banks by occasional river foods. There is a sharp hiatus in bone deposits where megafauna do not occur in sediments overlying the Katipiri or in dune field deposits. Small amounts of bone can be found in what I have termed 'intra-lake' areas that lie in desert areas between major rivers, usually on exposures around small playas and eroding sediments in the Tirari and southern Simpson Deserts. These probably indicate animals that followed smaller streams that fed into larger main channels linking them with small lakes or swampy areas and which are now buried as palaeochannels.

Central Australian Palaeolakes and Their Environments

Palaeoshorelines and ancient beaches around the modern lake testify to the existence of megaLake Eyre during MIS 5e. That lake included Lakes Gregory, Blanche, Callabonna and Frome to the southeast. Nothing short of a regular and intense monsoon climate in the catchment as well as over the lake with lower evaporation rates than those of today could have produced such a large filling (Webb, 2010). Modern monsoons have variable season length and rainfall intensity that cover 400,000 km^2 of northern Australia (Suppiah, 1992) but these go nowhere near creating such large lakes today. Other central Australian lakes experienced similar megalake phases against a background of fluctuating mid-late Quaternary wetting and drying (Nanson et al., 1992; Webb, 2008). Evidence from Lake Gregory, sandwiched between the southern Kimberley and Great Sandy Desert, as well as from Lake Woods in northern central Australia, indicates megalake episodes during MIS7 and MIS9. A wet event during MIS 5b is recorded at Lakes Lewis, Gregory and Woods together with a full flood down the Finke River. Strzelecki Creek, which fills Lakes Callabonna and Frome, also flowed at that time, moving water on through the Warrawoocarra Channel to Lake Eyre. That also occurred in late MIS 5d, a time when glacial temperatures were below −4.0°C (Bowler et al., 1998; Cohen et al., 2012; English et al., 2001; Nanson et al., 1995; Nanson et al., 2008). A smaller lake filling than that of MIS 5e occurred in MIS 5a. Recent investigation of prominent palaeoshorelines on the western side of the 3000-km^2 Lake Mackay, 1000 km northwest of Lake Eyre, reflect a similar event there during MIS 5e as well as at Gregory, Woods and Lewis, and is further evidence of a comparatively rapid return to wetter times following the MIS 6 glacial. It also marks a peak warm period *ca.* 128 ka, when EPICA data show temperatures reached 5°C higher than today. The Mackay evidence suggests a very strong over-lake weather system delivering enhanced monsoon rains at 22°S because its small catchment seems too small to deliver enough inflow to form palaeoshorelines. Lake Lewis, to the east, is much smaller than Mackay with a maximum filling area of around 1000 km^2, but the MIS 5e event is not nearly as clear there; it should be, however, because it is almost on the same latitude as Lake Mackay. It is difficult to believe a monsoon strong enough to fill Mackay is not registered at either Lake Lewis or Lake Woods, 350 km and 600 km to the east, respectively. However, what may be reflected is a reduced intensity that far east or perhaps more work is required.

The complexity of events rears its head once again and it is not so easy to relate them all as a uniform phenomenon, nor does it make sense against some periods of glacial–interglacial cycling. But in all this climate and environmental reconstruction, where are the megafauna? While Lake Mackay experienced high lake levels, the total lack of any megafauna fossils there seems to show the presence of water does not naturally mean the presence of megafauna which, on the face of it, is peculiar. The same thing is reflected at Lakes Gregory and Lewis. Perhaps Lake Mackay's position put it beyond the reach of these animals, but why? Animals wander and move so why would they not eventually find this lake? These questions may be answered in the next chapter but for now it is worth remembering that they need a way to get there. In the meantime we have seen how *Genyornis* reached Lake Lewis but went no farther.

Although palaeoshorelines like those of Lake Eyre may not be featured on all central Australian playas, their presence on several lakes testifies to high fill levels during MIS 5e, which at present seem to originate from temporally and spatially staggered intra-continental rainfall. It seems the 125 ka Eyre filling was probably caused by a climate system vastly different from that of today. While it seems logical to assume a mega-monsoon cause, a multiple and stochastic system might be a better explanation of what fired up all Eyre's drainage systems. Whatever the weather system was, it was more active, complex, and regular, and of longer duration than any around now and it penetrated far south into the continent. Examination of *Genyornis* egg shell isotopes indicates that the monsoon terminated during MIS 5a (Johnson et al., 1999). There is little doubt that glacial–interglacial climate patterns were much more complex than mere cold-dry/warm-wetter swings; hence, environmental interpretations using the bare cycles themselves are only a starting point and not the definitive picture of climate or its interpretation during the Quaternary. I suspect, however, that environmental interpretations resulting from interglacials are much more complex to achieve or predict than the effect of deep glacial cycles. In Chapter 7 we saw that not all interglacials were the same; most were short and not very strong, while a few were the opposite. Glacials, however, seem more structured and basically long, deep and cold. We cannot be sure, however, that the last interglacial wet, which took place during much warmer world temperatures than occur today, occurred in previous interglacials. They indicate they were not as warm as MIS 5e and that some were cooler than the present interglacial, although MIS 11 had similar characteristics to MIS 5. This would suggest also that each of the past interglacials probably has to be interpreted on its own merits. That will not be a surprise but doing it will always be a challenge, and the farther we go back the harder it becomes as data becomes more uncertain and interpretations become coarser.

In MIS 5e the abundance of water around Lake Eyre initiated high channel flow and over-bank flooding and encouraged gallery forest growth along palaeochannels and their associated back swamps and wetlands. Remnant scroll bars and oxbow lake features are embedded in the Katipiri Formation. These are also visible in aerial views which show them threading across smaller playa lakes (mentioned earlier), indicating a complex series of anastomosing channels and associated swamp niches. They are there as a measure of the broad rainfall and water flow throughout the region, bringing various mosaics of vegetation and plant diversity. It made the Simpson Desert more like a Simpson Savannah with wetlands, lakes and animals. Constant flows would have formed a wide variety of microenvironments in flood outs and around lakes as well as along smaller rivers and main channels. Fossil pollen from the region is limited but studies of lake sediments indicate many species found in the region today after good rainfall probably grew during the last interglacial. These include Poaceae (grasses), Asteraceae (daisies) and Chenopodiaceae (herbaceous flowering plants), but few species growing outside the region today have been detected among the pollen (J. Luly, personal communication). Tree pollen is sparse but *Callitris* and *Casuarina* pollen have been found together with rhizomorphs and many examples of ancient tree trunk and branches. An *in situ* partially fossilised tree trunk (?*Ficus* sp) was found in Katipiri Formation sediments on Kallakoopah Creek, which stratigraphically dates to at least MIS 5 (Figure 8.5). The presence not only of large trees but several species, together with the fossil remains of arboreal animals such as possums (*Trichosurus vulpecula*) and koalas (*Phascolarctos stirtoni*), are present in Katipiri sediments along major palaeochannels. The dietary preferences of these animals for eucalypt leaves, fruit and blossoms underpin the need for suitable tree species to be present probably as gallery forest and out on savannah woodlands, perhaps around small lakes. Smaller animal species in upper Katipiri Formation sediments found on the Kallakoopah and around Tirari Desert playas include the Southern Brown Bandicoot (*Isoodon obesulus*), Eastern Hare-wallaby (*Lagorchestes leporides*, now extinct) and Pale Field-rat (*Rattus tunneyi*), all far outside their recent distribution areas, as is the Western Grey Kangaroo (*Macropus fuliginosus*). The presence in the region of the Greater Stick-nest Rat *(Leporillus conditor)*, Bilby *(Macrotis lagotis)*, Burrowing Bettong (*Bettongia lessueur*) and the Southern Hairy-nosed Wombat (*Lasiorhinus latifrons*) is not particularly surprising as the SLEB was within or close to their known distribution areas before their extinction or severe range contraction during the 19th and 20th centuries.

The distribution of animals around Lake Eyre shows they lived right across the megalake system but while many fossils are associated with the Katipiri Formation, that formation spans a considerable period of time, possibly back to MIS 11. Temporal distribution, therefore, is not possible at the moment although we do know some species probably lived during the MIS 5e megalake stage. That time provided an environment in which terrestrial and aquatic animals could live, and there are enough fossil representatives to reconstruct a palaeotrophic setting; we will look at that below. However, the lack of such evidence around other central Australian lakes suggests that similar colonisations did not take place so readily even with assumed suitable conditions for colonisation to take place. But perhaps regional conditions were not the same as those in the SLEB. Lake Eyre's major catchment channels might have made the difference by providing suitable and/or stable corridors delivering cubic kilometres of fresh water to the region. They also originated in the northeast where we know animal stocks lived. In the case of other central Australian lakes, there may have been no stocks at the end of any of corridor leading to them. While we have noted that the presence of water (in lakes) may not mean the presence of animals, perhaps we should add to that, that the existence of a suitable corridor did

Figure 8.5 An *in situ* trunk of fossil tree that was excavated from overlying Katipiri sediments (see Figure 8.3) of MIS 5 age on Kallakoopah Creek deep in the southern Simpson Desert (top). The wood is supersaturated with salts, turning the tissues black (bottom image). *Source:* Photograph by Steve Webb.

not mean animals travelled along it. Ease of colonisation of Lake Eyre was not surprising nor was it remarkable given enough time to bring animals in. But, as we have seen in the previous chapter, the Lake Eyre megafauna guild is really a reflection of that to the south rather than the northeast from where its catchment rivers begin. What we may be seeing, therefore, is that water to form a suitable environment for Lake Eyre's animals came from one place while the animals themselves originated elsewhere. One group that must have come along with the water, however, were the aquatic species and they depended not on the time taken to fill the lake but on the time taken to assemble the required trophic chains to support them. The key to understanding megafauna extinction is to try and understand the nitty-gritty of the world they lived in. They lived around megaLake Eyre, drinking its water and wallowing in its streams, so what was that world like?

MegaLake Eyre Palaeoaquatic Ecology

Like islands, a lake's size dictates the complexity, population size and biological diversity of its fauna (MacDonald, 2003). The SLEB fossil record reflects that rule, showing a widespread and complex population living in and around megaLake Eyre. It, therefore, begs questions concerning lake formation time and the time required for trophic development of the complex biological system to take place before a permanent fauna consisting of large terrestrial and aquatic animals could populate the area. I have outlined a detailed model to show how this might have taken place elsewhere (Webb, 2010) so will not go into details here, but it is worth examining the fauna and what it represents in terms of the complexity of the environment around the lake.

Crocodiles

Pallimnarchos lived in the main channels that fed into the SLEB, which is the farthest inland it penetrated. They are not found on the smaller intra-desert sites, suggesting they kept to the larger waterways, the lake and associated swamps. Crocodiles were at the top of the lake's trophic pyramid and for them to be present, suitable habitats were necessary, similar to those inhabited by modern estuarine or salt water crocodiles (*Crocodilus porosus*). They need replenishment of water, nesting sites, containing sedges and cane grass (*Phragmites* sp.), rafts of vegetation and other aquatic plants as well as large plant species for refuge; there is no reason to believe *Pallimnarchos* required anything less unless it used burrows like the Nile crocodile (*Crocodilus niloticus*). Large isolated billabongs and waterways in the SLEB probably supported them during drier periods but for continuous occupation, any periods of drought would have to be short lived with regular rainfall high enough to trigger river flows to flush the system. Extended drought would require recolonisation each time from northeast Queensland, where *Pallimnarchos* originated. Like salt water crocodiles their diet must have included insects and crabs along river banks, things particularly favoured by juveniles. Crocodiles have a varied diet including fish, small mammals, birds and turtles in freshwater environments together with lizards and snakes. The likelihood that *Pallimnarchos* depended on *Diprotodon* steaks or any other terrestrial megafauna is slim even though they may have been taken on occasion. The best chance for *Pallimnarchos* to catch a *Diprotodon* was when it was trapped on a steep or muddy river bank or lake edge, which was always possible with such a large animal. Proof of this are the bogged leg bones found in position in Katipiri muds and the 10-mm-diameter crush pits penetrating deep

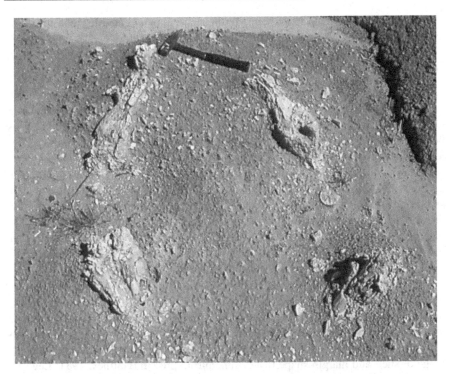

Figure 8.6 The upper and lower front limb bones of a *Diprotodon* bogged in Katipiri sediments near the Diamantina Channel close to Lake Eyre. The animal seems to die standing up, the body eventually falling forward as decomposition sets in.
Source: Photograph by Steve Webb.

through cortical bone into *Diprotodon* long bones found in the area (Figures 7–9 and 8.6). Carrion is also taken by crocs and they will move several hundred metres to collect a carcase and drag it back to the water. While crocodiles are at the top of the aquatic food chain, they usually only feed once per week on average.

Crocodile remains have been found in all the main areas of the megalake system. Population densities for estuarine crocodiles in the Northern Territory range from 0.5 to 1.5 km^2 per animal depending on the type of ecosystem (Harvey & Hill, 2003; Messel & Vorlicek, 1989), but it is not unusual to see several cruising in a 300-km stretch of river or billabong. Using these figures to estimate the Lake Eyre crocodile population during the lake's time as a 32,000-km^2 area suggests that a large crocodile population could have been present. However, with such long exit corridors stretching to the northeast, the chances of heavy losses during extended drought and arid phases were high. Animals could easily become stranded in isolated pools strung along the long, thin corridor. It is not known when *Pallimnarchus* finally became extinct in the SLEB but their habitats disappeared during MIS 5e and that was probably when they retreated for the last time because conditions were probably never as good after this. Appreciating the ecological requirements required by crocodiles in the SLEB in itself points to the rich biological tapestry that must have existed there during MIS 5e and very wet periods before that.

Turtles: Large and Otherwise

An environment that supports large crocodiles naturally supported turtles; indeed, turtles are a natural part of a crocodile's menu and tooth gouges in fossil shell fragments testify to that. A wide range of fossil turtles lived in the region, the largest being *Meiolania platyceps*, represented by horn cores and vertebrae, although these are not common by any means. The presence of *Meiolania* in the Burdekin catchment next door to the LEB catchment suggests it could have followed waterway corridors to Lake Eyre from northeast Queensland. It may not have been as aquatic as other turtle species, however, because of its size and morphology, including elephant-like feet rather than the webbed feet of other turtles. That suggests it was more like the giant Galapagos tortoises (*Geochelone elephantopus*) that have a largely terrestrial habitat (Gaffney, 1985). That adaptation would certainly have helped it make the journey inland, moving between basins even if very slowly.

Australia's modern aridity has forced modern turtles to the continental edges but the SLEB fossil evidence shows them living far inland. Carapace and plastron fragments are plentiful in Katipiri sediments across the region and the range of marginal and costal shell patterns and shapes indicate a wide range of species of various sizes. Unfortunately these are not readily identifiable to particular species because of the range of shell and margin shape that occurs during growth (Gaffney, 1981; Legler & Georges, 1993). Nevertheless, the range of shapes shows many species lived in the region. Turtles do not migrate so those in the SLEB must have been endemic forms during lake full event. They also hibernate but not for extended periods of drought, so recolonisation of the system must have taken place following any extended dry time (Wyneken et al., 2008).

Large Fish

Pallimnarchus also had a wide choice of large fish. Deep-water lake deposits along major channels indicate substantial water columns and lake arms that expanded and pushed up mid-late Quaternary palaeovalleys, particularly to the north and east of the present lake (Figure 8.7). Clays associated with deep-water formation are obvious signposts of aquatic environments in which large fish could grow and breed. Large fossil fish vertebrae found in assemblages indicate how big they grew. Calculations of live weights using vertebral diameter indicate fish of between 26–69 kg lived in the lake and rivers, demonstrating a mature and complex lake ecology with some specimens living to a great age. The size of the fish suggests two prime species: Golden Perch or yellow belly (*Macquaria ambigua*), and Murray Cod (*Maccullochella peelii peelii*). Perch can reach 75 cm and 23 kg but more regularly weigh around 5 kg. Their diet includes yabbies, shrimps, insects, molluscs and small fish. The record weight for a Murray Cod is 113 kg and measured 183 cm, however, a normal upper limit is around 125 cm and 45 kg. Murray Cod are top predators with a diet similar to Perch but they can consume bigger prey such as frogs, water

Figure 8.7 Truncated deep-water lacustrine Katipiri deposits at Punkrakadarinna on the Warburton River. They show a very thick deposit of lake bed clays that formed beneath an arm of megaLake Eyre that stretched northeast up the Warburton palaeovalley. Such thick deposits and their buckled form indicates substantial size and water depth of megaLake Eyre that presented ample opportunity for large fish and other large aquatic species to thrive.
Source: Photograph by Steve Webb.

fowl, small mammals and turtles as well as other reptiles. Both species favour river snags, require water temperatures above 23°C to spawn and can survive drought and migrate 1000 km.

Lungfish (*Neoceratodus* sp.) tooth pates are also found in fossil assemblages around Lake Eyre, some indicating large specimens. Today lungfish live only in a small area of the Burnett in central eastern Queensland. They reach 1.5 km and 40 kg and are capable of living in stagnant water during drought, although they generally require deep pools, slow-moving water and vegetation along river banks (Kemp, 1991). The presence of Lungfish, Perch and Murray Cod would indicate a secure trophic system that included a broad variety of producers in order for them to thrive. They also added further variety to the menu of *Pallimnarchus* and smaller sizes would also feed other predators. MegaLake Eyre's fossil fish remains are further evidence of the complex and rich environmental structure which formed in the region during MIS 5e.

Animal Palaeodemography in the SLEB

In Chapter 7 some observations were made regarding Lake Eyre basin fauna, but it is worth looking at this group again in more detail. The trophic description above suggests food and water were not a problem for megafauna during times like the MIS 5e interglacial only these were few and far between. Travel across open country between channels required soaks, swamps, water holes, small lakes and ephemeral streams as well as fodder. Water was an important ingredient in attracting animals but some species were probably adapted to semi-arid conditions as discussed earlier. Sometimes interglacials meant large channels, topped-up lakes and plentiful shrubs and grasses out on the plains. Meandering, abraded systems joining main river corridors made animal movement easier as well as never putting them far from water. Range-sharing was possible through differential targeting of food resources, with the broader adapted diprotodontids browsing lower levels and grazing while the largest macropods (*Procoptodon* and large sthenurids) browsed on taller shrubs and trees. But interglacial 'good' times were exceptionally short compared to glacials which could reduce or eliminate regional populations through extreme environmental degradation. The cyclic to and fro of animal groups in and out of the region is shown by the way they can be observed in assemblages above and below the sedimentary hiatus that marks glacial maxima. Megafauna die and are removed from the region at these times but those that escaped probably relocated to areas nearer the coast to repopulate again when environmental conditions improved, although the repetition of such events must have taken its toll each time, forcing slow attrition on populations over time. It must always be remembered that these were not overnight events but long-term trends.

Genyornis egg shell is found almost everywhere from Kallakoopah Creek in the southern Simpson Desert to Spencer Gulf on the south coast, west to Woomera and 400 km east to the adjacent Murray–Darling basin in New South Wales. Its widespread distribution and density of scatters represent thousands of nesting sites and is a very convincing evidence of a highly successful species. It is reasonable to assume that it probably moved into the SLEB from the south because there is little trace of it north of the Simpson Desert, although it may have moved northeast using now-defunct palaeochannels, eventually reaching Lake Lewis (see above). An expected natural movement of terrestrial species into the southern SLEB would be from the north following the long river corridors. *Pallimnarchus* and *Meiolania* must have come in that way but the separation of most northeastern species, generally located around and over the Great Divide, from those in the SLEB, may suggest a barrier for many. One exception seems to be *Procoptodon pusio*, which is found in Queensland but not in South Australia or anywhere to the west. The overwhelming number of SLEB species are also found in the south, strongly indicating they came from or were channelled north from there (see Table 7.1). The distribution of species located in the SLEB itself also reflects this to some extent. Moving from north to south, 11 species occur on the Kallakoopah, 12 on the Warburton and 17 on Cooper Creek, suggesting a greater species variety in the southern end of the SLEB, with most confined to the Warburton and Cooper channels; the majority of species avoided areas away from main water sources. *Diprotodon, Genyornis, P. goliah* and *Megalania* are found along all three channels and in intradesert areas while *Zygomaturus, T. carnifex, P. gigas, S. andersoni* and, not surprisingly, *Pallimnarchus* are common only to main channels. *P. rapha* and '*P*'. *gilli* are found in intra-desert areas and the

Kallakoopah, but *P. rapha* also lived on the Cooper, suggesting that it was another ubiquitous species although it has not been found on the Warburton. This might also be the case for *'P'. gilli* which has not been found on either the Cooper or Warburton channels. In contrast, *P. brehus* and *S. tindalei* have only been identified on these two, and *P. anak*, *'P'. browneorum*, *'Sim'. orientalis*, *S. stirlingi*, and *Sim. orientalis* appear only on Cooper Creek. Diprotodon's wide range in the region again suggests a broadly adapted animal able to occupy most niches and move into open areas between and away from waterways. While having adaptive flexibility had its advantages, but it also put them in greater danger when abrupt climate change took place because, as such, they were probably some of the last animals to leave an area undergoing change and they became stranded.

Trophic Growth, Faunal Colonisation and Collapse

The length of enviro-climatic optimums is very difficult to measure accurately but we can see evidence for them in places like the SLEB. The fossil assemblages there provide vivid evidence of the existence of a complex and varied trophic pyramid during interglacials. I have talked about the movement of animals to and fro in response to glacial–interglacial as well as stadial–interstadial changes. But how did that actually take place? What were the sorts of processes required to draw animals into central Australia after one of these extreme events? One way of approaching the question is to examine the only good example of how that happened. We can try and assemble the Lake Eyre interglacial environment and how such megalakes formed trophic systems in a bootstrap process from extreme aridity or, in other words, nothing. That process eventually attracted a complex set of animals to the lake, forming an equally complex trophic system.

Systems ecology has long recognised that trophic pyramids are more complex than simple levels of stacked animal groups. While it remains a useful concept to present a basic picture of community relationships, in reality such pyramids reflect complex food webs consisting of subtle relationships between different organisms, with energy going in various directions, not necessarily vertically from simple producers to carnivore consumers on top. Therefore, in order to appreciate community structure and function and its response to change and disturbance, it is necessary to understand how food webs evolved (MacDonald, 2003). However, gathering enough data to build a basic food web in living communities is a considerable task; to do so for extinct animal communities and long-changed environments is probably impossible. The fossil evidence presented here, therefore, is about as complete as one can expect from a fossil record (although the microscopic fossil record is yet to be studied) but there are some inferences that can be made even though the complete SLEB community of creatures is still unknown. It shows, for example, a complex ecosystem, a system that did not assemble overnight or that could develop without a suitable consistent climatic environment that lasted hundreds, if not thousands, of years.

The southern part of the modern Lake Eyre marks the lowest point on the continent at around −30 m. During a megalake, the depth varied from −10 to +10 m and it consisted of almost pure freshwater in over 90% of its upper levels in most places. With much higher lake levels, water depth salinity must have been pushed to the bottom and with flushing of large amounts of water through the system, the lake became almost saline free at times. It took time to grow a full mega-Eyre eco-system with its own unique constellation of ecological attributes, but trophic growth to completion almost certainly lagged behind megalake formation. Lakes can fill comparatively rapidly but constructing complex ecosystems takes time. Filling brought enormous amounts of water through the system, flushing salt and present-ing the initial catalyst for lake occupation by larger complex animals. Once they arrived, they had to be fed but without food they would not arrive. So, the first step was the blossoming of planktonic organisms and plant colonisers at the edge of the lake. The lake's 3500-km coast grew by expansion with mounting organic detritus. Phytoplankton and algae, as part of that detritus, fed micro-fauna (zooplankton) and small consumers which, in turn, supported larval fish and other varieties of small invertebrates. Long convoluted shorelines enhanced the amount of food available for smaller varieties of creature, forming an important basis for the influx of larger animals. Other lake fauna eventually included shellfish, crustaceans and molluscs and then large fish, turtles and crocodiles – a fairly complete biological community. Vegetation growth around lake shores formed environments necessary for croco-dile nesting and shelter and ranges for other large terrestrial and semi-aquatic spe-cies. Building biodiversity was a stepwise process that gradually assembled trophic networks supporting large colonisers. Lake growth over decades likely meant a substantial lag time between the initial colonist-producers and the arrival of larger species, but climatic conditions variously limited or enhanced the colonisation process.

Following the 1974 Lake Eyre flood, 80 species of water birds, 47 fish species and nine species each of zooplankton and phytoplankton were recorded in the lake (Morton et al., 1995). Pelicans (*Pelecanus conspicillatus*), cormorants (*Phalacrocorax* sp.), sea birds (gulls and terns), stilts, plovers and dotterels (*Charadrius* sp.) rapidly moved in and small fish such as hardy head (*Craterocephalus* sp.), smelt (*Retropinna* sp.) and yellow belly were the first aquatic species to arrive via river systems (Glover, 1989). Similar colonists were probably the first to arrive during MIS 5e, with birds initiating passive dispersal of seeds from upper catchment areas and small fish arriv-ing through active dispersal via catchment channels. Jump dispersal is a quick method of colonisation over large distances by a carrier agent like birds, and disjunct popu-lations (inoculation centres) set up before they grow then coalesce to form larger and more permanent communities. Animals and plants probably spread from refuge areas that survived during MIS 6 out near the coasts, and establishment, dispersal and growth of vegetation within the catchment probably occurred at different rates in dif-ferent areas.

Each biological element followed its own pattern of dispersal and popula-tion growth depending on carrying capacity and other environmental constraints (MacDonald, 2003). Initially, range expansion in an inoculation centre is slow because of the small colonising population. It then expands as a response to growing lake size, carrying capacity growth and niche development. Exponential population growth would follow given a burgeoning carrying capacity till that growth exceeded the resource base. Coalescence of inoculation centres and jump dispersal from them elsewhere then expanded range limits. Once diversity was in place, animal populations could spread throughout the lake system. The key to biodiversity is the permanence of water, the maintenance of the resource base and the maturing of various ecosystems not particularly lake size alone. A 50-km^2 lake could have the same species range as one necessarily 200 km^2; the latter would just have more of each species, similar to islands. Endemic populations could form with lake permanence but this required aquatic and terrestrial vegetation mosaics that, for example, supported crocodile breeding habitats and provided environments for other aquatic species; these are examples of mature ecosystems. Such systems also build conditions for sympatric associations among a range of animals, birds and fish. Aquatic fauna should not necessarily be seen in isolation from terrestrial species because the ecosystems they require needed regular local rainfall to produce savannah, gallery forest, back swamps, thickets, woodland areas and a wide array of intermixed niches, all of which are determinants of the types of creatures that inhabit them. There is little doubt aquatic fauna arrived first, followed by terrestrial forms taking much longer to arrive. For habitation success they also needed the right environmental conditions including good feed along the corridor they took to reach the region. We are also dealing with remnant populations; survivors of the previous Ice Age whose populations were not only reduced, certain species were probably lost from the range previously represented and they would not be coming back.

Studies of recolonisation following the last glaciation (MIS2), as well as more recent examples, show that the rate at which plants and animals return is extremely variable. Sixty years after the Krakatau eruption, 300 plant and 30 bird species were established on the island, while palynological evidence suggests North American trees spread during the early Holocene at 80–400 m/a or 1000 km in 2500–12,500 years (MacDonald, 2003). Some took several thousand years to repopulate their previous habitats following glacial retreat, with North American oaks recolonising at a rate of 380–500 m/a (Lévêque & Mounolou, 2003). European oaks (*Quercus* sp.), elms (*Ulmus* sp.), hazels (*Carylus* sp.) and lindens (*Tilia* sp.) took almost 6000 years to completely recolonise Europe, thus demonstrating considerable variety among colonising vegetation (Anderson et al., 2007). Doubling times range from 31 to 1100 years depending on the genera, but even within genera there are vast differences, e.g. pines (*Pinus* sp.). In Australia we did not see the retreat of ice but rather the retreat of sand which, I suspect, would have been a slower and more difficult process, and which left a more difficult environmental substrate to reclaim by vegetation. Development of a full trophic system around Lake Eyre, therefore, probably took centuries (?400–1000 years) to establish rather than decades, although some elements may have been in place comparatively quickly. Colonisation also relied on optimum climatic conditions that were maintained and that kept lakes full but optimum conditions may have lasted for only a comparatively short period within an interglacial and probably did not occur at all in some interglacials. That seems most likely when the graphs in Appendix 2 are viewed. I mentioned before that we have no real idea how long it took to form palaeoshorelines but, equally, we do not know what they really indicate in terms of lake-full time; that may have been very short but long enough to form a beach ridge. Interglacial onset was no guarantee that optimum conditions would occur or that when they did they would last for any great length of time, and that would have had implications for the degree of trophic growth in the time available and the complexity it would reach.

The ability of different plants and trees to colonise varies. Add to that the competition between some and we can see that complete vegetation mosaics probably took a longer time to establish than we might expect. If the process had a head start, with some species already in place, similar to the Lake Eyre flood event of 1974, then it may have begun quickly. Alternatively, if it required a bootstrap process on a scoured landscape such as that left over from MIS 6 then the primary base would have taken much longer to establish. So the 3–5000 year climb out of MIS 6 into MIS 5e, for example, would not be counted as time in which the interglacial was all rosy. It may have taken much longer than that just to bring the region back to a point where the introduction of greater rainfall or runoff into the SLEB began to have an effect on the environment – that is, of course, if the climate system that produced massive runoff and rainfall had itself come into existence by then. Whatever the situation, at their peak river systems must have achieved a biotic complexity akin to the rivers and swamps of Australia's present wet tropics, but for how long is unknown. These were optimum times for megafauna populations but I ask again, how long did they have it that way and how many of the 52 species were still around to enjoy it during MIS 5e? So far only 22 species are known from the SLEB and they are possibly spread across 100,000–300,000 years, the Katipiri times.

Interglacials meant range expansion and an opportunity for megafauna populations to strengthen replacing lost numbers. But these episodes were very short compared to glacials and were probably even shorter as times of respite for the megafauna than their rather artificial time frames suggest. For those species, present conditions were probably good enough for some local regrowth but it is less likely that they recovered enough to achieve anywhere near their full original population size before the good times were over. Indeed, the more likely scenario is one where MIS 5e just brought a lull in the continuing downward spiral of the overall population. Why? Because interglacials attracted animals into places like the SLEB, only to vastly reduce their numbers with glacial onset. In some cases this probably meant total species collapse or regional extinctions across the board and this had been going on over many cycles. The importance of considering all changes associated with glacial cycling without glossing over the smaller intra-cyclic reversals is fundamental to this issue and that was the point of detailing them in the previous chapter. All of them counted in the overall stress burden these animals underwent because any stadial–interstadial or smaller intra-cycle Dansgaard–Oeschger event embedded in the main cycles would have affected their biogeographic circumstances, sometimes bringing a little relief and at other times making things worse. Such switching would have been so important for pushing climate and environments this way and that. Contrasting ecological variation began 2 million years earlier and it grew in intensity over that time; perhaps the last glacial–interglacial cycle brought a set of circumstances that finally reduced Australia's megafauna (what was left of it) to non-viable fragments unable to withstand any further enviro-climatic stress that might have come from the arrival of humans as well as the MIS 2–4 glacials.

The presence of megaLake Eyre indicates climatic conditions that were far wetter than anything experienced in recent times or recorded in Australia at any time since. This suggests that a very different climatic forcing pattern operated during MIS 5e that maintained regular over-lake precipitation at both Lakes Eyre and Mackay. Accurate alignment of filling times with other lakes is not really possible, although that of Lake Mackay may have been close. Other lake fillings were more staggered but why is unclear; perhaps it is a case of the few dates we have for beach ridge formation need to be re-examined or more dates obtained. The possibility of giant monsoon combinations of variable intensity and other poorly understood climate-forcing mechanisms may play a part in these results. The actual filling time required may have been comparatively short but constant inflow was required for years to build the largest lakes. The terrestrial and biotic complexity that assembled from this is not visible at Lake Mackay or other central Australian lakes. While Mackay experienced great expansion and may have even begun constructing a basic aquatic ecosystem similar to that of Lake Eyre, it did not continue as far as we know. Why that was so is not clear; however, the lack of terrestrial and aquatic megafauna as well as large fish was probably due to the absence of a suitable corridor to bring them in. Animals would have needed a very long corridor to reach Lake Mackay from areas where we know they did live. At present Lake Mackay is a lone sentinel in the middle of Australia that perhaps signals the lack of megafauna in vast central Australia. There are more remnant lakes of Quaternary age to be surveyed, but I would suggest that megafauna will also be absent there.

The construction of basic palaeotrophic systems during the megalake filling may have taken place comparatively quickly but full trophic development to a level that supported large complex animal populations with equally complex relationships and ecologies and associated vegetation probably took much longer. That environment and the food chains required to support such biodiversity are suggestive of those found today in Australia's wetland tropics. For those familiar with the present-day environment of Lake Eyre, which is in an interglacial, the picture described here for MIS 5e presents a climatic and environmental juxtaposition almost too incredible to contemplate.

What we have in the SLEB is a good example of how megafauna lived and responded to glacial cycling and the associated super-drying of the continent. But we do not have the full range of species living there nor do we have any real idea of when the species we do have actually lived there. What I mean by that is while a range of megafauna species are present, that range could be spread over a very long time within an interglacial and/or as occasional visitors. Crocodiles do not necessarily reflect the terrestrial picture; their presence signals only that there was enough food available in the megalake systems and they did not necessarily depend on land-based animals for sustenance. Although there is so much more to learn from the region, particularly placing species in their final temporal context, at this point it stands as the only example so far of how megafauna lived in central Australia and what species were involved. The fossil and geological record needs to be studied further and in great detail together with other lakes, and much will come from that. But from work carried out elsewhere on fossil lake systems it is obvious that, one, megafauna were not present on other systems and, two, megafauna responded to glacial onset by abandoning central Australia or perishing there and that must have taken place at similar times in other areas around Australia. Vast and extreme environmental changes pushed these creatures out of the Lake Eyre region with retreat to the south. But it seems logical that each time that occurred, fewer animals actually escaped, leaving the main population increasingly depauperate of species and smaller overall. In the next chapter I outline why this was probably the case.

9 Australia's Megafauna Extinction Drivers

In 1967, Robert MacArthur and Edward O. Wilson published *The Theory of Island Biogeography*. It subsequently became a cornerstone in our understanding not just of island biogeography but also biogeography anywhere. The authors proposed that there was an orderly relationship between island size and animal and bird population size – or, more precisely, the size of an area and the number of species that occupy it. It was an idea developed from their research focusing on island animal communities in the early 1960s but Robert MacArthur in particular wanted to quantify the idea mathematically thereby developing a law that governed the number and types of animals present on a given island or living in a specific area. Their work followed that of Darlington, who suggested that different-sized populations (particularly those living on islands) followed a species–area relationship, and that this could be calculated: for example, dividing an area by 10 divides the number of species of fauna residing there by two. In short, the smaller the area the fewer the animals you find there – a logical conclusion, but one that is amplified when we talk about large animals.

While the work of MacArthur and Wilson was based on oceanic islands, I believe it is equally applicable to the Australian continent during glacial cycles. Australia is, of course, an island continent but during glacials it became a physically larger but habitably smaller and divided continent. Areas became isolated as severe enviroclimatic conditions eliminated and moved habitable areas towards the continental fringes and that occurred as the pre-existing deserts pushed outwards. In many places they reached the shoreline, even recently exposed shorelines, dividing the continent into a series of patches simulating a continental archipelago of partially liveable islands divided by a sea of sand. They provided a barrier to both east–west and north–south travel. Thus, sea level fluctuation and desert expansion would have been predominant environmental determinants in the movement and survival of megafauna populations. The overall megafauna population would have reflected this geographical pattern of fragmentation, with various groups sheltering in refugia of various sizes from small outliers to much larger regions such as Australia's southeast.

Consider the turmoil of events outlined in the previous two chapters as glaciations moved through their cycles. First, there was the main glacial cycle up to 60,000 years long, often an order of magnitude longer than interglacials. Second, there were the stadial and interstadial cycles, consisting of rapid climatic reversals with their accompanying environmental changes within the main cycles and, third, there were the many smaller intra-cycle climate switches that also brought considerable change. All these added up to cumulative stressing of animal populations over time. Then there was the nature of major cycle transition. In the vast majority of cases, that was

Corridors to Extinction and the Australian Megafauna. DOI: http://dx.doi.org/10.1016/B978-0-12-407790-4.00009-4

marked by a series of abrupt climate switches superimposed on a steady decline in temperature from warm to cold conditions, sometimes deep cold, and vice versa. The biogeographic consequences for animal populations arising from this regime of constant climate change involving wide temperature swings, with tens of millennia under full glacial conditions, must have impacted markedly on animal populations around the world but no less in Australia because of the massive impact of the aridity accompanying glaciations which covered much of the continent. Their biogeographic circumstances became severely affected and lasted for tens of thousands of years, resulting in reduced and redistributed populations, much reduced food resources and the elimination of suitable rangelands. Interglacials brought more placid conditions, but even they were often punctuated by sudden temperature excursions back to Ice Age conditions. Many interglacials were not long and most did not reach modern temperature levels, meaning most would not have brought much relief from the harsher conditions of glacial Australia.

Introducing the Drivers

> *Animals and plants are adapted to specific climate zones, and they can survive only when they are within those zones.*
>
> *(Hansen, 2009: 145).*

The eminent climate scientist James Hansen was underwriting the fact that animals will be affected more than most by future climate change. We know that because we can see it happening. We have solid evidence that is widely accepted concerning animal responses to environmental and climatic changes, particularly when such changes alter their habitats. In the last 150 years much has been learnt about how animal and plant communities work and what makes them fall apart and go extinct. It is a good time, therefore to look at some of these and apply the principles that lie behind them to the circumstances of Australia's megafauna during the Quaternary.

Ironically, much of our recent knowledge of the natural world has come from investigating remnant populations and those near extinction. It seems that only then is there an imperative to find out how they work. Many animal species have quietly shrunk in range and numbers through changes to their ecology, nutritional circumstances, encroachment by humans, hunting, pollution and many more subtle habitat changes that follow from these, as well as some that we really had no idea would make any difference to them till it was too late or we bothered to find out. Humans have also a dubious history of poor relationships and exploitation of animals to look at and hopefully learn from, and I will say more about that in the next chapter. Many species have gone extinct on our watch, making what some would call the six or seventh mass extinction, depending on how many previous ones you want to recognise. The count is becoming dubious because where one ends and the next begins is becoming somewhat blurred, especially in the later Holocene and what is now called the Anthropocene. It is also almost impossible to know where and when the last individual of any species goes extinct – we just know we have not seen them for some time and then realise the reason is that they are no more.

Did the last Tasmanian Tiger die in Hobart Zoo in 1936, or was there another that lived years beyond only to die alone, lacking a mate or another companion, somewhere in the Tasmanian wilderness? (Hopefully, they are still there....) Beside deliberate eradication of certain species, like the *Thylacine* and passenger pigeon (*Ectopistes migratorius*), or the Great Auk (*Pinguinus impennis*), which was hunted down to the last two individuals that were unceremoniously clubbed to death on Eldey Island in 1844, we understand now that we could have been better keepers of the planet's animals. After all we have the big brain that can understand the world around us – remember that *Homo sapiens* means 'wise man'! However, in 1844 we did not understand what affected and governed animals in terms of their survival, and what makes them go extinct, in the way we do today. There was no environmental education as there is today. Nor did class matter. The illiterate men who clubbed the Great Auk to death were no worse than the public school-educated Sahib who, riding his large elephant, shot tigers willy-nilly. Neither knew a jot about the natural world they were ravaging.

In the past century there have been hundreds of scientific studies undertaken to understand our planetary biology but it has been slow – some would argue, too slow. While that information has been growing, animals have been disappearing. But out of these dubious circumstances, episodes of happenstance and much determined field work, we have slowly become aware since 1844 of the many rules that govern how animals live, what they need, their co-dependence and why they tick, such as the species–area effect and population equilibrium models already visited in this book, among other things. The accumulated knowledge arising from those studies, which have looked at how animals go extinct, has forged a holistic view of the ties between animals and their environments. They are known collectively as biogeography. Biogeography has become a major key in understanding our environments and how the organisms of our planet interact, live, survive, adapt and perish in them. It has developed a wide set of criteria or rules that can be applied to the animal and plant worlds which unlock and explain the secrets of individual species survival and their ability to thrive in different biological and geographical settings. That study also teaches us how animals so easily go extinct. Moreover, we have learnt that there are many factors to consider when assessing the fragile nature of the environment/animal balance required to keep things in equilibrium so that life can continue on so many levels and in so many forms.

It has emerged that this equilibrium comprises a delicate and complex brew of natural factors, or *drivers*, that act for and against the status quo of species on individual, community and population levels and which operate separately or in combination. These factors have to be considered in any discussion of the status of a species, past or present. The *driver* list presented below is only a guide to some of those factors. Another consideration is how many are naturally or reciprocally associated. But what is important is that they apply just as naturally to extinct animals as to those living today. For the purposes of this book I have called the factors that cause extinction *biogeographic extinction drivers* or BEDs. They were briefly introduced in Chapter 4 but they are extremely important to any discussion of how Ice Ages affected Australian and any other megafauna: they are all driven in themselves by climate and climatic change, but they have to be tempered by individual continental circumstances. They also have to be considered in concert as to how they might have affected the chances of megafauna survival during glacial cycles as much as they are considered essential to assessing the chances of survival for animals today in the face of development, deforestation or CO_2-driven climate change.

Biogeographic Extinction Drivers

1. Long-term climatic and environmental change
2. Short-term abrupt climate and environmental change
3. The depth of broad environmental or focused niche change
4. The speed and extent of niche and general environmental change
5. Reduced or lost food and water supply (multiple causes)
6. Food chain breakdown (multiple causes)
7. Trophic collapse (multiple causes)
8. Large animal reproduction: slow, single births, longer gestation time and infant stage
9. Larger animals require greater food and water
10. Larger animals require big ranges
11. Larger animals are slow to adapt
12. Larger animals are vulnerable to inter-specific competition, particularly with smaller varieties that adapt more quickly and thrive in the same niche
13. Species replacement of one by another (invasion and migration)
14. Intensification of competition
15. Failure to compete at the same level as competitors
16. Failure to move against or recognise reducing environmental conditions
17. Rare or endangered status – small or single population or small isolated patches
18. Reduced population size, locally, regionally or continent-wide (especially big animals)
19. Population reduction at genera and individual species level
20. Continental population contraction to isolated regions/areas
21. Large animals are susceptible to regional population contraction to small areas
22. Population contraction at the genera and species level
23. Confinement of a single and/or small population
24. A small but scattered population
25. Isolation from members of the same species (disjunction)
26. Limited breeding potential and opportunity to replace numbers or grow a population.
27. Vulnerability to predation
28. Animals that are highly specialised, particularly larger animals
29. Genetic limitations arising from founder effect, isolation and genetic drift ('blind alleys')
30. Overspecialisation (species with complex, behavioural, physiological and morphological adaptations; isolated species or those occupying higher trophic levels are more likely to go extinct with general or regional environmental and/or climatic change)
31. Top predators are more likely to go extinct than omnivores
32. Collapsing predator–prey chain: extinction of prey species with reduced favoured prey
33. Predation of limited stocks – *boom and bust*
34. Pathogenic outbreak or newly introduced diseases
35. Reaching adaptive limits (body size becomes a burden)

It is important to recognise that these drivers do not, in most cases, act on their own but in combination, in concert or in sequence, as one can precipitate another. There may be more BEDs than those listed on the previous page but these should take us where we need to go. Before looking at them further, it is worth taking a closer look at the climatic mechanisms and environmental changes that 'drove' many of them.

One prominent effect of glaciations was how they effectively shifted biomes elsewhere. In the case of the Northern Hemisphere land masses, Tundra replaced Boreal forest across Asia, Boreal forest moved south and desert replaced savannahs in a number of areas. In Southeast Asia, rainforest shifted south onto the exposed Sunda shelf to be replaced by savannah and open woodlands in northern rainforest areas. Animals move with their favourite ecosystems (remember the Hansen quotation above), so in both regions animals moved with range shifts following their respective biomes and favoured niches. Consequently, the Southeast Asian 'extinctions' were not so much *extinction per se* but on many occasions *regional extinctions*, the disappearance of animals from one area and re-emergence in another; we look at that in the next chapter. Such movement is not new and we saw similar movements around the world throughout the Tertiary in response to environmental and geographical alterations. The roll-on effect is species dispersal, assortment, and redistribution. Movement may not be quite that easy, however, with some species finding that their alternative environment is not as complex, large or as suitable as they might require. In those cases they are affected one way or another, usually with a reduction in numbers that, in some cases, reduces them to dangerously low population levels, a point we might today call *at risk* or *endangered*. In Table 7.1, the large range of species found only once or twice could certainly have this label hung on them. Radical climatic shifts happened at glacial end-points where sea levels rose quickly, engulfing previously exposed shelves. Shallow shelves formed the biggest exposed land masses, extending out hundreds of kilometres from interglacial shorelines, and that occurred in Southeast Asia and northern Australia in places. But while that exposed millions of square kilometres of additional land, such shelves also experienced the speediest inundation at glacial terminations.

The speed of rising sea levels at the end of each glacial episode is a tricky question in terms of how environments responded. How fast, for example, was vegetation catch-up on the mainland as continental shorelines approached? We experienced this conundrum in the previous chapter with megalake formation time. The best tools we have for estimating environmental reformation, whether round a lake or a continental edge, are studies of vegetation catch-up in lowland and highland forests in response to global warming (Bertrand et al., 2011). They show that the former responds more slowly to rising temperatures than the latter, which is interesting because the Australian biota is more like lowland than highland forest communities. That suggests that coming out of an Ice Age quickly, as is evident from the glacial graphs in Appendix 2, would cause a lag between new vegetation communities and the loss of continental shelf. Without replacement of niches and suitable habitats, animals moving off drowning shelves would have been squeezed between safety and the sea as the shoreline approached. Animal populations must have lived and flourished on the

Sunda shelf for more than 60,000 years at a time among the rainforest and savannahs and around large lakes that formed on them. Suddenly, all this would go into
reverse with interglacial onset as landmass reduction occurred, eliminating millions
of square kilometres of fully established environment. We should not forget that the
Appendix 2 graphs starkly show that glacial termination was a considerably faster
process than glacial onset. So, how many BEDs listed on page 220 were operating
during these enormous geographic changes taking place during Ice Ages? With these
factors in mind, we can turn to Australia and what was happening here.

Australian Enviroclimatic Change During Glacial Cycles

Today, the Australian continent covers 7,500,000 km². The mainland extends
2500 km north to south, or 3000 km to the tip of Tasmania, and it stretches from
temperate to tropical latitudes 10–45°. Consequently, it supports a wide variety of
environments from Alpine in Tasmania and the Snowy Mountains to sub-Equatorial
in the far north. During glacials it became Sahul, a continent up to one-third bigger (10,000,000 km²) which joined with Papua New Guinea when sea levels dropped
below about −15 m and adding Tasmania when they reached −54 m. Fluctuating sea
levels during the mid-late Quaternary exposed Australia's continental shelf to some
extent for almost 95% of that time and maximally between 30% and 50%. With the
exception of the Snowy Mountains and parts of southwestern Tasmania, Australia
remained ice free during glacials, making its Ice Age experience very different
from that of most continents, particularly North America and Eurasia. Rather than
the development of glaciers, expanding permafrost, tundra and snow fields on those
continents, Australia's overwhelming changes were increased aridity and expanding
deserts which virtually took over the continent. The mechanism behind this was the
expansion of pre-existing central and western deserts (Williams et al., 1998). Africa
responded similarly, but the difference between the two continents was that Africa
straddled the Equator, dividing its continental deserts – but I will return to that later.

During glaciations, temperatures in Australia ranged from 3°C to 5°C below those
of today but they varied around the continent so that in the Snowy Mountains they
were 9°C below (Anderson et al., 2007). It is difficult to precisely align Antarctic
ice core-derived temperatures with Australia's or indeed with any of the associated
environmental changes here. We do know, however, that Antarctic temperatures
correspond to widespread environmental shifts expressed in many forms around
the world. We also know they were extensive, in many cases extreme, with radical
changes altering biome structures everywhere. Some biomes, for example, were, in
effect, shifted wholesale towards lower latitudes, particularly those of the Northern
Hemisphere, and that is widely documented. One piece of evidence for this comes
from the marked increase in atmospheric methane accompanying interglacial
resumption recorded in Antarctic ice cores. These signals indicate exposure of land
after glacial retreat, permafrost melt forming wetlands and the changing position of
the inter-tropical convergence zone (Loulergue et al., 2008). The correlation of the
methane signal with glacial cycling, therefore, is a direct proxy of enviroclimatic

changes altering vegetation structures and landscapes. Those signals also largely originate from changes occurring in the northern hemisphere on the other side of the globe from Antarctica, where the ice core temperatures were derived. The proximity of Australia to Antarctica, therefore, probably makes the Antarctic data equally or even more relevant as a pointer to environmental shifts taking place here. It is reasonable to assume that the rapid temperature switching recorded in ice cores drilled 3000 km from Australia can be used to indicate equally rapid environmental changes that were taking place here and which were particularly associated with temperature changes.

Glaciations cause drier conditions around the world but this was taken to extremes here, particularly in central Australia. There, persistent strong winds brought widespread transportation of sand and soils from mobilised dunes and dune building. They also brought about widespread desiccation and deflation, all of which promoted desert expansion (Croke et al., 1999; Magee, 1997; Magee & Miller, 1998; Magee et al., 1995, 2004; Nanson et al., 1998, 2008). Major dune building phases have been dated to 256, 173, 112 and 56 ka (Johnson, 2004). Aridity brought drying of inland waterways, streams and wetlands, with reduction of trees and shrubs and shrinkage of forests, and these conditions prevailed over 85% of the last 1 million years. All Australian records seem to show a trend towards a drier and more variable climate during the last 350,000 years, with heightened aridity in the late-Quaternary (Pack et al., 2003). At that time the Tanami, Great Victoria, Great Sandy, Little Sandy and Gibson deserts, an area of 1,068,000 km², expanded towards the coast and in some cases reached it. Desert expanded south to the Nullarbor Plain, west to Australia's northwestern coast, north into the Kimberley, into Arnhem Land at the Top End of the Northern Territory and onto the exposed continental shelf. Central Australia resembled the Sahara, devoid of vegetation. That would have resulted in massive environmental change across the vast majority of Australia's landmass, changing, eliminating, shifting or severely affecting many habitats. Evidence from the northwest indicates a change from C3 woodland to C4 grassland plants, suggesting a reduction in total rainfall and increased seasonal rainfall consistent with a drying climate even in an area that today is within the summer monsoon boundary. At Lake Lewis 600 km to the southeast, and also in the southern limit of monsoon, widespread dune formation took place after MIS5 (English et al., 2001). Simpson Desert expansion probably began 300,000 years ago (Wasson, 1989), although it had probably expanded and contracted to various degrees many times before that. Other evidence shows reduced catchment rains and channel flow to the SLEB during the last interglacial (DeVogel et al., 2004), underpinning similar results at Lakes Amadeus (Chen et al., 1993) and Woods (Bowler et al., 2001), which also experienced reduced rainfall in the late-Quaternary, particularly in the monsoonal catchment. Further north at Magela Creek in Arnhem Land (13°S), a tributary of the East Alligator River there is evidence of decreasing monsoonal activity and a drying climate in the late-Quaternary although it lies well within today's monsoon belt (Nanson et al., 1993). The Nullarbor and central and northwestern Australian coastline would have been sparsely vegetated then and in some areas the desert extended to the high water line.

In the southeast, the Little Desert in western Victoria extended almost to the coast where stabilised, vegetated dunes are seen today. Similar dunes also occur in northern New South Wales near the Queensland border 5 km from the modern coastline. Dune building also took place along the large river systems of the southeast such as the Murrumbidgee and Darling Rivers. Another proxy indicator of aridity across eastern Australia is dust; records show sustained increases in particle levels beginning *ca.* 350 ka (Dekker De et al., 2010; Hesse, 1994; Hesse & McTainsh, 2003). Excessive expansion of the arid centre of the continent has been seen in palaeoenvironmental changes in southeastern Queensland, including retraction of vine thicket and increasing grasses not associated with anthropogenic activity, accompanied by reduction in megafauna species variety (Price, 2006; Price & Sobbe, 2005; Price & Webb, 2006). The coastward spread of aridity is also signalled further north in central eastern Queensland near Rockhampton. Excavations at Etna Cave show replacement of rainforest adapted by arid adapted animal species *ca.* 180 ka corresponding to the major dune building phase of MIS6 (Hocknull, 2005; Hocknull et al., 2007). Faunal changes go hand-in-hand with major vegetation alteration, which included reduced rainforest species and increased open woodland, eucalypt and grass species, suggesting a stepwise transition of increasing savannah as more arid environments moved in and replaced them. Similar changes have been noted in the ODP 820 ocean drill core taken off the northeastern Queensland coast near Cairns. Three signals corresponding to a reduction in rainforest plant species, a concomitant increase in arid-adapted eucalypts and an increase in charcoal particles suggest widespread biome changes from rainforest to more arid conditions beginning *ca.* 160 ka, again in the depth of the MIS6 glaciation. Initially thought to be anthropogenic in origin, these are now more parsimoniously linked to an increased fire regime that accompanied the dry climates of a glaciation where vegetation still existed, as was the case in the tropics. It seems that during glacial periods, Australia's arid centre moved out to severely squeeze the coastal strip so that in some places coastal environments were reduced or disappeared, leaving desert to meet ocean. These sorts of conditions strongly suggest that a similar pattern took place elsewhere on the Australian coastline with the resulting effect that it was reduced to unconnected patches or refugia in some cases. These conditions, however, cannot be considered unusual and they probably occurred throughout the mid-late Quaternary during glacial cycles.

Accumulating evidence is showing more and more massive climatic stresses on late Quaternary megafauna. An example is work on a well preserved and unabriaded assemblage from the Darling Downs 500 km south of Rockhampton (Price et al., 2011). Results there show a stepwise disappearance of species in the region taking place for 75,000 years before the supposed arrival time of humans in Australia. *Phascolonus gigas*, *Palaorchestes azeal* and *Zygomaturus trilobus* had all disappeared from the region by 122±22 ka; *Diprotodon optatum*, *Thylacoleo carnifex*, *Troposodon minor*, *Protemnodon roechus* and *Macropus agilis siva* had all gone by 107±18 ka; and *Protemnodon brehus*, *Protemnodon anak* and *Megalania prisca* had disappeared by 83±10 ka. The above dates coincide with MIS5e, c and a, respectively. These data seem to be another indication of megafauna disappearance over time and when conditions were much better than during full glaciations. In other words, animals are out on the Downs when climate is warmer and environmental conditions are more liveable and conducive to expansion possibly from areas nearer the coast. They are also taking place in a part of the best-stocked region of Australia. That area was probably one of the last refuges for many of the largest marsupials in Queensland because of its enviroclimatic history. Each time the region is colonised, however, there are fewer species able to recolonise, and those found in this latest piece of research were probably some of the last.

While Australia's environment at modern temperatures is well known, conditions that occurred during the last interglacial have not necessarily been repeated in the present one and that was observed in the previous chapter. Moreover, we have seen that each of the last eight interglacials had a different interglacial temperature profile from the other, with MIS5e and MIS11 being closest to the present. Others were shorter and/or much cooler, few reaching modern temperatures. Therefore, extrapolation between today and past interglacials is not always a practical exercise except to compare those that were similar at some time in their cycle. It is from these that some general inferences are possible concerning the existence of certain ecological structures that changed biogeographic circumstances. Interglacials were generally wetter periods with more regular rainfall in the interior but this may not have been the case for all. Inland river flow and lake fillings were sometimes extensive but at other times stream flows happened in short intensive bursts driven by interglacial climate anomalies not apparent in the present, i.e. something strongly reflected in SLEB sediments of MIS5e and on other large lakes in central and northern Australia. Smaller temperature shifts within larger glacial cycles must also have brought sudden and disruptive environmental changes, albeit not necessarily extensive, by shifting temperatures up and down full glacial–interglacial cycles and the record indicates they occurred in rapid succession. There is now a growing body of evidence showing spreading aridity during the mid-late Quaternary and that can be viewed in the following references: Callen and Nanson (1992); Nanson et al. (1992), (1993), (1998); Chen et al. (1993); Magee et al. (1995); Magee (1997); Magee and Miller (1998); Bowler et al. (1998), (2001); Jablonski and Whitfort (1999); Johnson et al. (1999); Wang et al. (1999); Kershaw and Whitlock (2000); Moss and Kershaw (2000); English et al. (2001); Kershaw et al. (2003a,b); Pack et al. (2003); DeVogel et al. (2004); Cohen et al. (2012). The trend is summed up by Kershaw and co-workers:

...all Australian records...show some trend towards drier climatic conditions superimposed on glacial cyclicity through the late Quaternary (Kershaw et al., 2003a: 1277). They also suggest that: *superimposed on this...is a trend towards drier and/or more variable climates within the last 350,000 years* (Ibid, 2003a: 1271).

Other research has concluded there are a series of '...*Australian records that indicate increased continental aridity in the Late Quaternary*...' (Pack et al., 2003:629). The mounting evidence for extreme aridity across the Australian continent during the mid-late Quaternary now forces us to look at the consequences of such change and what effects they had on the megafauna themselves and the resources they relied on.

From Centre to Edge

Modern floral distributions have prompted some to suggest that eucalypt domination fragmented Australia's animal communities, indicating a driver of increased aridity, environmental change and vegetation change.

> *The evolutionary history of Australia's vertebrate communities makes no sense at all in the context of modern floral distributions. What were these once widespread and apparently very productive plant communities now "masked" by eucalypt domination? The answer is obvious, if not overly simple: they are represented by the fire-sensitive communities that still exist in fire-shadows and as remnant associations growing under optimal conditions, especially in parts of the Northern Territory, Queensland and Western Australia and elsewhere....*
>   *(Murray & Vickers-Rich, 2004: 300–301).*

The long evolutionary history of the megafauna was not only based on their ability to adapt to a drying continent but must also have been intimately tied to a symbiotic relationship between them, plant communities and the maintenance of favourable niches. Seed dispersal through frugivorous browsing on medium-sized shrubs must have been one of these. Dromonorthids, the larger kangaroos (Sthenurines) and the diprotodontids were among those that probably helped maintain certain ecosystems such as vine thickets, which included numerous other plant species bearing edible fruits, berries and seeds, through faecal dispersal (Ibid, 2004). With increased aridity during the last 300–400,000 years or more, continental drying must have also become a forcing process, together with an elevated natural fire frequency, both of which affected these floral communities. It is obvious that glacials brought massive vegetation changes to Australia on the local, regional and continental levels as desert rolled out of the centre and stayed there for up to 60,000 years. But that drove a cascade of changes to plant distribution and composition that favoured fire-adapted species and grasses at the expense of fire-sensitive communities, which withdrew to remnant patches and fire-shadows. In turn, grasses would have succumbed to strengthening aridity together with the last few shrubs. They gave way to ever-increasing and broadening sandy patches loosened soils that coalesced with underpinning soil movement and thus promoted desertification. A good indicator of this process is dust. It has been shown that the quantity of dust increased by 25 times during glacials, particularly facilitated by coupling of climates over the Antarctic and lower-latitude continents (Dekker De et al., 2010; Hesse, 1994; Hesse & McTainsh, 1999; Lambert et al., 2008). The late-Quaternary dust records verify this by showing that during glacial conditions soil loss was extensive as a reflection of root systems, grasses and other soil stabilising plants dying. Something very similar to this occurred in Southeast Asia with the impact of glacial cycles on environments there:

The environmental changes of the Pleistocene are correlated with a series of profound changes in the diversity and distribution of mammals in South-East Asia. Indeed, the first consequence of the geological and climatic changes was the size reduction of the sub-tropical and tropical zones.

(Tougard and Montuire, 2006: 136).

Others have also reiterated that environmental and biogeographical changes accompanied glacial cycling in Southeast Asia:

Southeast Asia has undergone dramatic climatic and geographical changes during the Pleistocene. [Glaciations] are associated with repeated shrinking and expansion of rainforests and correlative spreading and shrinking of savannas...pine woodlands and savanna grasslands expanded and replaced the rain forest over large areas of Southeast Asia during the glacial periods.

(Chaimanee, 2007: 3190).

These changes in Southeast Asia were also reflected in Australia but to a much greater extent, which would have had a particularly negative impact on browsers. Their populations contracted to isolated pockets of plant communities on which they relied. As a result, their numbers would be reduced to unsustainable levels, greatly limiting the part they once played in seed dispersal. That negative feedback would reduce further recolonisation attempts by negatively affected plant species. The reduction and loss of plants then impacted animals that relied on them, resulting in reduced niche frequency: fewer animals could be fed and supported, thus reducing individual community size as well as overall population size. Over tens of thousands of years during glacials, such processes must have particularly affected megafauna by changing and reducing population composition and dispersal across the continent and by species contraction, therefore affecting broad population reduction over a long period while smaller, grazing macropods flourished in these circumstances. These events underscore the disadvantages megafauna browsers faced during glacials, compounded by the erratic nature of repetitive stadial–interstadial and other climatic reversals buried within each glacial. This process began in earnest in the early-Quaternary and if we take an average of eight reversals of one type or another in each Ice Age then the above process did not happen just once but at least 800 times during the Quaternary.

At present, Australia's 10 main deserts total 1.37 million square kilometre or 17.3% of the continent. During glacial maxima with lowered sea levels, desert covered variously 70–90% of the 7.5–10 million square kilometres of continental mass, depending on low stand level. Basically, Australia's arid and semi-arid regions, which today cover about 70–75% of the continent, combined into pure desert. Millions of square kilometres of rangelands were lost, driving megafauna to the continental edges, although not all of the edge was habitable as previously noted. As favoured plant communities retreated before the spreading desert, animals that relied on them did the same. Different species required different minimum areas for support depending on their size, behavioural patterns and adaptive qualities, but big browsers living in small patches would not survive in the medium term. Even though large areas of the centre and the west may not have been inhabited by megafauna, those living in semi-arid regions must have moved closer to available water and that meant coastward, while others living in isolated pockets and along drying river systems disappeared or were stranded. Those processes alone reduced the overall megafauna population as well as affecting those living precariously as specialised species focused on small niches and patches. Patches experienced vertical and lateral niche shrinkage, reduction in vegetation variety and fragmentation, all of which compounded problems of population maintenance and survival, particularly among small populations. Larger areas were less susceptible but would have shrunk.

One effect of fragmentation is that while animal populations can exist, the resulting fragments do not necessarily represent places where future population recovery can take place when conditions change. Indeed, patch fragments were more likely to have led to localised extinctions, particularly during extreme, long-term glacials. Megafauna living in small patches or isolated groups would be eliminated as their patches shrank to nothing and blew away, eliminating the animals and causing overall population reduction. Smaller animals were less vulnerable and would survive longer in a given patch size. Also, small patches could support greater numbers of small species and their faster reproduction rates aided them in maintaining and recovering their population numbers. Broadly adapted, generalist animals had the best chance of survival during glacials, but larger, specialist animals with longer life spans were less able to cope (see below). Problems arising from glacial onset did not finish there because for megafauna moving to the edge of the continent there was another problem: *equilibration*.

Turnover, Equilibrium and Recovery

The graphs in Appendix 2 show slow glacial onset so when environmental change took place, animals would have slowly contracted to better areas. Their movement from a degrading to a more stable environment was not, however, a simple rule for rescue. MacArthur and Wilson's 1967 *turnover* explanation for balancing oceanic island populations is just as useful for describing isolated continental patch populations suffering habitat contraction during Ice Ages. *Turnover* takes place when a given carrying capacity can support only a certain number of animals and resident groups suffer or become reduced when migrating animals move in, although an actual reduction occurs for both groups. The slow movement of animals into better environments means they join those already in place and all target the same foods. A natural balance ensues when the burgeoning effects of population increase are circumvented by overexploitation of limited resource. There are two possible outcomes to this: one is the displacement of the endemic group by the new arrivals; the other, and more likely, is the blending of both groups, putting them in *disequilibrium*, but only for a while because that situation is not sustainable. In the medium or long term, larger animals are most at risk in this situation, particularly where populations are at unsustainable sizes. A given environment usually has a finite carrying capacity that will only support a certain number of animals. The outcome is that the food supply will be overexploited, triggering population collapse. The population bulge then *equilibrates* back to the natural carrying capacity of the region which, as before, has little room for more animals, leaving the same number of animals occupying the area. The net result is that the migration fails to rescue animals in terms of maintaining the population size. Instead, there is a net population reduction. Movement of animals to peripheral areas, therefore, is not a recipe for survival of all: rather, it means the loss of animals through a much-reduced carrying capacity, and this must have taken place hundreds of times. It also has to be remembered that a given carrying capacity was slowly reduced as glacials deepened. While the environmental consequences of glacials are obvious, the effects of smaller climatic fluctuations within glacials and the abrupt stadial events within interglacials are more difficult to ascertain, but similar outcomes must have occurred even if in a more limited form.

There is a 'rescue' factor in this story and that lies in the degree of habitat restoration that is possible following glacial termination, something essential for any sort of megafauna population recovery. Individual areas would have experienced various degrees of recuperation during interglacials and interstadials, but no two were the same and population recovery was not the same. Animals could come back from the brink, with some actually expanding back into old refuges, but if they did not have the opportunity to recover during these short time frames before the next onset of aridity then they were already experiencing a lower population entering the next Ice Age than that which entered the previous one. The whole process was an inexorable reduction to various degrees in different regions: the smaller the initial population the quicker it reached an extinction threshold, decreasing ever more rapidly as glacial cycles proceeded. It would have been a slow process, taking place over many glacials and other climatic reversals each time, variably reducing individual species living in different regions and reducing to some degree the overall megafauna population. Losses through each cycle varied depending on the circumstances of a particular species or population and the strength of the enviroclimatic changes involved. Those variables affected some species more than others. It was an inexorable process, moving forward little by little, sometimes rapidly, other times very slowly, even with slight improvement in numbers on occasion during warmer, more stable interglacials. Figure 9.1 has been constructed to demonstrate how this process acted variously on four different species over a number of glacial cycles. The process was not, however, a lineal one. As the population fell, so the downward trend became exponential and that applied particularly to bigger rather than smaller animals. Under these processes, the latter were, in a way, being separated from their giant cousins. Smaller marsupials could survive on small patches, with their greater reproductive frequencies and on smaller amounts of food.

Generally, interglacials meant restoration of inland environments and a chance for megafauna to reoccupy their previous habitats. But could megafauna populations be restored to pre-arid levels in the short time slots available during interglacials and interstadials? In most cases both these events were short, some very short, and often did not reach average interglacial temperatures, thus preventing inland environments from reverting to their previous optimum levels of restoration and occupation. This also produced a recovery that varied with the length of each interglacial and the environmental conditions it brought. In other words, the continent remained little

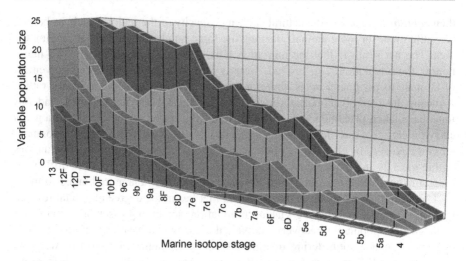

Figure 9.1 A differential megafauna extinction process using four theoretical species (gray shades) of different populations sizes, variable population histories and staggered terminal outcomes during the mid-late Quaternary. What is missing is the spatial component that whereby different species histories look different in various regions.

improved during many interglacials, which suggests severe limitations on population recovery. Interstadials were even shorter and even more limited in this respect. Glacial onset then cut short any environmental recovery and the same process of inland abandonment reduced populations and the cycle began all over again. Thus the overall Australian megafauna population was slowly reduced during the repeated glacial cycling of the Quaternary in a staggered manner, with each species facing its own problems. Some fared better than others but all experienced a reducing population to some degree, slow at first but increasing over hundreds of thousands of years. But there is more.

Megafauna Reproduction

As impressive as large animals look, they are vulnerable. While isolated groups are naturally predisposed to extinction, isolated groups of large animals are even more vulnerable. They need large ranges and plenty of food and water and this is reviewed below. But they also have one particular Achilles heel: their reproduction rate. During the last 50–100 years we have seen the world's largest animals suffer population reductions as well as extinction of some species. Besides natural environmental changes, humans have expanded into their territory, hunted them for one reason or another, burnt their habitats, logged, clear-felled and generally destroyed their ranges, all of which have played a large part in these reductions, pushing some species to the brink of extinction and others beyond. Behind the inability of large animals to resist a comparatively rapid slide towards extinction is their slow reproduction rates, something that does not affect rats and rabbits or small marsupials. So we can be fairly certain that it was not easy for megafauna to breed up quickly from any population reduction, collapse or other set back.

All the large animals have long gestation times; they usually have single births and the infant has a lengthy nursery time with its mother (Holliday, 2005). We are well aware of the 640-day gestation time of an elephant but let's choose something smaller – a rhino, for example, which is similar in size to a *Diprotodon*, has a gestation time of 16 months, a birth interval of two to three years, and the calf remains with its mother for two to four years. It is, therefore, a long time before the infant reaches maturity, defends itself against predators and joins the reproductive capabilities of the group. An infant animal also attracts predators and in so doing draws the mother indirectly into conflict with them. Moreover, a young animal is less likely to keep up with the group, making it a vulnerable straggler, easily picked off; it tires before older animals do and as a small animal, it is more vulnerable to fatigue, drought conditions and limited food supplies. For large animals the loss of an infant measured against the time to produce one makes it an extremely important loss when trying to maintain population numbers. Times of environmental hardship during glacials would also have militated against reproduction when nutrition was at a low level or difficult to access. All larger megafauna were probably prone to similar pressures and drawbacks in their reproductive histories, especially long gestation periods. Of course, the proper environment and nutritional intake are vital for females to bear young, and so the kinds of massive environmental changes described for glacial cycles would have impacted on that process, and on the young themselves. As physically powerful and fearsome as these animals might have looked, they were as vulnerable as a rhino!

Large modern kangaroos have a gestation period of around 33 days; the young 'joey' (pouch young) then suckles in the pouch for up to nine months, after which it leaves, although it continues suckling visits for another three months. Kangaroos are also fertile while the pouch young are suckling; as soon one leaves the pouch, another is born and takes its place. Kangaroos are also able to adjust their breeding conveyor belt to food availability and environmental circumstances: breeding down during drought and breeding up when more plentiful times return. Megafauna recovery from glacials was probably very slow, in tune with their big-animal reproductive capabilities as well as the difficulties they faced from other constraints outlined above. Small populations are also subject to a critical reproductive rate dictated by reduced numbers of reproductive females. Eventually, reduced reproduction cannot attain the minimum replacement rate and the group is no longer viable. This process is underpinned when reproduction rates are naturally slow – and megafauna reproduction must have been slow. The chances of recovery increased with the length of the interglacial, but not necessarily for all populations or for the population as a whole.

Gender balance was critical because the loss of females compounded recovery problems. Gender imbalance in local populations could have arisen following adult losses. Isolation, population disjunction and small group size were some of the factors that could severely reduce reproductive capabilities. Finding a mate is difficult when groups are separated by closed or long, difficult corridors or they occupy small isolated and shrunken patches. Population collapse is prevented only by the existence of a minimum number of fertile animals for a particular species. An irregular demographic profile in which a group is dominated by very old or very young animals also prevents a collective push towards recovery. Moreover, isolated groups would experience the lack of new arrivals that would reduce species turnover to zero, with net losses greater than replacements, resulting in *non-equilibrium* (Brown, 1971, 1978). Isolation of small groups and individuals also makes them vulnerable to predation. A couple of adult *Megalania* or *Thylacoleo* are far less troubled by an isolated individual or a young animal than by a large herd, and small group attrition is a real possibility as individuals are slowly taken by predators over time. Of course, the carnivores do not get off Scot-free, particularly those also occupying a small patch, because they experience boom and then bust as their prey is eliminated.

The speed at which habitat restoration happened at glacial termination would be another factor at play in population recovery but that probably did not occur very often in the late-Quaternary. The implications of BEDs made it difficult for any megafauna population, either as a whole or in terms of individual species, to recover from a severe reduction in numbers. It was like turning round the proverbial ocean liner, but before the 'liner' in this case could complete the turn during an interglacial, glacial or stadial intervention slowly pushed it back on the same course again, taking it further along the road of population reduction. Full population recovery required a long period of environmental stability at optimum conditions and that took place increasingly rarely as the Quaternary proceeded.

Patch Size

Animals require different range sizes to support them depending on the species, behavioural patterns, specialisation and adaptive qualities. Small patches do not usually support large browsers, although they can be sufficient for small animals. Even large animals living in larger patches are vulnerable because the patch size will determine how many can live there, and that may not be very many; thus the population faces a number of problems. One of these is patch fragmentation, which involves the vertical and lateral niche shrinkage and reduced niche variety mentioned earlier. This compounds problems of population maintenance, particularly among already small populations. Generally, megafauna needed large patches but if they became fragmented so would their population. Small populations of large animals may continue for a while but small patches are not places where large species thrive in the long term. A large fragmented patch, therefore, is really a number of small patches and is, therefore, just as likely to lead to localised extinctions across the fragments, but that also depends on the number of fragments and their size. Smaller animals were less vulnerable to patch shrinkage and fragmentation, surviving longer in a given patch size than larger animals, and their faster reproduction rates helped them maintain their numbers. Then again, broadly adapted, generalist animals survived better under these circumstances than specialist animals, especially those with long reproduction times. On the wider continent, the relationship between the number of species and the area they occupied must have been significantly affected during glacials when species–area ratios increased as more animals tried to live in smaller patches with reduced food and water supplies. But glacials were slow onset so population reductions would also have been slow, almost imperceptible. The broader environment would also have confined larger species to particular areas, suggesting that populations were never large and had limited distribution. Facing a world of increasing and rapid enviroclimatic change during the mid-late Quaternary would have brought all the above factors into play with increasing impetus, mostly impacting large species.

We have seen how spreading aridity closed corridors, long-narrow ones first and then short-narrow types. Corridor closure could fragment habitats and trap and isolate animals. Conscious movement is not considered here, because it implies immediacy; instead, movement to the continental periphery was an extremely slow shift of groups within the retreating hem of comparative abundance. New habitats opened as those of central Australia closed. In coastal areas with a shallow continental shelf, exposure of new land may have taken place as much as 40 cm annually as it is estimated to have done on the shallow Southeast Asia shelf. The emergence of the shelf, albeit slow, would have helped relieve population pressure along the coastal strip and in country back from the coast and remembering the principle of equilibrium. Continental shelf environments also lasted tens of thousands of years, which allowed the growth of communities of mature vegetation to be established fairly quickly as exposure may have occurred around 20 km per millennium (Mackay, 2009). Unfortunately, we may never know how these were used by megafauna or what favoured niches those environments may have produced even though it was such a large and almost permanent part of Australia's geography during much of the Quaternary. But as we saw in Chapter 7, there were the dangers of isolation and drowning when sea levels rose.

Megafauna biogeography must have differed radically between glacial and nonglacial phases. It responded in different ways and to different degrees at various times in different parts of the continent. This would have led to biogeographic differentials that affected some species more than others. Cumulative effects resulting from such differentials are likely to have forced the repositioning of animals and the sorting species composition as well as redistributing and eliminating species. An example of this is the westward and southern expansion of the Western Desert, engulfing much of the west Australian coast and the Nullarbor Plain. That isolated the southwest and effectively separated the continent east from west for most of its width and north from south except for a slender east coast link. The habitat fragmentation, niche shrinkage and megafauna population fracturing that followed would have been regulated by distance effects between refuges. But smaller niches and refuges disappeared at these times, together with any resident megafauna. In some cases, critical colony size would be reached, reducing animal populations to non-viable levels with differential extinction rates and changes to species distribution and breeding patterns, another element reducing the overall continental population. The effects of isolation and inadequate gene flow through smaller and smaller populations could then compound these problems. Small, isolated populations suffer genetic drift and founder effect, leading to expression of harmful alleles over the medium to long term and that would occur with each glacial–interglacial cycle.

Australia's Glacial Biogeography and Extinction Drivers

It is difficult to imagine with any real feeling the radical and widespread changes that accompanied a world temperature drop of between 7.0°C and 9.0°C. But this is an underestimate because some interglacials were warmer than the present one, sometimes up to 5.0°C warmer. That makes a net maximum glacial–interglacial difference of up to 14.0°C, a larger change than that from the Eocene Maximum to modern temperatures. The difference is that that took millions of years, whereas glacial temperature drops took only a few thousand. While some animal groups in Australia may not have been disadvantaged by these events, the overwhelming case must have been that many were, particularly those living on the edges of main groupings and others who had formed isolated pockets in various places. Any significant planetary cooling or warming has an effect on biota, as we are constantly being reminded today. It alters terrestrial biological communities and their trophic connections and changes oceanic temperatures with consequences for marine communities. Therefore, knowing the details of climate change, particularly over the last 1 million years, and not just the big events, has a profound bearing on how we should view and interpret what happened to the megafauna populations that experienced those changes.

The spread of aridity reduced food supplies and that meant the loss of large animals through a mixture of starvation and natural attrition due to range contraction and pasture loss, as we have seen above. Large animals require greater food and water resources than smaller varieties. They also require bigger ranges for manoeuvrability and resource gathering, this is particularly so for grazers. The browser also requires enough shrubs and small trees which are the first resources to suffer under Ice Age conditions. Rainforests were transformed into open woodland, woodland gave way to savannah and savannah shifted to desert. Australian biomes largely moved north and those in the southeast contracted coastward. Unlike Southeast Asia, the finite island continent of Australia, and its limited continental shelf in the south and east, imposed limits on how far biomes could move.

A differential occurs between biomes as they 'move', which affects the number and types of fauna that live in them. Generally, small species inhabit rainforests and large animals graze on savannahs and in open woodland. The high-rise of the rainforest provides different levels of food resources and shelter, whereas these are limited on the flat sprawl of the savannah and even more so in deserts. Increased aridity meant major loss of biomes containing variety. Desert dwellers live in an environmental extreme and are not put out very much if their home becomes savannah. Savannah is not hard to put up with; there is always somewhere to live and it is more benign than desert. But that flexibility is not generally present in animals that live in open or closed woodland. They cannot cope so easily when their habitat loses most of its trees or becomes desert, and that is what happened during glacials as narrow strips of savannah were squeezed between encroaching desert and coasts, particularly where little release onto an exposed continental shelf was possible.

It would have been difficult for larger megafauna to adapt to encroaching desert in this way, although they would be forgiven for thinking they had adapted enough to a drying continent. But when deserts virtually took over the continent for the majority of the mid-late Quaternary, it was difficult for them to maintain their numbers as before. So, the reduction of woodland/savannah in the face of spreading deserts would have been a significant driver of megafauna population reduction. In contrast, forest-dwelling species disappear if their forest disappears because they have no fall-back environment to enter. They move with a moving forest or their population contracts; if forest lands suffer net reduction, so do the animals living in them. Once again, a population that has shrunk, this time with its forest, may have reached a level where it has become too small to recover. A similar process could have occurred to savannah dwellers. Large animals favour woodland and savannah and so did many megafauna species, as far as we can tell, so any change in these environments affected them. They are also usually slower to adapt, thus vulnerable to inter-specific competition, particularly with those adapting more quickly in the same niche. The best chance to save the population in this case is to link up with other populations of the same species. But that requires corridors, and those had closed or were closing.

While widespread environmental changes affected virtually the whole of the inland, they were not necessarily uniform and probably occurred on different scales in different regions. Certainly, many megafauna niches disappeared during glacials. Temperature perturbations within glacials must have pulled environments this way and that, creating confusing patterns and resulting in added stressing of animal populations. Environmental change is almost never uniform on the continental scale: some areas will change radically while others experience only minimal change. These can be used as refuges for shifting or dwindling groups of animals but they are not permanent places of safety – and animals also have to reach them. There is then the element of the geographic extent of environmental alteration, but equally important is the speed at which it takes place. Speed of change has to be measured against standard climate change over thousands of years as well as that occurring through abrupt climate change. Slow change offers time for adaptive behaviours to develop but rapid environmental change over hundreds of years or even decades does not. Drought and good times are the Yin and Yang of nature's clock in the present interglacial. We measure those in seasons or by 'good years' and 'bad years'. Sometimes in Australia droughts can last a decade or more but they always come to an end. They do not reflect to one degree the extremely long glacial changes and their accompanying conditions. Modern climatic change does not add up to that encountered in Ice Ages. A decade of drought is undetectable against the changes

Table 9.1 Two Possible Lines of Consequence for Australian Megafauna Population Reduction During 20 Mid-Late Quaternary Glacial Cycles

Event	Consequence 1	Consequence 2
Aridity spreads out of Australia's centre.	• Channels dry, other water sources disappear, food stocks crash.	• Small refuge areas and habitats disappear. Larger refuges become smaller.
Adjacent environments become marginal, a trend graduated with increasingly better browsing to the coast.	• Main corridors close, isolating various parts of the continent. • Corridor closure also creates environmentally marooned islands with sub-populations. • Adaptive species favoured. • Inland patches disappear and peripherally placed patches shrink.	• Animal populations are fragmented and isolated as corridors close. • Usual movement patterns cease with onset of disjunction. • Australia becomes a series of isolated regions forming a continental dry-land archipelago with pockets of isolated animals.
Animals occupying the continental centre die or migrate out.	• Migration confined to escaping species or those that can live on the edge of the spreading arid zone. • Most large species do not move far with macropods comprising the bulk of those that do.	• Some species disappear locally. • Other animals become tethered or trapped in areas of increasing aridity. • Carnivore populations experience boom and bust, suffering drastic reductions or become locally extinct.
Animals enter territory already stocked with the same species. Overpopulation and increased competition.	• Overstocking of fauna occurs in areas becoming marginal. • Local population collapse due to overcrowding and overexploitation of carrying capacity. • Animal losses within and across different species.	• Losses among megafauna that move into marginal areas, others displace endemic groups or move on. • Animals under severe demographic, behavioural and breeding pressure become vulnerable to losses. • Species already 'endangered' slide closer to or are overcome by extinction locally. • Overall population size is reduced.
Aridity coinciding with oceanic regression and shelf exposure. **Aridity not coinciding with oceanic regression and shelf exposure.**	• Animals move onto exposed shelf depending on the size of megafauna populations living near regressing shorelines. • Animals trapped between spreading aridity and coasts in narrow strips of favoured environment and isolated niches. • Population concentration leads to reduced numbers of animal.	• Slight population recovery with expanding shelf and new emerging ecosystems (a slow process). • No further migration. • Overall population reduced from time of previous occupation of the centre.

(Continued)

Table 9.1 (Continued)

Event	Consequence 1	Consequence 2
Continental shelf transgression with or without alleviation of aridity in the centre. The possibility of rapid sea level transgression.	• Animals living on the shelf are confined on shelf islands that form as sea levels rise. • Inter-island corridors disappear as sea rises together with links to the mainland. • Animals trapped on shrinking islands that then become submerged.	• Loss of all fauna inhabiting continental shelf with the exception of the few who moved before links to the mainland disappeared. • Population numbers drop further.

that took world climates through a giant roller coaster of temperature swings over hundreds of thousands of years. They were extreme and ponderous and the scenario painted above is admittedly only a stab at what it must have really been like during those times. Every time world temperatures plummeted Australia was largely turned into a desert; for up to 69,000 years as in MIS8. The speed of temperature change varied from the slow slide of glacial onset that took tens of thousands of years to reach glacial maximum to end-point changes that took only a few thousand years to warm the world 9°C. During the last 800,000 years the world spent over 85% of the time in temperatures below of those today and it is fairly safe to say that this brought a multitude of radical stress events to megafauna populations (Table 9.1). But the fluctuating temperature spikes within glacials and glacial onsets, noted interminably in Chapter 6, were rapid events occurring over a much shorter time, perhaps a few hundred years or even less. Warming periods within glacials were also rapid, with a return to interglacial conditions, usually taking a few thousand years to move nearly 10°C. While appearing reasonably unimportant against the main cycle, these intra-cyclic anomalies may have contributed vital thresholds of continued habitation in many areas of the continent, bringing the environment back to normal, or almost so, but in short bursts. These would have provided animals with a chance for opportunistic foray back into old habitats, only to trap them once more as the situation just as quickly reversed again. A sequence of such events served to further wear down populations through corridor closure and patch shrinkage time and time again through cumulative, widespread, varied but significant impacts on megafauna. It is often the small changes that tip species into oblivion – the stepwise continuation of negative trends that in themselves are not particularly dangerous but which cumulatively pose special conditions and limits on even the most adaptable creatures.

Differential spacing of fauna usually reflects a spread of suitable habitats, its preferred ecology and niche availability. At its most environmentally benevolent, the Quaternary Australian continent probably imposed constant changes on megafauna distribution. The environmental changeability of the mid-late Quaternary, even in optimum times, probably limited the size of those populations in relation to the amount of available living space. At its worst, suitable habitats were not spread out but confined, very confined in some areas such as southwestern Western Australia and probably in the north of the continent; it is possible that only the southeast was to any degree constant. The rest of the continent severely limited the size of population that could inhabit it. Thus the size of the continent does not reflect the size of the megafauna population that was sparsely distributed outside the southeastern and east coast regions for most of the time and was probably only moderately plentiful there.

Late-Quaternary palaeoenvironmental change cannot be ignored in any future discussion of megafaunal extinction because the severity of biogeographical changes that accompanied glacial cycles was probably significant, with variation during every glacial cycle. The result was increasing biogeographic stressing among the various species living in different regions of the continent. Increasing aridity would also mean large-scale shrinkage of benign environments and their replacement with harsher ones, demanding greater adaptive qualities and favouring smaller animals. Any negative climatic fluctuation changed refuge environments so that some always teetered on the edge of oblivion. In turn, the balance is determined by the marginality of the particular area: those with plenty of fodder/water supported more animals than those with less, but the refuge area could not increase its capacity even by 10% until more favourable environmental shifts occurred at glacial termination, when habitat expansion was again possible but they did not always eventuate.

Modern Australian Extinctions

At the end of this chapter, it is worth reminding ourselves of Australia's largest living 'megafauna', which consists solely of a few kangaroos, and others who have gone extinct since European colonisation in 1788. The largest animals are listed below together with their distribution and maximum weight for male individuals (Strahan, 2000). Note that five are close to or above the megafauna weight threshold weight of 50 kg.

- **Red Kangaroo** (*Macropus rufus*) **66–90 kg**. Very common, found in much of semi-arid and arid Australia west of the Great Divide. Absent in most of Victoria, Eyre Peninsula, southwestern Australia, Top End and Kimberley north of Broome.
- **Eastern Grey** (*Macropus giganteus*) **66 kg**. Common in eastern two-thirds of Queensland, NSW and Victoria.
- **Western Grey** (*Macropus fuliginosus*) **54 kg**. Found in western NSW, Victoria and across the Nullarbor to southwestern Australia.
- **Antilopine Wallaroo** (*Macropus antilopinus*) **49 kg**. A tropical animal found in the Northern Kimberley, Northern Territory's Top End and Cape York.
- **Common Wallaroo** 'Euro' (*Macropus robustus*) **47 kg**. Found all over Australia except Victoria, Nullarbor and western two-thirds of Cape York.
- **Agile Wallaby** (*Macropus agilis*, sub-species *agilis, nigrescens and jardini*) **27 kg**. Northern Kimberley, Top End, coastal Gulf country and Cape York.
- **Black Wallaroo** (*Macropus bernardus*) **~22 kg**. Found in a small patch in southern Arnhem Land.
- **Black-striped Wallaby** (*Macropus dorsalis*) **20 kg**. Occupies tropical Queensland to mid-coast NSW and inland to the Great Divide.

Table 9.2 Marsupial Species Once Common on the Mainland but Now Confined to
Offshore Islands

Common Name	Scientific Name	Last Seen on Mainland	Original Location	Original Range
Western Barred Bandicoot	*Perameles bougainville*	–	WA to NSW	Extensive
Burrowing Bettong	*Bettongia lesueur*	1940	Inland Aust	Extensive
Rufous Hare-wallaby	*Lagorchestes hirsutus*	1950s	Inland Aust	Extensive
Banded Hare-wallaby	*Lagostrophus fasciatus*	–	S & SW WA	Two limited patches
Greater Stick-nest Rat	*Leporillus conditor*	–	WA to W NSW	Extensive
Shark Bay Mouse	*Pseudomys fieldi*	1980s	Inland Aust	Extensive

The Common Wallaroo (*Macropus robustus*), the two greys and the red kangaroos
are the only extant, large and widely distributed Australian marsupials, with smaller
wallabies and wallaroos scattered regionally. What is worth noting is that none
of these animals is found everywhere. The red is arguably the most common, with
the others largely inhabiting eastern and southern Australia (that is familiar). Other
animals are found in patches with two tropical animals. Other macropods are much
smaller, with most wallabies weighing 10 kg or less. Smaller species, such as the
Bilby (*Macrotis lagotis*), common Brushtail Possum (*Trichosurus vulpecular*) and
Western Quoll (*Dasyurus geoffroii*), were widespread, but have severely contracted
since European colonisation. Five others, the Western Barred Bandicoot (*Perameles
bougainville*), Rufous Hare-wallaby (*Lagorchestes hirsutus*), Burrowing Bettong
(*Bettongia lesueur*), Greater Stick-nest Rat (*Leporillus conditor*) and Shark Bay
Mouse (*Pseudomys fieldi*), now live only as relic populations on small islands off the
western Australian coast (Table 9.2). All had extensive ranges throughout inland and
central Australia at European arrival (Strahan, 2000). Reasons for their drastic range
contraction, as well as the total extinction of another 17 species since 1788, include
the introduction of cats, foxes, rats, rabbits and domestic livestock and the cessation
of Aboriginal mosaic burning, among other possibilities (Flannery & Schouten, 2001;
Strahan, 2000) (Table 9.3). Not one of these extinctions, however, can be properly
traced to a single known cause; few suggested causes are substantiated to any degree
and none applies to all disappearances – and we have been here to record them! At
least half the species were already rare, had extremely limited ranges or were delicately balanced in their particular environment like some mouse species. Many were
animals whose demographic circumstances made them vulnerable to *any* change
in their habitat, like the Toolache Wallaby (*Macropus greyi*) which occupied a very
small region along the southern section of the South Australian–Victorian border.
They are unlikely to have survived even minor change in their life patterns, which is
what seems to have happened to the Toolache after settlers moved in. Animals like
the Black Wallaroo listed above could easily come under a similar threat because of
their limited range. Furthermore, it cannot be confirmed, or rejected, that any of these

Table 9.3 Marsupial Extinctions on the Australian Mainland since 1788 in Historical Sequence

Common Name	Weight	Scientific Name	Last Seen	Original Location	Range Size	Reason for Extinction
Darling Downs Hopping-mouse	?50–100 g	*Notomys mordax*	1840s	Inland SE, Qld	Extremely limited	Unknown
Big-eared Hopping Mouse	?50–100 g	*Notomys macrotis*	1843	SW	Extremely limited	Unknown – cats, agriculture
White-footed Rabbit Rat	200 g	*Conilurus albipes*	1845	SE (Q to SA)	Broad	Cessation of firing, rats (disease), cats
Gould's Mouse	50 g	*Pseudomys gouldii*	1857	Eastern Inland Aust	Broad	Cessation of firing, rats (disease), cats gone before foxes arrived
Broad-faced Potoroo	?1.5 kg (est.)	*Potorous platyops*	1875	SW Aust	Three small patches	Unknown, rare when European arrived
Eastern Hare-wallaby	1.5 kg (est.)	*Lagorchestes leporoides*	1889	SE Inland Aust	Broad	Unknown – before intensive settlement
Short-tailed Hopping-mouse	100 g	*Notomys amplus*	1896	CA	Extensive	Unknown
Great Hopping-mouse	?50–100 g	*Notomys* sp.	?1900	Flinders SA	Extremely limited	Unknown
Basalt Plains Mouse	?~100g	*Pseudomys* sp.	?1900	SW Vic	Extremely limited	Sheep grazing on tussock grass
Long-tailed Hopping-mouse	100 g	*Notomys longicaudatus*	1901	Inland Aust	Extensive	Unknown
Pig-footed Bandicoot	200 g	*Chaeropus ecaudatus*	1901	Inland Aust	Extensive	Unknown
Lesser Stick-nest Rat	150 g	*Leporillus apicalis*	1933	Southern, CA	Extensive	Unknown
Desert Rat-kangaroo	900 g (ave.)	*Caloprymnus campestris*	1935	CA	Very limited	Unknown, never abundant
Toolache Wallaby	15kg	*Macropus greyi*	1939	SE SA & SW Victoria	Extremely limited	Hunting, land clearance
Desert Bandicoot	800 g (est.)	*Perameles eremiana*	1943	NW SA & NW WA	Extensive	Cessation of firing
Lesser Bilby	360 g (ave.)	*Macrotis leucura*	1950s	CA	Two patches	Unknown
Crescent Nailtail Wallaby	3.5kg	*Onychogalea lunata*	1956	West and CA	Extensive	Unknown – fox!
Central Hare-wallaby	1.5 kg	*Lagorchestes asomatus*	1960	CA	Broad	Unknown

CA - Central Australia; NWSA - Northwestern South Australia; SW - Southwestern Australia; Vic - Victoria; SE - Southeastern Australia.

species would have survived with or without colonisation by non-Aboriginal people and their domestic animals. It seems obvious that some species were probably going extinct as part of the natural course of events among animals.

It is clear that megafauna would have been in circumstances no different from those faced by our modern fauna. We know some were geographically restricted; others are rare in the fossil record. Still others would have been confined to dwindling or rapidly changing environments to which any further negative alteration from climate change would have put them in danger and eliminate them altogether. Others may have been overspecialised, while the inability to continuously or quickly adapt throughout the largest animals was detrimental to their survival in such a turbulent period. Isolation and population fragmentation compounded most or all of these problems. These situations are typical among animal populations anywhere and it must have been so for the natural world of the megafauna. In other words, many were poised at the edge of extinction as the result of many factors. Like our modern fauna, some were always rare and these were more vulnerable than most. Others were more common or very common and groups were composed of a range of species that variously fitted into all these categories, as much a part of their natural pattern as any other animal population. These factors worked variously, in different ways, at different times and across large expanses of time. All megafauna must have had a complex relationship with the environment, including a variable ratio of species composition in any given population, and these must be reflected to some extent in the dispersal of species as they are found in the fossil record. I have used the term *climatic change* many times in this book not because it is a buzz phrase at the moment but because it happened hundreds of times in varying frequency and severity during the Quaternary. The degree of climatic change had to affect to some extent any animal that experienced it. Some succumbed more than others to such changes but there can be no doubt that climatic changes of the kind that took place during glacial–interglacial cycling must have had a severe effect on many groups and that large animals were the most vulnerable. Fluctuating aridity made for a tapestry of contrasting and complex situations across Australia in both spatial and temporal dimensions for at least the last 1 million years but the details of that tapestry are now lost. Nevertheless, basic relationships drawn from extant species and standard biogeographic and ecological principles are extremely useful if megafauna extinction is to be understood more deeply.

One argument against environmental stress as a cause of Australia's megafauna extinctions is that those animals lived through previous glacials. This is an extremely simplistic argument because it does not take into account the complexities of living through glacials and the resulting changes to the lives and biogeography of the animals. It implies that glacial cycling could have had no effect whatsoever on megafauna populations and that that they and their lives were static. For this reason I reject the argument as invalid. We have seen that megafauna were not static, and could not have been, in the same way that other animals are not static – things happen to animals when environments and climates change. While some did live through previous Ice Ages, many died and populations crashed or at least experienced downturns. They must have also experienced difficulties in maintaining their populations, let alone increasing them again. Enviroclimatic attrition worked its way through various species at different rates and at different times. It wore these large animals down and pushed them to their limitations. In some cases, further adaptation may not have been possible while in others, animals could not keep up with the replacement rates required to save them. We know this must have been the case by just referring to well-known natural laws that apply to all animals. Once again, these may not have worked on mythological beasts, a status that the venerated megafauna have sometimes been elevated. The above argument says that none of this happened to megafauna. It was an argument used by archaeologists, not biologists, zoologists or biogeographers, and I have to say that it has often been the case that in the past palaeontologists have concentrated more on taxonomy and systematics than on the actual lives and biological circumstances of their fossil species.

10 Megafauna and Humans in Southeast Asia and Australia

A Wish in a Cave

Could the artists at Lascaux, Chauvet, Altamira and other western European palaeolithic art sites have been using memory or stories to paint across their cave walls those magnificent images of animals that went extinct, or did they know them? What drove people to paint buffalo, horses with spots and rhinos with very long horns in such beautiful colour and vivid movement? Are they early 'films' as some have suggested? The images are much more than just paintings: they seem to be a dedication. It has been suggested that they represent increase ceremonies; that by painting an animal's image it makes it more common and easier to hunt or even brings it back from oblivion. The coloured frescoes have been painted with perspective and life blood running in the veins of these animals: they are living and breathing, moving and dying. Certainly, some images, such as those of the speared cave lions (*Panthera leo*) in the Diverticule of the Felines, Lascaux, seem to depict real episodes of hunting or perhaps defensive encounters against powerful marauding predators. Were the amazing depictions of the vertical movement of Woolly Rhino (*Coelodonta antiquitatis*) heads brandishing their long horns and the drawings of cute spotted horses mere tales of extinct animals brought to life from memory or had the painters seen them that day?

Painting may have been carried out by shamans, schooled in the practices of magic, mysticism and sorcery handed down from ancestors. But artists are artists and these were the best. Shamans are not necessarily good at painting even if they know a thing or two about magic and the afterlife. Perhaps shamans acted as advisors to the artists? We know that drawing and painting images may have been an integral part of ritual and ceremony for thousands of years. In that time it has not been unusual to see that much of that art has been centred on the power or influence of animals, animal spirits or the spiritual connections people had with them so it could have been a collective effort mingling the combined talents of artists and magic men or woman. But perhaps it was quite the opposite. The reasons for painting such images may have had nothing to do with present-day interpretations offered by archaeologists, anthropologists or palaeopsychologists who have speculated as to why people took such trouble to enter and deeply penetrate deep, dank, dark passages and decorate them.

Light was provided in the pitch-dark and very cold tunnels by torches or burning sticks probably covered with animal fat because the flame had to last. Perhaps bundles of these torches were taken in and slowly ignited, one after the other, with one or two left along the tunnel to light the way back out of the labyrinths when painting

Corridors to Extinction and the Australian Megafauna. DOI: http://dx.doi.org/10.1016/B978-0-12-407790-4.00010-0

was finished. Torches would slowly burn down and sputter out, so the artists had to keep that in mind as they worked. The darkness must have frightened them as they entered the tunnels and slid on the slippery floors soaked with water dripping from damp walls and ceilings. They wound their way along the slim corridors which, for them, contained creatures and unfriendly spirits that lurked in the dark cracks, crevasses and corners along the tunnel. The drive to paint must have been strong and the effort made to carry it out suggests it must have been for some special purpose because if they just wanted to paint they could have done so outside. So they stood in the darkness, braved the damp and eeriness and created galleries of unique, beautiful and triumphant art, perhaps to muffled chanting combined with ritual and ceremony of some kind. Were the megafauna animals they painted worth it? The artists and those who supported them must have thought so because that's all they painted! Megafauna were important; at the very least they were food, warmth and props for skin tents and in Ice Age conditions, you would not want to run out of such commodities. So the object became one of supporting the animals with magic and ritual.

So why paint such beautifully composed and crafted images of everyday animals – or were they everyday animals anymore? We do know that Europe lost its cave bear (*Ursus spelaeus*), hyena (*Crocuta crocuta*), cave lion (*Panthera spelaea*) and woolly rhino between 40 and 25 ka (Stuart & Lister, 2012). Chauvet Cave is an Aurignacian site dated between 32 and 30,000 years ago, so perhaps the artists there were documenting dwindling or species loss. Perhaps the animals had become so rare that people believed (or hoped) that by painting them they could make them return. They might have realised they had lost or were losing species we now call European megafauna, just as we now understand that we have lost and continue to lose our animals today. Let's take it a little farther: did they know what was causing the disappearance or slow decline of any particular species or did people see how rare some were becoming for no reason that was apparent to them? Perhaps not: that's where the magic of painting, ritual or both came in and the possibility of increase from depiction. Did they twig that it might be overhunting or, more likely, something else causing the disappearances? Extinction is often a silent predator – no noise, blustering, screaming or unimaginable catastrophe; just a silent whimper or sigh of the lone sentinel of the last animal of its species, unknown and unseen. Who has ever seen the last animal of any species? We have. But if the paintings are a kind of representation of the concern or realisation by early Europeans of the loss of their megafauna, it does not look like 'blitzkrieg' to me.

A recent report detailed the particularly rare event of documenting the last of a species. That was the killing of the last Javan Rhinoceros in Vietnam. The animal was found with its horn removed, a victim of the senseless poaching for a relatively small part of its anatomy – its long horn, so delicately painted by the palaeolithic artists of Europe, and so coveted by ignorant people today. Ironically, the ignorant and superstitious minds of those who today trade in animal parts to concoct cures and aphrodisiacs are something one might expect of 'cave men', not today's 'modern', educated, technologically advanced humans. In the rhino's death we see the elimination of an animal that saw off the Ice Ages and survived two particularly severe and long-term wars, the second of which included widespread use of highly toxic defoliant. No, it took a notional whim by those who would want to make money from others believing in a twisted logic and magic to eliminate these animals from the region – that's 'blitzkrieg'.

In the natural flow of biology and evolution we know for a fact that the path to extinction is often subtle, quiet and misleadingly sublime. A given range of species contains animals that are more vulnerable or specialised than others, while a few teeter on a knife edge of existence. This is fine till the balance is slightly tilted one way or another – but seeing the tilt happen is often impossible. Some animals are certainly less able to adapt and cope with extreme or drastic climatic or environmental change than others, but when they already vulnerable even small changes to their world can have dire consequences for them. Not knowing the predisposition that led to the extinction of a particular species allows us to choose the most convenient excuse for their demise even if it may not be correct. The megafauna extinctions have been written about *ad nauseam*, together with suggestions as to why and when the group disappeared. I believe each continent had its own relatively unique set of circumstances that brought about these extinctions, perhaps with humans playing a differential role in them in various regions. Large extinctions, as we have seen, happen in pulses joined by slender threads made up of lesser events and the temporal beat of background extinction. The Earth has not experienced a full mass extinction since the K–T event, although significant but smaller extinctions took place after it in the Tertiary as well as before it in the Jurassic and Triassic. Animal turnover is a natural part of evolution and that continues to one level or another; some suggest we are overdue for a big extinction event because these are part of the natural cycle of life. Others point to the loss of species over the last 500 years or so and suggest we are undergoing the 'sixth mass extinction' now – or perhaps it is the seventh or eighth, but who is really counting? – because now we are seeing the nitty-gritty in an extinction process and we are unable to put a pin between the last and the next or, in most cases, really understand what causes them. There is also another factor, and one very close to the 'appeal fatigue' that is spoken of when referring to the constant appeal for money to help with human mass disasters, famines and other calamities, and that is the fact that we are always reading or hearing of another species in trouble, endangered or close to extinction. So the currency becomes devalued; people take less and less notice of these reports and warnings, and the response become more sluggish – not because people do not care but because they have heard countless similar appeals before and feel that in such a tumult they cannot really help.

There may have been an instance in the past when one species caused the extinction of another, perhaps through one hunting out another or one outcompeting another for the same resources or to take over their ecological stage. But the recent extinctions are just that: they are one animal (*Homo sapiens*) competing against the rest, but it has been and continues to be an unfair fight with the number of victims measuring between hundreds and millions. Extinctions in Western Europe during the last 1000 years included medium to large animals such as wolves, bears, beavers, as well as the drastic reduction of others such as bison (*Bison bonasus*). Elsewhere, specific species like the Great Auk (*Pinguinus impennis*), the Moas (*Megalapteryx* sp.), Steller's Sea Cow (*Hydrodamalis gigas*), the Dodo (*Raphus cucullatus*), Bluebuck

(*Hippotragus leucophaeus*), Thylacine (*Thylacinus cynocephalus*) and Quagga (*Equus quagga*), to name only a few, have become rallying flags of the destruction wrought by human hunting. Some of the largest animals that ever lived, indeed, one of the largest ever, almost succumbed before a moratorium saved them from being harpooned to oblivion. The losses are joined by dozens of other small mammals and birds around the world, victims one way or another of territorial invasion, introduction of disease and non-native species, logging, clearing for farming and grazing, plantation development, expansion of human communities, human vanity for adornment with animal skins and exotic bird feather cloaks and hats, quack pharmacy, primitive potions and the human population explosion in general. Plants are yet another story.

Not all our recent losses are due to a deliberate policy to drive animals to extinction. The mere fact of the world's exploration over the last six centuries, particularly the discovery of long-isolated islands and the vulnerability of species inhabiting them, was one reason for extinction. Species living on islands succumbed just because they lived in an extremely delicate balance within a remote and finely tuned ecosystem, and any interference with that, however small, caused them to disappear. Others were just rare. Nature has its rare animals apart from those that are made so by human influences. The fragile nature of island species and their highly specialised lifestyles in small niches always predisposed them to extinction by the smallest of changes to their natural balances but these did not arrive till humans set foot on their lost worlds. That is not to say, however, that spectacular climate or environmental change to their various homelands would not have achieved the same results had they occurred. Imagine if there were no humans in the present interglacial: all those species that I have listed above would still be here. Then an Ice Age occurs, and depending on where they live, all delicately balanced species would disappear along with many others that were already endangered, rare, highly specialised or were confined to one or two populations occupying small ranges and unable to cope with such massive changes. Perhaps the palaeolithic painters were making a pictorial wish to reverse such a situation. The point I am make here is that the destruction of many animal species by recent human activities as we explored, farmed, built cities and generally expanded across the world, has been used as a yardstick to explain the megafauna extinctions of the past: they call it simply 'blitzkrieg'.

Blitzkrieg is a small word with blood-freezing connotations for those Europeans who saw the German mechanised army of the Second World War arrive on their doorstep, particularly the Polish people. The term means 'lightning war', denoting a quick strike seemingly out of nowhere. Initially, it was overwhelmingly believed that the Australian megafauna extinctions were caused by a 'blitzkrieg' type of event: the arrival of humans on the continent who then swiftly hunted the unfortunate creatures to extinction, probably in 1000 years or less. The idea was first mooted by Paul Martin, who had used the term as far back as 1967. His 'overkill' hypothesis was initially proposed to explain the demise of the North American megafauna which seemed to coincide with the arrival of Clovis paleoIndian people. The use of 'blitzkrieg' was deliberate and emphasised the swift and devastating mass killing of all the largest, and only the largest, animals. It should perhaps have been called 'overblitzkrieg' because it indicated an *over*powering of all forms of these creatures by *over*whelming numbers of people needing to be fed, so that they *over*hunted them and killed even more than they needed. Lastly, they *over*ran all continents and did the same thing there. In terms of the basic archaeology and palaeontological evidence of the time, this idea was plausible because of the assumed coincidental timing of the event with the arrival of people and also because the natural history and biogeography of the megafauna were largely overlooked. Anthropogenic 'blitzkrieg' continues to be offered by some as the best explanation for the end of the megafauna and has become the term to explain the 'sudden' demise of the largest of the Ice Age animals *anywhere*; it dominated thought on the subject for several decades wherever such extinctions took place. In Australia, landscape burning has been added to the armoury of the first arrivals. The first people here, it has been proposed, introduced widespread continental burning, thus negatively changing the landscape in a way that disadvantaged browsing megafauna by instigating the loss of trees and shrubs and encouraging the spread of grasses. That argument has been softened slightly to suggest that the practice of landscape burning might have had an indirect rather than a direct effect on megafauna by landscape modification (eliminating large areas of forest, particularly in the north of the continent) and by changing shrub and woodlands to grasslands and even changing rainfall patterns across the continent.

In Australia, however, the argument is taking on a new balance between death by human hand and climatic and environmental change. I do not intend to review this literature because it has been undertaken elsewhere but I include the following references for those wishing to pick up the nuances of the arguments: Brook and Bowman (2002); Choquenot and Bowman (1998); Field and Fullager (2001); Flannery (1990), (1994); Horton and Wright (1981); Johnson (2002), (2006); Johnson and Prideaux (2004); MacPhee (1999); Martin and Guilday (1967); Martin and Klein (1984); Meltzer (2004); Miller et al. (1999); Mulvaney and Kamminga (1999); Murray (1984), (1991); Rich et al. (1985); Roberts et al. (2001); Trueman et al. (2005); Webb (1998); Whitney-Smith (1996); Wright (1986); Wroe et al. (2004); Barnosky et al. (2004); Webb (2008–2010).

Megafauna Extinctions in Southeast Asia

Although rarely undertaken, discussion of Australia's megafauna extinctions really has to begin outside Australia to examine a very important aspect of the whole blitzkrieg debate. Blitzkrieg in Australia has assumed humans arrived just prior to the extinctions and does not address the logical conclusion that those capable of carrying out such a mass extinction must have left a trail of similar destruction in their wake throughout areas of Southeast Asia on their way to Australia. It is an aspect of the debate that has attracted little attention here, although I have no idea why that is so. So what is the evidence for human impact on the Southeast Asian megafauna? If humans impacted the Australian megafauna they must have done so on their way here as they passed through Southeast Asia – or were they just waiting to kill the Australian megafauna?

It is worth beginning this section with a couple of quotes referring to Southeast Asian megafauna losses:

> *... few taxonomic groups disappeared from SE Asia or became extinct during the Pleistocene, although some that did include proboscidians (Stegodon and Palaeoloxodon), and carnivorans (Crocuta and Pachycrocuta) and hippopotomus (Hexaprotodon).*
>
> (*Louys & Meijaard, 2010:2*).

Here is another:

> *Southeast Asia has undergone dramatic climatic and geographical changes during the Pleistocene. It is supposed to have been significantly drier and cooler with more seasonal climates during glacial periods...[which were] associated with repeated shrinking and expansion of rainforests and correlative spreading and shrinking of savannas...pine woodlands and savanna grasslands expanded and replaced the rain forests over large areas of Southeast Asia during glacial periods.*
>
> (*Chaimanee, 2007: 3190*).

Among the largest animals inhabiting Southeast Asia today are elephants, rhinos, tapirs, tigers, cattle, pigs and bears, all of which are megafauna and have always lived somewhere there. At the beginning of the Quaternary, there were around 45 genera, 11 of which contain the above animals. What seems to have happened is that the region lost 34 genera over 2 million years, far more than Australia lost in the same period. The losses, however, were staggered temporally and spatially across an enormous area that is now divided into modern geopolitical countries. Table 10.1 lists lost genera and the apparent times they went extinct, and Table 10.2 divides them by the name of the modern country in which they have been found. In a few examples we see the same genera go extinct at different times, which indicates either two separate species within that genera going extinct or the same species disappearing in two different geographical regions or countries. The reason for noting these in this way is that many Quaternary megafauna extinctions in Southeast Asia took place in different regions at different times due to environmental changes brought on by glacial cycling (Ibid, 2007). In addition to the quoted references above, other useful authors writing about the late-Quaternary situation of the Southeast Asia megafauna include Vos de (2007), Louys et al. (2007), Tougard (2001), Tougard and Montuire (2006) and Stuart and Lister (2012).

Southeast Asian extinctions included several *Stegodon* elephants, a hyena (*Crocuta* sp.), and the giant anthropoid ape *Gigantopithecus*. Other species suffered reductions including pandas (*Ailuropoda*) and Orang Utans (*Pongo*). Palaeontological evidence from regional fossil assemblages strongly suggests that extinctions were temporally staggered and regionally variable and were almost certainly linked to world temperature reduction, sea level change, land bridge formation and loss, and the environmental and biogeographic consequences of some or all of these. As the above quotes show, glacials brought various changes to biomes including effectively moving them. Biotic consequences included the opening up and

Table 10.1 Staggered Quaternary Genera Extinctions in Southeast Asia

Genera (n = 35)	Early-Quaternary	Middle-Quaternary	Late-Quaternary
Acinonyx	1		
Cervavitus	1		
Crocuta		3	1
Dicerorhinus		1	
Dicoryphochoerus	1	1	
Dorcabune	1		
Elaphodus		1	
Eostyloceros	1		
Epileptobos		1	
Equus	2		
Gazella	2		
Gigantopithecus		2	
Gomphotherium	1		
Hemibos	1		
Hesperotherium	1		
Hexaprotodon	1	1	1
Hipparion	1		
Megalovis	1	1	
Megantereon	1		
Megatapirus		2	2
Melodon		1	
Merycopotamus	1		
Metacervulus	1		
Pachycrocuta		1	
Palaeloxodon		3	2
Paracervalus	1		
Potamochoerus	1		3
Proboselaphus		1	
Procapreolus	1		
Proleptobos	1		
Sinomastodon	1	1	
Spiroceros		1	
Stegalophodon	1		2
Stegodon	1	4	1
Ursus			1
Regional Extinctions	*25*	*25*	*13*

Animals recorded in fossil deposits in China, Burma, Java, Thailand, Malaysia, Borneo, Laos, Cambodia and Vietnam. Multiple numbers or repeat extinctions in temporal columns indicate extinction of different species in same genera or the same species in different parts of Southeast Asia (see also Table 10.2).
Source: After Chaimanee (2007) and Louys et al., (2007).

spread of savannahs where rainforest had previously existed and reduction of tropical habitats in northern areas of Southeast Asia. In effect, biomes moved south onto the newly exposed continental mass of Sunda. Biome loss and transfer and habitat reduction – that sounds familiar!

Table 10.2 Regional Distribution of Megafaunal Extinctions Across Southeast Asia (Refer Also to Table 10.1)

Genera (n = 35)	Early-Quaternary (China, Burma, Java)			Middle-Quaternary (China, Burma, Java, Thailand, Malaysia, LCV)						Late-Quaternary (China, Java, LCV, Borneo)			
	C	B	J	C	B	J	T	M	LCV	C	J	LCV	B
Acinonyx	/												
Cervavitus	/												
Crocuta					/		/			/			
Dicerorhinus				/					/				
Dicoryphochoerus				/									
Dorcabune	/												
Elaphodus									/				
Eostyloceros	/												
Epileptobos						/							
Equus	/	/											
Gazella	/	/											
Gigantopithecus									/				
Gomphotherium				/									
Hemibos		/											
Hesperotherium	/												
Hexaprotodon	/	/						/					
Hipparion		/									/		
Megalovis	/			/									

Table 10.2 (Continued)

Genera (n = 35)	Early-Quaternary (China, Burma, Java)	Middle-Quaternary (China, Burma, Java, Thailand, Malaysia, LCV)	Late-Quaternary (China, Java, LCV, Borneo)
Megantereon	/		/
Megatapirus		/	
Melodon	/		
Merycopotamus	/		
Metacervulus	/	/	
Pachycrocuta		/	/
Palaeoloxodon		/	/
Paracervus	/	/	/
Potamochoerus		/	
Proboselaphus		/	
Procapreolus	/		
Proleptobos	/	/	
Sinomastodon	/		/
Spiroceros		/	/
Stegalophodon	/	/	/
Stegodon	/	/	/
Ursus			/
Total Extinctions	**14** 10 1	**7** 3 2 2 2 9	**3** 3 3 5 1

LCV – Laos, Cambodia and Vietnam.

/ – Complete extinction of genera in region or across Southeast Asia.

Southern Southeast Asia became Sunda during times of low sea level, comprising all land and exposed continental shelf below 15° north. Wallace's Line marks its southeastern boundary across which lies Wallacia, comprising Australia, Tasmania and Melanesia. During glacial maxima, lowered sea levels exposed a vast subcontinent encompassing the Gulf of Thailand and the Java Sea as well as extending out into the South China and Banda Seas. The emergence of the Sunda shelf probably occurred fairly quickly because of its shallow slope angle, similar to that off northwest Australia. Shoreline advance during lowering sea levels has been estimated at around 40 cm per week or 20.8 km in a thousand years (Mackay, 2009). If we take an average 20,000-year descent into an Ice Age (refer to graphs in Appendix 2) at the above rate, that would produce a 416-km-wide shelf, given a shallow and fixed shelf slope angle. Following shelf exposure dispersal and distribution of various plant communities would take place but probably not at a speed required to replace those being lost due to the changes occurring inland. Eventually, given the lengthy nature of glacial cycles, mature ecosystems would have been established across the shelf, a welcome event for animals migrating south onto the shelf that offered them the opportunity to recapitulate initial population losses. Looking at the process in reverse, as we did in the previous chapter, and by taking an average 4000-year glacial termination period (see Appendix 2), sea levels would advance at 2 m per week! Others have recorded a half metre per year sea level rise for a 300-year period between 14.6 and 14.3 post-LGM on the Sunda Shelf (Hanebuth et al., 2000). With such a rapid return to warmer temperatures and high sea levels at glacial terminations, the speed of shoreline advance would have quickly submerged the mature glacial environments, such as rain forest, that had flourished for thousands of years. Such relatively rapid destruction of coastal environments by a 100-km annual shoreline advance would not have provided enough time for the formation of replacement ecosystems inland, something also noted in the case of the inundation of the Australia continental shelf in the previous chapter. The 'lag' of vegetation type would impose stress on megafauna populations while those trapped on islands either became isolated or were drowned. For example, Orang Utans became isolated on islands after high sea levels returned. That saved some because they became extinct on the mainland: their remains in archaeological sites in Vietnam show that they were hunted there in the late-Pleistocene in a similar manner and time to those found in Eurasia, as discussed in Pushkina and Raia's work quoted above (Tougard & Montuire, 2006).

The amount of land exposed and the shape of the prehistoric coastline would have changed constantly as sea levels fluctuated during glacial cycles. Corresponding changes in the size and distribution of coastal rainforest and other ecosystems across the region would have followed as the shoreline moved back and forward. Temperatures were at least 5°C lower in tropical lowland rainforests during glacial events and such reductions were enough to severely alter rainforest ecology in some areas; in others, it was reduced, moved further south, became isolated in pockets or disappeared entirely. Such biogeographic change in vegetation and biome structure would reflect on animal populations with reduction of numbers and redistribution among rainforest dwellers in the same way it probably did among the marsupial giants of Australia. As a result, cases of extinction were largely local, caused by range reduction through Ice Age environmental changes which transferred biomes to other regions; those animals that could followed. In fact, climatic cooling in the northern hemisphere during the Quaternary sent faunas southward while in Australia it sent them to the coasts. It was at this point that the mechanics of biogeography then shifted into gear, moving and changing animal populations in ways explained in Chapter 9. Results for animal populations in Southeast Asia included attrition and reduced populations in species such as the giant anthropoid ape *Gigantopithecus*, which may not have been large in the first place, and the hyaena (*Crocuta* sp.). Others, such as the panda (*Ailuropoda* sp.) and Eldi's Deer (*Cervus eldi*) also suffered reduction in range size and distribution. Pandas, however, had reached their peak in the mid-Quaternary and have been declining since then to reach their present endangered levels (Huang, 1993). But the consequences for animal populations in Southeast Asia were far less severe than they were for those living in Australia because it was not an island that changed into a desert.

Stone tools found in the Nihewan Basin of northern China date to 1.35 Ma but the layers below this show evidence of the processing of animal tissues around 1.7 Ma, particularly marrow processing (Fleagle, 2010). Animals found on the site include horse (*Equus sanmeniensis*), deer (*Cervus* sp.), elephant (*Elephas* sp.), bovid (*Gazella* sp.) and woolly rhinoceros (*Coelodonta antiquitatis*). As we were meat eaters from before 2 million years ago, that evidence is not really remarkable. In addition, we know *Homo erectus* lived in China and Java for at least 1.5 million years but while there was an extremely long shared history between them and various large animal species, humans seemingly had little or no impact on the fauna until our population began to rise appreciably after the Last Glacial Maximum. There is no direct evidence of the hunting to extinction of any animal species by humans before that time. With recent discoveries in southern Siberia at the Denisova Cave in the Altai region, however, we now have a clearer picture of human settlement in the wider region (Derevianko & Shunkov, 2009). Subsequent genetic work using human skeletal remains from the cave has shown that the 'Denisovans' living in the cave spread their DNA widely, particularly in an easterly and southeasterly direction extending as far as Papua New Guinea (Abi-Rached et al., 2011). Genetic work has shown Denisovans were related to Neanderthals, both of which shared a common ancestor in the European archaic *Homo sapien* people, represented by *Homo heidelbergensis*, which was derived from *H. erectus*. That suggests Denisovans were a break-away group which moved east through Eurasia and entered southern Siberia. The genetic evidence suggests the groups probably divided before 350 ka, indicating that the Denisovans were a successful and viable group at least till their occupation of the Denisova Cave *ca.* 55 ka. Neanderthals living in western Eurasia spread as far south as Israel, north to Byzovaya near the Arctic Circle (Slimak et al., 2011) and east to Iran and Kazakhstan. A toe bone resembling those of the Iranian Neanderthals has also been found in Denisova Cave, although dating somewhat younger than the Denisovan occupation, that suggests Neanderthal movement across the region, probably from Iran through Kazakhstan to the Altai region, denying a static situation as far as older populations of humans were concerned during the late-Quaternary. The picture now presents a far more dynamic and complicated picture than we had previously suspected. Apart from being a significant find in terms of the story of regional human evolution, the presence of Denisovans also shows this was not an empty, quiet region with little going on in terms of human movement, settlement and hunting. The implications go further, showing the dynamics apparent in the Altai extend much farther throughout Sunda and down into Sahul. It is important to note also that both the Denisovans and Neanderthals lived with the regional megafauna from the mid-Quaternary onwards without detriment to the large animals as far as we can tell.

The spread of Denisovan DNA through Southeast Asia occurred through migrations but it was probably carried further south and into Sahul after they mixed with others travelling through the region. Inter-breeding between modern humans who came through sometime *ca.* 70–80ka and Denisovan or Denisovan/Anatomically Modern Human mixes then moved across Sahul. Those genes now included modern human DNA and it was those people who moved through the region to Sahul. When that took place is not yet known, although it could have been around the same time. Once again, we have dynamic movement through the region of a people who were not necessarily archaic, or may have been partly so, and who came into contact with a wide variety of large animals that remained unaffected in terms of being hunted to extinction for no other reason than they had always lived together. It is clear that since *H. erectus*, animals were hunted but not to extinction. The evidence so far shows that the movement of modern humans through Southeast Asia had very little if any effect on the fauna although an association between humans and animals existed for hundreds of thousands of years including large animals. While that is quite natural, any association in the archaeological record means little with regard to the hunting-out of a species and driving it to extinction unless massive kill sites are obvious. Indeed, the fact that humans and megafauna lived together for tens of thousands of years without such sites is a strong indicator that the association was not one of faunal attrition or massacre. There seems to be little difference whether the people we are talking about were Anatomically Modern Humans or more archaic forms that had lived in eastern Asia for hundreds of thousands of years, or a mixture of both: humans had nothing to do with the megafauna extinctions recognised in the region. One thing we can rely on is that any people moving through Southeast Asia were well aware of large animals and in the long continuum of migration would naturally not have been surprised to find similar sizes and types of beast wherever they travelled, even in Australia. The only surprise may have been to find animals with its baby's head poking out from a hole in their stomachs. The point here is that Southeast Asia's megafauna was left virtually untouched by humans till the increased hunting of the late-Pleistocene post-LGM, which even then was only minor.

To people they were just ordinary animals, so it is no surprise to find the occasional piece of evidence of human–megafauna interaction in a few Southeast Asian archaeological sites because they had been living there for a long, long time. But more intensive hunting only gained momentum from MIS2 onwards, probably as a result of human population growth, as may have been the case in Eurasia, with more intensive settlement and improvements in weapon technology and hunting techniques. Deliberate targeting of single species is not something that is commonly seen in late-Quaternary people, who had a wide choice of protein from which to choose, and nothing even close to a mass kill site has been found in Southeast Asia. In other words, people travelling through the region might have met large animals but they did not feel the need to exterminate them. So far then, late-Quaternary kill sites in Southeast Asia – nil.

Southeast Asia today boasts many large animals but Australia does not. That in itself is a statement with regard to the zoological history of the two regions. Southeast Asia also bore changes associated with glacial cycling but in its case they were probably less intense overall; in addition, the open-ended geography of the Asian continent allowed it to effectively extend its land mass towards the Equator and facilitate the movement of habitats south in response to Ice Age-driven environmental change and lowering sea levels. Environmental shifts to the south at these times saw megafauna following them that led to some local extinctions while the same species popped up elsewhere as they followed their preferred ecosystems. Southeast Asian animals could move in many directions. They were never troubled as they were in Australia by widespread desertification that buried their continent under sand, dust storms and drought and eradicated millions of square kilometres of vegetation and rangelands, squashing them into corners, fragmenting groups and populations, isolating and trapping them against narrow coastal regions. Had that occurred it would have led to the elimination of far more species in Southeast Asia than was actually the case. It seems Southeast Asia was left virtually untouched by the passing of human migrations but not by glacial cycling. Indeed, it seems to have been the palaeoenvironmental and sea level changes associated with glacials that were the main culprits for the minor extinctions that took place there, as animals were trapped in their old habitats which changed around them; this seems to be the consensus of others as well.

After discussing Southeast Asian megafauna it is worth briefly looking at the emerging picture of megafauna extinctions in Eurasia because, after all, it was the conjoined land mass north and west of Southeast Asia. First, we'll begin with a quote:

> *People became increasingly able to hunt abundant prey species, many of which, however, are still living. Palaeolithic populations of large mammals of the "mammoth-steppe" were already suffering from the deterioration and contraction to the north of their preferred habitat, while humans appeared to show little interest in the now-extinct species, even when a conservative archaeological approach was used that should have favored finding human influence on extinct fauna... Our findings are consistent with the climatic explanation of the late Pleistocene extinctions in Eurasia.*
> *(Pushkina & Raia, 2008:777).*

That work represents only some of the research into reasons for the megafaunal disappearance over the great land mass of Europe and Siberia. Below are some statistics relating to Eurasia megafauna, followed by some conclusions drawn from the work that has been carried out there. During the last 100,000 years, Eurasia lost 35% of its 49 large species of animals considered megafauna (>44 kg). As usual, the largest animals were most affected and losses in that group included three species of elephant (*Palaeoloxodon antiquus, Palaeoloxodon naumanni* and *Mammathus primigenius*), two rhinos (*Stephanorhinus hemitoechus* and *Coelodonta antiquitatis*), a hippo (*Hippopotamus amphibious*) and an array of other large animals weighing over 500 kg (Stuart & Lister, 2012). Extinctions were taking place before modern humans arrived in Eurasia and when they did, the overlap with large animals was extensive – at least 30,000 years. Some recent research has focused on the extinction of the woolly rhinoceros (*C. antiquitatis*) which reflects similar stories for other species. Extensive dating of over 290 specimens clearly shows these animals went extinct in a staggered pattern both temporally and spatially, which reflects the Southeast Asian extinction pattern (Ibid, 2012). The main conclusion drawn from the work seems to show clearly that the demise of the woolly rhino was triggered by enviroclimatic changes which brought about alterations to the distribution and general availability of its preferred fodder. That included the spread of shrubs and trees at the expense of grasses and herbs which, as a grazer, the woolly rhino preferred. The rhino's general pattern of disappearance was from west to east over a time period of around 20,000 years, with the last animals living in eastern Siberia and going extinct by 14.1 ka. It is also pointed out that other Eurasian megafauna went extinct in a staggered pattern stretching across the last Ice Age and with different responses to the environmental changes that were taking place.

What is fascinating about the picture emerging in Eurasia is the differential pattern of responses by different species to the LGM and the enviroclimatic changes it brought. Certainly, they were not uniform across the continent and this was reflected in the extinction patterns of the animals. It seems the woolly mammoth and the giant deer (*Megaloceros giganteus*) were two species that reacted differently, although they suffered range contraction and restriction in small refugia '*where they remained for several millennia before extinction*' (Ibid, 2012:15). That sounds familiar. Some examples of Eurasia's staggered extinctions include the narrow-nosed rhino (*S. hemitoechus*) and giant deer (*Sinomegaceros yabei*) ca. 45 ka, spotted hyaena (*Crocuta crocuta*) ca. 31 ka, cave bear (*Ursus spelaeus*) ca. 28.5 ka, cave lion (*Panthera spelaea*) ca. 14 ka and the woolly mammoth (*M. primagenius*) at 10.7 ka, steppe bison (*Bison priscus*) at 9.8 ka and giant deer (*M. giganteus*) at 7.7 ka, an almost 40,000-year time span. Hardly a blitzkrieg.

Because We Cause Extinction Now, We Did So in the Past....

Megafauna extinction arguments that pivot on anthropogenic factors have become mired in parallel arguments surrounding early modern humans, their approach to the environment and the historical perspective. Today's Aboriginal Australians feel blaming them is insulting to a people who have had custodianship of the land for 50,000 years or more and they point to the fact that no other species went extinct under that stewardship except the mainland *Thylacine* around 4000 years ago – although that had more to do with the dogs brought by new arrivals than the arrivals themselves. Moreover, why would they kill off a perfectly good food supply? They did not terminate any other item on their menu. Looking after the land and its animals not only feeds them but it is also a central tenet of their philosophy of living which is deeply embedded in the Dreaming, the ancestral spirits that often take the form of animals and the totemic links that people have with those animals. On the historical side, they also believe that blaming them for the megafauna extinctions is just another 'white fella' argument that helps subvert 'black fella' claims to ownership of traditional lands in Australia by pointing to the 'fact' that they did not take care of the land and its creatures as they claim. They also point out that it was 'white fella' animals that have caused the extinctions or at least the considerable reduction of many endemic Australian species. Therefore, the 'blitzkrieg' argument has moved to a political level and one that is despised by Aboriginal people.

While the blitzkrieg argument was a convenient and simple explanation for the megafauna extinctions, it also seemed to fit human arrival times. But it ignored the fact that animals go extinct as a natural phenomenon, for many reasons and sometimes even in a staggered manner. They do not need humans to help. It ignored the fact that the megafauna extinctions were yet another one of those episodes, particularly in the face of data-free blitzkrieg arguments. As we have seen, the reasons for extinction are usually extraordinarily complicated and not always obvious in the archaeological record. The blitzkrieg idea also completely ignores the palaeoenvironmental maelstrom the world passed through in the Quaternary, which took a toll on Australia's megafauna as it incrementally sapped the population as Australia moved from one glacial event to another. That toll was probably sporadic, regionally variable but it was steady and accumulated from the different biogeographic changes that governed animals then as they do now. It could also have gradually introduced extinction processes that imperceptibly and subtly removed species from the landscape over a long time.

It should not be overlooked that megafauna continued to thrive elsewhere and still do. These include buffalo of various types, elephants, giraffes, rhinos, tapirs, camels, large bears, various large felids and a whole range of other large species such as camels, Musk Ox, ungulates and some kangaroos that weigh well over the human-imposed 44–50 kg limit for a megafauna species. All these species were left perfectly unharmed by native peoples and an exponentially growing human population till the advent of trophy hunting and human inroads into their habitats. Inuit did not hunt the Polar Bear to extinction even though they did hunt it. It was and still is a dangerous pest to have around and that was particularly so when people lived in traditional ways out on icy wastes, but it has not been annihilated! Inuit incorporate these bears as a special part of their culture and mystical belief system and they carve and draw their images in a manner resminiscent of the Lascaux painters. The same goes for the elephant, rhino, tiger, lion, komodo dragons, large pythons, grizzly bear and many, many other large species, all dangerous and all still there, although the Romans virtually eradicated the Middle Eastern lion just to supply their circuses. It was not until the railroad was built across America that the buffalo almost went extinct from being hunted to feed the railroad workers. Before that, they could be counted in millions during Native American times even though *they* hunted them.

The anthropogenic argument usually extrapolates both our present and recent historical abysmal environmental record into the distant past. In doing so, it does not allow for the differences that existed between late-Quaternary people and those living during the last 300 years or so. For example, the argument is always backed up by the interminable example of the extermination of nine species of Moa by

Maori ancestors in just 170 years. Most readers of this book will be familiar with the story and with the argument that, by implication, because those early inhabitants carried out the Moa exterminations, other native people living 40–100,000 years earlier must have exterminated Australia's megafauna. That example usually overlooks the extenuating circumstances of those Maori ancestors and their continued habitation in a land depauperate of terrestrial protein, with the exception of Moas, their eggs and a few scrawny parrots. The Maori had also arrived not on a comparatively small, isolated country but on two small islands or patches without a connecting corridor that together comprise the same area as the SLEB (265,000 km²). But the two main islands of New Zealand are, individually, comparatively small and, therefore, make hunting-out an easier proposition than on a continent. The distribution of an estimated 158,000 Moas on both patches meant a divided population of finite size because of range size, carrying capacity and no escape. Like flightless bird species on other islands throughout the world, the arrival of humans on each patch was far more dangerous for the Moa (who suffered from extreme naiveté) than if the two islands had been joined to make a larger range. Male birds were smaller than females, some only a fifth of the weight and a third of the height, and one turkey-sized species had a back height of only 0.5 m and weighed 17 kg. Because of their size, targeting male birds would have been optional and an excellent way of rapidly reducing the population. Also, those Maori hunters probably had more advanced weaponry if we are to take construction of large, ocean-going watercraft as a yardstick. There were more people involved than the first Australians could muster and they were more concentrated. Using the New Zealand example, therefore, seems as logical as using Dutch sailors and their shipboard animal companions and the Mauritius Dodo extermination as a suitable example of anthropogenic reason for Australia's megafauna extinction.

Nevertheless, while extinction is a natural function of evolution, it can be argued that an anthropogenic role in it is not. In Australia, evidence is building that the last megafauna disappeared quite sometime after humans arrived. Before that, the argument had been that they were all gone by 40,000 years ago (Gillespie & Brook, 2006; O'Çonnell and Allen, 2004; Miller et al., 1999; Roberts et al., 2001;). Dates obtained from the Malukananja rock shelter in western Arnhem Land show people had landed on Australia's shores shortly around 62 ka and an optical date of 53 – 60 ka has been found for sites in Deaf Adder Gorge 70 km to the south (Roberts et al., 1994, 1998). Sustained anthropogenic burning in northeastern Queensland has also been blamed for vegetation changes observed at Lynch's Crater *ca.* 45 ka (Turney et al., 2001). In southern New South Wales the Willandra Lakes have produced a number of archaeological dates in the 40–50,000-year range associated with human occupation, while in far southwestern Australia redating of sediments from the Devil's Lair cave deposits puts people in the region sometime between 50 and 55 ka (Turney et al., 2001). All these reports suggest that people arrived sometime between 55 and 60 ka, but are they the final word? In 1962, it was thought that people had only been in Australia for 16,000 years. Other examples of recent archaeological and palaeontological finds noted in this book are pertinent examples of how we might think we have the answer until suddenly new data turns the picture on its head.

It has previously been suggested that the final megafauna extinctions occurred a long time after the arrival of humans with some species lasting almost to the last glacial maximum (Trueman et al., 2005; Webb, 2006). At the time of writing, there is a reliable report of a well-dated site (25–28,000 ka) containing *Pallimnarchos*, *Diprotodon* and giant kangaroos (perhaps the largest yet found). If true that would be further evidence together with the art of a large temporal overlap between humans. Besides not knowing when humans arrived in Australia, we still do not know when the very last megafauna disappeared although it looks more and more as though it was close to the last glacial maximum. So, at present the overlap is at least 15–25,000 years but I would not be surprised to see this figure extended much further; a similar picture seems to be emerging for other continents. Such a large overlap also means humans here could have been imposing stresses on what was left of the megafauna over tens of thousands of years. But even if those stresses contributed to the final demise of these animals they must have been minimal to take so long to have an effect. That would be a more logical proposition and one that might implicate humans at a very low level in the final demise. We must also not forget that with dates for megafauna closing in on the last glacial maximum, stresses from that direction now have to be considered also. Again I point to the lack of archaeological evidence that associates humans and megafauna. Moreover, if that was so then the few remaining species were rare, with very small populations severely restricted in their habitats; there can be little doubt that animals in such circumstances would not last long and would eventually disappear in an exponential fashion. On the other hand, if people arrived between 45 and 50 ka and the enviroclimatic attrition argument is not correct, then they eliminated these animals so quickly it defies the imagination. It also begs the question of why the largest modern animals, such as the Red, and Eastern and Western Grey kangaroos and Wallaroo, were not also eliminated? I would also expect to see the skeletal results of such a massacre common in all early archaeological sites and on a massive scale: megafauna bones are difficult to hide! So, where are the piles of corpses representing a massacred continental population, particularly in the earliest sites?

For over 40 years archaeologists and others have scoured the fireplaces of Australia's earliest inhabitants, who lived along the edge of Lake Mungo and other Willandra Lakes. The Willandra has revealed campsites, middens and 45–35,000-year-old campfires with faunal and fish remains, stone artefacts and many human burials. In all that time, however, the only marsupials found as burnt remains of ancient dinners are small to medium-sized modern species. They include various small bettongs (*Bettongia lessueur*), bandicoots (*Macrotis lagotis, Perameles bougainville* and *Isoodon obesulus*), Quolls (*Dasyurus geofroyi.*), the Hairy-nosed Wombat (*Lasiorhinus kreffti*), possums (*Trichosurus vulpecula*), small wallabies (*Lagorchestes leporides, Lagostrophus* sp. and *Onychogalea* sp.) and the Red and Eastern Grey kangaroos. But even the bones of these larger species are not common in archaeological campsites and fireplaces. The burnt-out burrows of the Hairy-nosed Wombat are numerous, visible as fossilised tunnels standing out as darker tracks in the eroding sediments of the lake-shore areas. Burning or smoking out wombats was one of the hunting techniques used in the Willandra, particularly along the ancient beaches around the lakes. Many of these species have undergone severe range restriction or, as in the case of the Burrowing Bettong (*Bettongia lesueur*), have gone extinct on the Australian mainland during the last 200 years. A few megafauna species do occur in the Willandra region but they do not occur in the fireplaces and campsites of the earliest inhabitants. So, the only conclusions that can be drawn from the archaeological evidence from the region, and from one of Australia's oldest and certainly biggest and best-studied archaeological areas, are that either megafauna were not hunted or they did not exist when the first Australians arrived there, even though they did at one time. Therefore, while limited megafauna bone has been found in the region there is no evidence for a connection between them and humans. The same goes for all the other oldest Australian sites around the continent such as those on the Darling Lakes west of the Willandra Lakes. The megafauna are there but no evidence exists for a direct connection between them and humans. The only 'evidence' for the act of overkill in Australia is perceived evidence: the timing of the human arrival here and a few dates for some megafauna specimens but these are not the most recent dates for these creatures. I cannot but suggest that the increasing likelihood of megafauna overlapping with humans in Australia for thousands of years brings the blitzkrieg argument to a grinding halt.

There is no firm archaeological or palaeontological evidence in Australia either for a mass extinction event involving the whole range of species or for direct connection between *any* megafauna species and humans – except rock art. But what would it mean if they did exist? The odd cut mark or stone tool placed near a piece of megafauna bone also does not prove megafauna were killed *en masse* by humans; these may merely indicate scavenging, and a couple of megafauna specimens do not make a massacre. Archaeological evidence leaves little doubt that the earliest Australians preferred small animal varieties. There is also little doubt that if megafauna were around when humans arrived they were probably looked upon like any other animal: as fair game for a suitable meal. Remember, they are only another large animal in the same way as other animals were outside Australia. But I believe there would not have been any primary imperative to kill large animals. Like elsewhere, perhaps humans thought the easiest way to hunt was to focus on smaller animals, which caused fewer problems or dangers, and leave the larger ones alone (unless they were down and dying); smaller animals were much easier to capture and kill and served the dietary needs of the band. Larger animals were a waste of meat.

The only exception that points to a link between humans and megafauna in Australia seems to be burnt *Genyornis* egg shell fragments in campfires; either these were thrown there or they are the remnants of overcooked or exploded eggs. Some pieces of shell have been found with a small hole punctured in them and that may indicate a human technique for preventing egg explosions when placed in a fire. However, it is still not certain whether humans or animals were responsible for making the holes because there are no examples with both a hole and signs of burning. Even assuming these holes were made by humans, it does not prove that the *Genyornis'* extinction was caused solely by humans. It proves only that they predated the eggs. That was probably carried out by chasing or distracting the adult bird away from the nest. Apart from a few cut marks on a few pieces of megafauna bone, some of which may not be contemporary with the death of the animal, and a few stone tools found with their remains, evidence for human butchering or any other interaction with any megafauna species does not exist (Fillios et al., 2010; Horton & Wright, 1981; Vanderwal & Fullagar, 1989; Wyrwoll & Dortch, 1978).

One difficulty with solving the megafauna extinction issue is that we have little understanding of the final extinction time of individual species, of the number that existed when humans arrived (or at various times leading up to that, whenever that was) and of where the last animals were living; these problems will always fuel arguments. But those arguments will be settled only with a lot of field work to solve the problems. Also, we cannot be sure that the most recent date for any given species represents the last animal or group of animals and that may never be sorted out for any species. What would help is if we had very recent dates for all species that would at least place them in a time frame in which we could show they all existed at the time of human arrival or at some other time. That would be compelling evidence to support a wide range of species coexisting with humans here or at a particular threshold time before that. But we do not have that evidence now and I do not expect it will ever be found. We have mountains of fossil bones but, unfortunately, the prohibitive costs and the time required to date literally hundreds of bones and teeth means that most museum collections and other assemblages will probably remain largely undated. Another problem is that a large part of the continent has yet to be surveyed, although that is slowly being undertaken. One problem with such surveys is that there also seems to be large areas where Quaternary sediments either do not exist or are buried. This is particularly so in areas of gibber plains and desert dune fields and places where erosion and scouring have rendered landscapes back to Tertiary-age sediments. Preservation is the key and fossil remains in those types of environments are often lost or buried, so we may never know whether megafauna lived in those areas or not. All that can be done is to continue surveying sediments of the right type and age, and the fossil lakes and palaeochannels of central Australia are the most likely places they will be found.

The present picture shows a megafauna population that lived only in certain parts of the continent. It was divided into sections, each with its own combination of species that were in restricted distributions, some very limited in variety. That strongly suggests the megafauna extinction was not continent-wide but rather in a series of patches, each with its own combination of species. That also greatly reduces the likelihood of a blitzkrieg scenario with all species being wiped out across the continent. Accumulating data, some of which has been presented here, shows that Quaternary environmental change must have been a significant forcing mechanism of megafauna attrition for more than a million years, resulting in very few species reaching the accepted extinction threshold; those that did probably lived in isolated and geographically restricted, remnant populations containing few animals (Webb, 2006, 2008; Wroe & Field, 2006). Therefore, what we do know about megafauna distribution and the biogeographic effects of Quaternary glacials must be included in any future assessment of the overall extinction event. The anthropogenic argument in

Australia has come under criticism, not for the sake of being different, but due to our improved ability to examine more closely the enviroclimatic changes associated with glacial–interglacial cycling and the biogeographic consequences for the megafauna.

The overkill argument is based on one cause – the killing of all the megafauna through hunting – whereas the environmental and biogeographic argument is based on the many aspects of enviroclimatic stressing that took place during glacial–interglacial cycling over hundreds of millennia. We can now see them for the radical events they were and the vast changes they must have brought to the Australian continent as they did to other continents. This has been made possible by a range of palaeobiological work undertaken throughout the world as well as the extensive ice and oceanic core data which in itself has improved our understanding of the shape and complexity of glacial cycles. Moreover, wide-ranging field work in Australia over many years that comparatively few researchers care to undertake is slowly helping us unravel the effects of glacial climates on our continent. Examination of the effects of extreme Quaternary climate change on megafauna has been made possible by assembling various forms of empirical geological, palaeobiological and palaeontological data and applying the basic principles of climate science, biogeography and faunal demographics to marsupial megafauna as a real animal group rather than a stand-alone assemblage of 'special' animals. This is taking place not just in Australia but worldwide. Discussion of megafauna extinctions has now moved to include an enviroclimatic contribution with regional differences placing their position towards one or other end of the argument spectrum. In Australia the anthropogenic cause has always been somewhat convincing because of the supposed coincidental timing of extinctions with the first human arrivals but it has lacked a wider view that does not address the remaining uncertainty of *when* humans first arrived, how many landed and their capabilities, particularly the use of fire and landscape burning. That 'coincidence' is no longer correct, nor have the proponents of the argument been able to show a direct connection between megafauna and humans. Moreover, we still have little or no idea of the rate of arrivals, the way they explored the continent and how fast the continent was settled. Overlap between humans and megafauna is now more than likely but it probably did not include the full range of animals nor were megafauna living all over Australia as they once did. The earlier people arrived the longer that association existed, but there still remains no evidence of the implied massacre that took place in the blitzkrieg argument: no butchering or scavenging of animal carcases and no evidence of the charred remains of megafauna in 40,000 year old or any other campfires, only extant varieties. Kill sites in Australia – nil.

Fire, Humans and Megafauna

Australia is famous for its infamous bush fires. They have entered cities and run unchecked across thousands of square kilometres of densely inhabited countryside, destroying towns and homesteads and killing many people. A mention of disasters such as *Ash Wednesday* (1983), *Black Friday* (1939), *Black Saturday* (2009) as well as the equally ferocious *Tasmanian* (1967, 2013) and *Canberra* (2003) bush fires makes the naturally warm Aussie go cold. Today, most of the largest fires occur in southern, eastern and southwestern parts of the continent although large ones can burn out thousands of square kilometres in remote areas – but they are largely ignored because they do not usually threaten life and property. Fires in central and northern Australia are lit by people or ignited by lightning but part of the reason for the former is that fuel is kept in check by Aboriginal mosaic burning in many areas and undertaken in Autumn and Winter. However, it may well have been giant anvil-shaped smoke clouds billowing thousands of metres into the stratosphere from natural conflagrations that originally sign posted distant Australia encouraging the first people to make the crossing here. Those fires would have been sparked by lightning strikes on vast amounts of natural fuel build-up across northern Australia at the beginning of the monsoon season when temperatures began to rise and the rains had not yet arrived. Megafauna had experienced and survived hundreds of such fires right across the continent before humans arrived. That had been the only natural cause of fire before the arrival of humans and fires in unburned country would have been large and often out of control, making them very dangerous for animals. This would have been enhanced by the growth of highly combustible plant species which are generally found in almost all parts of the continent except rainforests. Both browsing and grazing animals would have been affected by fire, but in different ways for different species.

As people moved inland, frequent burning over a long time with regular rainfall would have opened up forested areas, promoting grasslands at the expense of closed wooded area and shrub lands. These changes favoured grazers, large and small, and may have increased their numbers. At the same time, browsers were negatively affected as the spread of grasses made large areas vulnerable to drought when grasses eventually died off and Australia's thin soils eroded or blew away. Fire-prone plants would be replaced by fire-tolerant types, which meant that not only were shrub lands shrinking but also their compositions were changing. Some shrubs may have disappeared through increased fire frequency, disadvantaging those animals dependent on them. In total, these changes could have had profound effects on those megafauna that were adapted to the previous environmental status quo but there probably were not many of them left. The tropics could have responded differently with wet sclerophyl and vine-thick areas less affected by fire and which encouraged animals into pockets. In Australia, the blitzkrieg argument incorporates anthropogenic burning of the landscape as a deleterious impact on megafauna. So if the firing of the land by the new arrivals severely impacted megafauna it was because it was of a type and frequency not experienced by them before and/or it was frequent enough to change their habitat in a negative way that particularly affected larger species. So, was regular anthropogenic burning a contributory factor in the megafauna extinctions?

Ethnographically, Aboriginal firing of the land brings many benefits. First, it clears the land of unwanted spinifex (*Triodea* sp.), a highly flammable and prickly-tipped grass. That makes cross-country travel safer and easier, particularly in central Australia. The act of burning also drives out game that can then be hunted. Burning just before rain and storms arrive in spring and summer also removes the ignition potential from lightning. Fast-moving accidental bush fires move fast and pose extreme danger for those on foot. Following the fire, ash is deposited on the ground which is then absorbed during follow-up rains. This natural fertiliser helps replenish nutrients and minerals in a soil that is generally lacking in them. The result is the production of new grass or 'pick', which then attracts fauna back into the area, so completing the ecological circle for the hunter–gatherer. Clearing ground in this way also reveals the holes and hiding places of lizards and snakes, some of which are cooked in the fire – the original fast-food takeaway. Perhaps the most important aspect of firing is that it encourages germination in many plants, shrubs and trees as well as releasing seeds that then take root and grow. So firing is food production for the future. There is also a spiritual side to it: by keeping the land clean, Aboriginal people are looking after their country. This shows the ancestral spirits they are worthy of keeping it because it was those ancestral spirits that provided the land and gave it to them in the Dreaming.

So, what about the first arrivals? Did they have such natural technology? For firing to have affected the landscape and the megafauna, the new arrivals would have had to have this practice ready to go as they stepped off the boats. Firing would also have had to be applied widely and regularly to change the environment so radically and comparatively quickly. The process of burning would have begun in the north, probably on the continental shelf, and gradually moved down the coasts as well as inland, taking perhaps thousands of years to change the whole of the continental environment. Remember, however, that large areas of the inland were desert during the 50–70 ka window within which we suspect the new arrivals landed. For regular firing, there would also need to be a permanent population living in all parts of the continent. A small wave of arrivals would not have fitted that bill. Instead they were of a size and composition such that they would have moved on, firing once in each area they encountered; but that does not constitute *regular* burning, enough to change landscape. Population size is, therefore, important and I would not expect the earliest people here to have had a very large population, one that would cover the continent. Ah, I hear you say, but if the continent was largely arid then they probably did not live in the central parts and only a few small groups could have moved through the habitable parts of the continent, mainly along the continental edges. Ah-ha, I say in reply, then those regions were not fired because there was not enough vegetation to burn and sustain it and that would have been the case for a large part of Australia during the period from 20 to 80 ka. The megafauna population at that time was also reduced for all the reasons given in previous chapters. But did the first people bring a burning technology with them? To answer that we have to go back to Southeast Asia because we need to know who the arrivals actually were.

Evidence for a pre-MIS5 entry for humans has not been found in Australia except for some controversial evidence that I discussed earlier (Webb, 2006). However, if the first Australians arrived ~150 ka when sea levels were at their lowest, they were far too early to have been Anatomically Modern people. This suggests they must have been 'locals' from the Indonesian archipelago or from other parts of Southeast Asia; humans had been living in both areas for hundreds of thousands of years. There is no evidence from Java or anywhere else in Indonesia for the use of landscape burning or fire of any kind by its early endemic inhabitants (*H. erectus/ Homo soloensis*). Moreover, their extremely limited numbers and the unsophisticated nature of their cultural items such as basic cores, chopping tools, and basic flakes, together with a few Acheulian hand axes, were unlikely to have placed the threat of extinction on Javan animal populations (Simanjuntak et al., 2010). Moreover, they knew all about large animals such as tigers, elephants, rhinos and large *megalania*-like lizards. A later arrival of people perhaps *ca.* 70–80 ka may have seen the arrival of mixed genetic stocks of Anatomically Modern and Denisovan people, mentioned earlier in this chapter, but they have left no evidence of landscape burning in the regions they traversed before entering Australia. Furthermore, there is no evidence of humans using fire to change the landscape anywhere in the world at or before this time and that includes Southeast Asia. If the first people (whoever they were) used firing we should see such changes in the environments of our nearest neighbours, but we don't. There is no good evidence for widespread, regular burning in Southeast Asia or on the myriad islands between there and Australia as a result of the long history of humans in these regions. The ecology of tropical islands as well as the southeast Asian continent is also not conducive to widespread landscape firing in the way it has been practiced ethnographically in Australia. Fire is deleterious to monsoon rainforests which largely contain fire-sensitive plant communities. Food plants can be easily lost, together with a highly specialised rainforest fauna. Rainforest foods are not enhanced by fire: quite the contrary. They grow quite well without it and prosper through interaction with the diets of birds and animals that inhabit them. Human inhabitants of rainforests must have been well aware of the environmental sensitivity of their surroundings or else the long human habitation of such regions would not have been as successful as it was. Fire use would, over tens of thousands of years, have altered the sensitive structure of rainforest and the delicate balances that occur there and it would have left permanent scarring that would be easily observed today.

Although landscape burning has been useful in Australia, developing the technique probably took sometime, certainly with regard to the highly specialised way it is used by Aboriginal people. Whenever people arrived, it was during a time of lower sea levels than now. So if they brought firing technology with them the first place to be burnt would have been the continental shelf. If they stuck to the coasts, being unused to living in arid environments, then landscape burning may not have been used till they moved farther into areas that responded well to it and the right conditions existed. But from what I suggested, above I assume they did not have it and Australia's glacial environment meant that much of the continent would not have had enough vegetation to burn let alone start massive destructive fires; in other words, it was not a useful technique during Ice Ages or long periods of desertification. On the other hand, if fire was used as a method of clearing, the very small initial population might have produced equally small burn areas, although the odd conflagration could have developed on a windy day in a dry and previously unburnt landscape and that would have been as dangerous for humans as for any other animal large or small.

I would suggest that whenever humans arrived in Australia, they came without landscape firing practices. Those were developed over some thousands of years when Australia's early arrivals eventually occupied arid and semi-arid landscapes that lent themselves to burning on a broad scale. I suggest also that occupation of those landscapes took some considerable time because they were novel for a tropical people who were adapted to rainforests, islands, coasts and ocean living. Desert dwellers were produced in Australia and not imported. Once they had adapted culturally and to some degree physically to their new environment, then they began to understand how it worked and how useful fire was to manage it. The bulk of opinion today is that people did not arrive in Australia till 60 ka or slightly later. Regular landscape firing may have affected larger browsing animals much more so than grazers and smaller species, but occasional burns would have been of no more inconvenience than fires started naturally by lightning strikes.

There is one last consideration. It has been claimed that the world's human population went through a bottleneck after the time of the Toba volcano super-eruption in Sumatra *ca.* 71 ka (Ambrose, 1998, 2003). It was a time, apparently, when the world's population is said to have been driven below 10,000 souls and possibly as low as 2000 by this event. If this is true, and I have my doubts, the numbers arriving in Australia for the first time would have been extremely few in the 10,000 years following the eruption. I would, therefore, find it hard to accept that so few could do so much to so many species in such a short time. Moreover, what were the consequences of such a large explosion on megafauna here or anywhere during the 10 year 'nuclear winter' that is said to have followed (and whose time frame I also doubt)? The link between volcanic activity and extinctions is well charted in this book so it is worth considering whether the aftermath of the Toba explosion contributed to the reduction of megafauna in the region or on other continents.

Megafauna Extinction: Summarising Considerations

Today, we are constantly made aware that future climate change will affect our biosphere and is likely to cause the extinction of many animal groups. The predictions cite how animals will be affected by increasing drought in some places, the loss of sea ice in others, by warming oceans and even by sea level rise elsewhere that will inevitably result from global warming. Those claims are not frivolous; rather, they are based on sound climatological, ecological and biogeographic principles. Those principles have always operated not only now but also in the past. Why, then, are those same principles ignored when it comes to the megafauna extinctions and the severity of the Quaternary Ice Ages that they lived through? They brought about infinitely more severe conditions than those predicted for our near future, not once, but repeatedly for 2 million years.

Unified theories to explain the extinction of megafauna everywhere have diverted attention from explanations that address situations and extinction histories that likely took place on different continents. It seems to me this has happened to the universal megafauna extinction debate for far too long. The megafauna extinction debate has always needed to turn a corner, employing less emotion and rhetoric and more science and closer examination. Hopefully that is gradually taking place. The complex issue of extinction and its myriad causes can now be appreciated rather than ignored. Viewing all the evidence so far, I cannot see how the megafauna extinctions were caused by humans, at least not in the majority of cases and particularly in Australia. But I can see how enviroclimatic changes on a world scale and over hundreds of thousands of years could have caused them, and the cumulating data is now telling us that. One of the main reasons for that conclusion is that there is much more logical data to support an environmentally caused extinction than a human-induced; in Australia, there is not a scrap of empirical evidence for the latter. The megafauna debate has, generally speaking, focused narrowly on a universal explanation for an extinction which was initially understandable. Apart from the fact that unifying theories regarding prehistory are vanishingly rare in their accuracy, they also distort the differential biogeographic circumstances brought about by the glacial conditions of the Quaternary which are perfectly understandable and logical.

Beyond the megafauna debate, different reasons for animal extinctions on different continents have been adequately demonstrated time and time again through the Earth's long history, and many of those have been documented here. So much can happen in 1 or 2 million years and we are better able to see and examine the details of what went on in the last 1 million years than between 18 and 19 million years ago, for example. If we could see the details of climate variability at that distance, interpretations of marsupial adaptation, evolution and extinctions would probably change, possibly drastically, and mean so much more. We have the good fortune to be able to see and learn more about the Quaternary than we have been able to learn, arguably, about all time before that. So we now know much more about the conditions under which these animals laboured for a very long time although much more will be known in future. Our knowledge is increasing all the time and providing a much clearer picture of the environmental circumstances that resulted from the Age of Ice Ages. That accumulation of data increasingly makes it much less likely that a single cause and, in particular, an anthropogenic cause claimed these animals. Like most extinctions, it is now obvious that the whole thing is much more complicated than that.

Extinction today may been seen, both ethically and philosophically speaking, as undesirable primarily because humans are the main cause of it, by means that are often ancillary to our progress such as population pressure and development. Before those were factors, animals went extinct the way they should, for natural reasons that depended on their ability to maintain their populations. Population maintenance relies on numbers, gene pool size, reproductive speed, fecundity and ability to adapt and respond to environmental and climatic change. One animal may be able to do this where another cannot and large animals usually fall into the latter grouping. Smaller species recover far more quickly and easily from set backs than big ones because they need less, reproduce faster and have greater numbers of offspring at one time. On the other hand, large animals usually live longer, so if allowed to do so they can have multiple offspring over a long reproductive lifetime which maintains their population. However, the ability to live longer does not necessarily mean they make old bones or have the opportunities to mate and produce, nor does it guarantee optimum sex ratios, and the odds of all these factors kicking in become increasingly shorter as their populations become smaller, fragmented and isolated. They also become victim to founder effect and genetic drift. Also, large animals are often more specialised and require greater food and water resources than smaller ones. So a large species must have greater numbers to survive through a given set of negative circumstances than a smaller one, although they usually have fewer because of their vulnerability. Larger animals can also be slower to remove themselves from areas of sudden environmental change and they need bigger ranges, so range shrinkage affects them more than smaller animals, making them vulnerable. In Australia, range shrinkage was exceptional during ice ages. In their way, the megafauna browsers were specialised, requiring ample trees and shrubs. So probably few were able to cope with arid conditions because of the amount of food and water they required and the widespread reduction of browsing fodder. Therefore, size would certainly have mattered in the game of survival and extinction: 'the smaller the better' certainly applied during glacial conditions. It also matters in terms of geographic distribution, individual range and population size and composition, all of which rely on sex ratios and fertility or the ability to have offspring in the first place.

Evidence for human arrival in Australia is poised between 50 and 60 ka. However, using far more tenuous evidence described in an earlier chapter, as well as recent human fossil discoveries in Asia, it might go back to the penultimate glaciation *ca.* 150 ka. With such an ancient neighbourhood on our doorstep, I will always feel that that could be a distinct possibility. The main criticism of an earlier entry, however, is that no 'solid' supporting archaeological evidence has been found to support it. Now, I claim that megafauna did not live on half of the continent because there is no evidence for them there; this follows the same line of logic – no evidence, no occupation. Some might not accept that and use the old argument that absence of evidence is not necessarily evidence of absence in this case because some remote areas of the continent have not been thoroughly investigated. However, accepting a human presence using the same argument is thought to be totally unacceptable as a valid argument that humans could be here, we just haven't found them yet. You can't have it both ways. The contradiction is obvious but we must work with the evidence at hand. So the point made here is that at present it seems that humans arrived here between 50 and 60 ka and megafauna did not live on half the continent – or at least it seems that is the case, given the evidence we have on both counts.

So, the present data suggest that late-Quaternary Australian megafauna were sparsely distributed over only about two-fifths of the continent, with some expansion during interglacials. They were largely concentrated in the east and southeast but most species had limited ranges within those areas. A large section of central and western Australia had few, if any, animals. Species common to the LEB, southwestern Australia, the Nullarbor and the northwest strongly suggest that a few species were more widespread than others and that groups moved opportunistically into central areas on occasion. One reason for thinking this is the apparent lack of any megafauna in central Australia, with only *Diprotodon* and *Genyornis* living on the periphery. The fact that these two have been found in semi-arid areas suggests they could penetrate parts of the continent that maybe others could not. Some may have been semi-arid adapted while the majority were confined to temperate and well-watered environments; that may be the reason that the last of these animals probably lived in the north of the continent, if we go by the rock art record. It may be unremarkable, therefore, that these were possibly some of the last megafauna to go extinct although a few other remnant species probably lingered in the better-watered north and northeast. Their common occurrence in many parts of the continent together with a couple of other species is unlikely to be a coincidence but rather a demonstration of their adaptive strength. But the reproductive strategy of large marsupials was probably not conducive to maintaining their respective populations once their populations reached a certain level, became fragmented or were even randomly hunted, although we lack any evidence for that.

We know that there was a staggered loss of animals as background extinctions in Australia extending back over 1 million years in response to the pattern of changing environmental and climatic conditions that strengthened throughout the Quaternary. The large animals of the mid-late Quaternary such as *Bohrer* and *Euryzygoma* were no exception to this pattern and succumbed slowly but inevitably either because they were unable to adapt to stresses and conditions or because their populations were too small to sustain them. There are always animals that succumb in this way and that, hopefully, has been clearly demonstrated in prior extinctions documented in this book. The environmental circumstances met within Australia during the Quaternary do not necessarily coincide with conditions on other continents and there can be little doubt they were different. We coped with desert and drought; others coped with glaciers, tundra and freezing conditions. Moreover, the response of animals on other continents can also be expected to be different because of natural differentials in their adaptive capabilities, their individual population sizes, their biogeographic differences and the possibilities for migration away from deteriorating conditions: east to west in Eurasia, north to south in Southeast Asia. Therefore, the arguments laid out here for Australia's giant marsupials are not expected to serve as explanations for Quaternary extinctions elsewhere even though it is tempting to suggest that emerging evidence from Eurasia seems to be suggesting a similar tapering off of species over 20,000 years or so. Having said that, given the world climatic background, megafauna extinctions were certainly something that could be expected because they would have been felt by fauna no matter where they lived. We also have to appreciate proportions here. The megafauna extinctions were only one of hundreds of such events our planet's biological calendar and in reality it was a comparatively insignificant one at that.

In the face of the arguments put here, the question of the timing of Australia's megafauna extinction is perhaps not as important as it might appear because it seems almost certain that it was not a singular event spatially or temporally but rather a slow exponential reduction throughout the gambit of species. And it gained momentum as animals became fewer and more susceptible over time, with species going extinct at different times and in different places. The causes were differential species vulnerability; the particular circumstances in which animals found themselves, their geographic distribution and disposition, their abundance or rarity in the first place and the circumstances they found themselves in. It will remain vital to extinction arguments, however, that accurate terminal ages for all species be determined in order to fill out the picture properly. However, most people would agree that actually finding the last of any species is probably impossible. Moreover, the usual process of extinction is not a bang but a whimper, ending with a few stragglers that hang on somewhere or other. A few animals remain here and a couple there, the few infants born are not enough to replace losses and they too disappear or do not find mates for one reason or another – and the species has gone, *plip!* Anyway, we do not have to date the last specimen of each species; we just have to see when the most recent dates seem to cluster and peter out with no younger dates emerging. Unfortunately, that process will take sometime but hopefully such dates will be established in years to come.

Enough evidence of long periods of exceptional mid-late Quaternary drying in Australia exists to prompt further research to refine our understanding of the extent of this process. It is more and more likely that cumulative changes in climate leading to increased aridity imposed environmental pressures which affected megafauna deleteriously everywhere over the long term. Aridity eroded trophic levels, impacted vegetation, removed vast ecosystems and with it species variety, bringing about local extinctions, regional species withdrawal and population reduction across the continent. Glacial desertification pushed out to the continent edges, squashing remnant populations into smaller range areas and isolating others in continental island patches. Stress was imposed by patch isolation, corridor collapse and shrinking ranges, while other populations were pushed together, and all these forced animal reductions as habitats disappeared, pushing small populations together for a while; others disappeared. The absence of megafauna in the west and centre of the continent shows the magnified impact of these processes in arid and semi-arid regions. While animals can exist in fragmented habitat, they might not do it for long and, again, larger animals are less likely to thrive than small varieties under such circumstances. Also, they do not necessarily represent places where future population recovery can take place when conditions change. Indeed, they are more likely traps leading to localised extinction, particularly during extreme, long-term climatic events. Smaller animals were also less vulnerable to such change and would survive longer than larger ones in a given patch size; their faster reproduction rates also aided them in maintaining their population numbers. Broadly adapted, generalist animals had the best chance of survival during these times, but larger, specialist animals with longer life spans were less likely to cope. One argument against environmental stress as a cause of megafauna extinctions is their persistence through previous glacial events. But for the reasons detailed here it is becoming apparent that not all species did persist (see above) and those that did were just scraping through with various-sized reductions during every cycle.

What Happened to Australia's Megafauna? A Conclusion

As much as my enthusiasm has possibly done the opposite, I do not want the reader to think I believe what I have written in this book is *the* truth about the megafauna extinctions. The one thing I have learnt over the years is to be cautious; new evidence will always come along that can easily sink theories and take ideas in completely different directions. What I wanted to do in this book, however, was to examine the basic circumstances under which the megafauna lived for 1 or perhaps 2 million years. Those circumstances were dire in the long term and in themselves they provide a very broad, complex but natural explanation for a process of gradual extinction that grew slowly but ended exponentially. Glacial cycles had enormous consequences for Australia's biogeography and they cannot be ignored because to do so is unworthy of those who would wish to reach a better understanding of the issue. Moreover, they offer a much more logical reason for the demise of these animals than swift extermination by humans – for which there is no evidence. It is also silly to blame humans simply because it is popular to blame humans for *any* extinction or because the very fact that we cannot prove humans caused the extinctions somehow makes it likely that they did. We also know there is an overlap of perhaps 15,000 years or more between humans and megafauna in Australia, so the rolling blitzkrieg never happened here.

My aim has been to emphasise that we know a lot about why animals go extinct and that we watch the process everyday. I wanted to apply that knowledge to this issue. The reasons for the megafauna extinctions outlined in this book are not derived from *likely* reasons but from a large amount of *sound research* and the work of many in various disciplines that has taken tens of thousands of hours to collect, assemble and analyse and that relates to how animals live and die on our planet now and how they did so in the past. We know these tenets; we do not have to argue about them. The facts derived from that work clearly show us how animals live and thrive and disappear against fixed and changing determinants of geography, time, environmental circumstances and climatic change. The work shows also why animals live in some places and not others, why some number in thousands and others live in small vulnerable populations permanently perched on the edge of oblivion and how they can be pushed over with the slightest breath of change. Such studies now have an extremely broad horizon which takes in many different disciplines, some bunched under overarching titles such as *Population Biology* and *Biogeography*. The principles that have been derived from these (and I have mentioned only a few) are hard won, strong tenets of planetary zoogeography. One thing that has come out of the kinds of research that has contributed to the arguments presented here and which burns brightly in many conclusions about animal geography, behaviour and response to climate and environmental change is that the majority of extinctions are not always easily understood and they almost always have a multiplicity of causes.

At the end of the Tertiary world, climate entered a cyclic roller coaster that became more severe as we passed into the Quaternary; so began the Age of Ice Ages. Australia's marsupials made a brave effort to adapt to the gradual drying of the Australian continent that began 12 million years ago and they were successful for millions of years. The long, slow drying process provided plenty of time for various adaptations to take place and gigantism was one of those. When the Age of Ice began, therefore, Australia possessed a suite of large animals equipped to face an arid continent under the terms they were used to – but those terms did not include increasing aridity to the extreme and then cycling it. To add to megafauna's troubles, those cycles deepened as the world cooled, taking average temperatures even lower. There was also a trend that made each cycle more severe than the last till they formed long cycles of cold lasting tens of thousands of years. Glacial cycles increasingly changed most of Australia into one big desert. After 100 cycles, Ice Age severity affected more and more of the continent so that for over 85% of the time during the last 20 cycles probably 85% of Australia was a desert; the comparatively small remainder was where the fauna lived squashed into corners. Interglacial respite enabled remnant populations to recover, but not all interglacials were that kind and recovery became harder and harder to achieve after each glaciation. The last three prior to the LGM were particularly severe and long. While our well-adapted megafauna survived through much of the Quaternary, there were casualties along the way and that placed on them a pattern of attrition that rendered the population severely reduced as the world entered another Ice age 90,000 years ago: that was enough to finish off many species in most parts of the continent, leaving small, scattered and endangered groups in the southeast and possibly limited areas in the tropical north. The arrival of humans may have more unwittingly than not been the last straw for the broad backs of these spectacular animals but, so far, the proof of that straw has eluded us although the straw may have been very long. Perhaps we should not be saddened by the passing of the megafauna but rather surprised they made it as far as they did.

There are some aspects of the past which we will never know in detail. Nevertheless, the arguments offered and conclusions drawn are, I believe, logical, realistic and based on well-accepted scientific data. They leave out the emotive factor of peering at a massacre carried out by humans against animals around the world. Such a conclusion is more sensational, granted, but the arguments are unrealistic and the evidence for it, at least in Australia, is non-existent. Late-Quaternary palaeoenvironmental change cannot be ignored in any future discussion of megafaunal extinction and the evidence for it playing the major part in the extinction is not only convincing: but it also is mounting everyday. We have peered at a world that ushered in a new era, that of the Ice Ages. Going on the pattern so far we will have another one, perhaps not in the too distant future. When that occurs, it might prompt the question, Who or what killed the human race?

The Last Word... for Now

I undertook some field work in Arnhem Land in 2010 hoping to find *Genyornis* egg shell – none found. Arnhem Land is situated at the Top End of Australia's Northern Territory and comprises an area measuring 97,000 km^2. It is a country within a country. It is Aboriginal country, a place where many Aboriginal language groups live a traditional life, or as close as it gets these days. They hunt, fish, look after the country, carry out regular burning, perform ceremonies and rituals, collect bush foods, look after rock art sites: they are rangers and nature wardens. Some paint stories of their country on bark, on grave posts and, on occasion, in the cave systems and rock shelters of the 'stone country' just as they have for over 30,000 years. It is here that the vivid pictures of megafauna can be found and, no doubt, many more will come to light in the future. Most importantly, Aboriginal people try to keep their traditions and languages alive and ongoing into the future, and one way of doing that is teaching their younger people about it. Paintings are not just ceremonial; they relate histories of people, places, ancestral spirits and the country and they depict life, environment, culture and the wildlife of the country. Images of sailing boats, Macassan praus, battleships, biplanes and internal structures of steam ships can be seen painted in the typical X-ray style of the region. They mark the historic coming of others to the coast. The east coast now has a large bauxite mine feeding the thirsty industrialisation of the world while, only a few kilometres away, there are pristine beaches and natural bushland with all the creatures that should be there. Unfortunately, the beaches have huge scatters of plastic bottles, rubber shoes, rope, ripped fishing nets, lids, cups, broken toys and more, messages from that industrialising world that chucks rubbish in the sea as though it will swallow the lot: perhaps not *pristine* beaches after all, with turtles struggling in plastic netting....

For most Aboriginal people the ancient ancestral spirits still rule and make things happen in their lives and the world. Their spirits are of those special places that were created by giant beings so long ago in the Dreaming. They have special places, some sacred, others secret and known only by those who should or need to know. Places are revered, such as Yalangbara, the small headland where the two ancestral spirit *Djang'kawu* sisters came ashore in goanna form and brought the region to life long ago in the Dreaming. Aboriginal people live in their own communities partially protected by an intricate permit system to which intending visitors have to adhere. Permits are required to drive the one major dirt road that runs 700 km across the region from central Australia northeast to the coast. Different permits are required to enter communities and other permits are required for other purposes which, although understandable, can be frustrating for the visiting researcher who merely wants to carry out innocuous research and learn in such a fascinating landscape, much of which is largely untouched by the white colonial finger and which obviously contains a lot of secrets of Australia's past. The region revolves around the Arnhem Land

Escarpment or 'stone country', a granite edifice that goes back a couple of billion years. That is surrounded by tropical savannahs, billabongs, swamps, bird-filled wetlands, vine thickets, tropical forest and savannah and incised by deep gorges. Some who have worked there for years have told me that there are places not visited by white man and have even suggested that some secluded, deep gorges cut through the escarpment over tens of millions of years may not have been visited by anybody!

In 1849, the British introduced 20 banteng cattle (*Bos javanicus*) into the Coberg Peninsular on the northwest coast of Arnhem Land to feed the pioneers of a new settlement. In short time the settlement was abandoned due to the unpreparedness of the settlers, climate and disease. The cattle were set free and from then on multiplied and spread widely throughout the region as most introduced animals do. One highlight of my trip to Arnhem Land was to see up close these wild buffalo that often weigh between 600 and 800 kg. I did not expect to see them because we had heard they had been eradicated in many areas. Why? Because they are a pest and are known to kill and trample those who venture too close. They destroy, trample and break down vegetation as their herds blunder through swamps and wetlands; they also carry brucellosis, a disease dangerous to bovine domestic livestock and humans. So, for the last 100 years buffalo have been systematically hunted, initially for their skin and meat. Many very brave or very foolhardy men have charged alongside a large stampeding buffalo on equally brave horses, shooting the great beasts with Martini–Henry rifles and other armoury using thousands of rounds of ammunition. Better weaponry has been introduced over the years which now includes M16 automatic carbines and specially built, stripped-down and reinforced four-wheel drives. Other shooting parties head out in helicopters on well-planned and wide-sweeping eradication programs. One such foray was being carried out by the Dhimurru Aboriginal Corporation as I left Arnhem Land. But the animals are still there! And I saw many more than I expected to. Ironically, while they are a pest in Australia they have become rare outside it, and this irony is something that now has to be taken into account. Rather than total eradication, conservation and confinement of these animals may now have to take over to preserve the species.

Even with sophisticated weaponry and deliberate acts of hunting to feed busy tanning and meat factories during the 20th century, with programs designed to eradicate them, and despite these creatures having only single offspring and living in a rather confined area in relation to the rest of Australia, the cattle are still there! They are truly megafauna beasts and the history of the past 100 years shows how difficult it is to remove them, even using a deliberate and well-planned act aided by automatic weapons of great power and systematic hunting by professionals whose only job is to kill them. It would be even more difficult, I would assume, with a bunch of spears in the hands of a few groups of human hunters on the ground with many species to choose from. Arguments about megafauna being eradicated to extinction by humans have never rung true with me and the story of the buffalo of the Northern Territory reinforce that. The deliberate killing of food sources till they no longer exist is not logical on a continent that had so many species to choose from and probably had a small human population. Smaller animals are so much easier to catch; they are less dangerous; they replace themselves comparatively quickly; and people relied on maintaining their food resources. However, if today the desert came to meet the buffalo of Arnhem Land they would die in their thousands; if desertification lasted for thousands of years at a time they would go extinct.

We will always argue about dates of this fossil site and that. We will also never date the last of the megafauna because we will not find it. So, some arguments will go on and on and there will always be disputes about what we find because those who do not agree with the findings will reject them however logical they may be and so we will go on and on, circling our intellectual roundabout once again. But perhaps that is what science and the search to understand our planet is all about. Oh, by the way, the natural hunters of Arnhem Land who have been there for over 50,000 years don't go near buffalo, at least not without a high powered rifle, and I don't blame them.

Forty Considerations About Why Australia's Megafauna Went Extinct

1. *At present, megafauna demography strongly suggests a limited population and species distribution outside southeastern Australia. Large areas of the continent probably remained unoccupied by them even during interglacials.*
2. *Only a fraction of the total megafauna species range is found in areas outside southeastern Australia.*
3. *Only 34 of the 52 megafauna species can readily be allocated to that category using weight (>50 kg).*
4. *Some megafauna species were probably always rare and others had extremely limited ranges, making them vulnerable to any climatic or environmental change – particularly repetitive forms.*
5. *Late Tertiary gross environmental changes across the continent took Australia from a heavily forested (rainforest) environment, through open woodland and savannah (megafauna time) to periods where 70–90% of the continent became desert in the Quaternary. Body size increase as an adaptive strength became maladaptive as Quaternary glacial cycles intensified and desertification expanded.*
6. *Glacial conditions imposed limitations on megafauna adaptive capabilities.*
7. *Extensive drying of the central continent and desert expansion during glacials gradually reduced the overall megafauna population.*
8. *Glacials caused drastic reduction of standing water as well as viable grazing and browsing opportunities for over 80% of the mid-late Quaternary.*
9. *Steady development of extreme and long glacial phases across the Quaternary subjected megafauna to regular population contraction and collapse during glacials.*
10. *Continental glacial and interglacial environmental change particularly affected small and isolated populations living in patches.*
11. *Constant climatic change over at least 1 million years brought adaptive pressures that could not be tolerated by many megafauna.*
12. *Mid-Late Quaternary Glacials were deep and long with short or very brief interglacials.*
13. *The prevailing adaptive environment for megafauna was glacial conditions that extended for at least 85% of the last 1 million years.*
14. *Interglacial or interstadial events provided only brief respite for megafauna recuperation and some interglacials afforded little or no recuperative time or appropriate conditions.*

15. *Isolation, disjunction and tethering were some of the mechanisms that eliminated local groups and reduced the overall animal population through incremental losses.*
16. *Animals were trapped by closing corridors and contracting refuges, a process that added to incremental losses.*
17. *Irregular and differential population reductions occurred across all 52 megafauna species considered here.*
18. *Sheltering populations inhabiting the continental shelf were eliminated by island tethering that became inundated.*
19. *The population was incrementally reduced during every glacial event.*
20. *Interglacials did not always afford environmental relief due to their limited length and cool to cold (continuing desert) status in some cases.*
21. *There was limited opportunity to elevate numbers to previous levels during climatic amelioration (interglacials) because large animals are more vulnerable to short-term enviroclimatic change than smaller species and their reproductive strategies are cumbersome.*
22. *Glacial–interglacial cycling placed various populations and/or species on an extinction spiral with regional variations.*
23. *A wide range of background extinction processes were operating (see the list of Biological Extinction Drivers in Chapter 9).*
24. *Normal population recovery is a slow process for large animals even over the long term and the southeast acted as a source area for repopulation of some areas by some species.*
25. *Megafauna species were going extinct even before MIS 5.*
26. *There is no evidence for impact of archaic human populations on animal populations in surrounding regions for hundreds of thousands of years prior to the megafauna extinctions.*
27. *People encountered large animals in Southeast Asia for at least 1.5 million years but did not exterminate them and that is the case for other regions.*
28. *There is no evidence that late Quaternary human migrations through Southeast Asia killed off megafauna species.*
29. *Asian megafauna overlapped with modern humans by tens of thousands of years and no kill sites found have been found.*
30. *Southeast Asian extinctions were a product of environmental change rather than anthropogenic interference.*
31. *There is no evidence for overkill in regions surrounding Australia.*
32. *Southeast Asian extinctions were caused by environmental change.*
33. *Australia's climate continued change to a general greater aridity and desertification during the Quaternary.*
34. *Australian fossil and rock art evidence now shows megafauna and humans probably overlapped by at least 15,000 to 20,000 years possibly even much, much longer if earlier landing occurred.*
35. *With the possible exception of Genyornis eggs, there is no evidence for direct contact (hunting) between humans and megafauna in Australia.*
36. *Megafauna populations were the only remnants when humans arrived.*
37. *There is no consensus on extinction timing of any megafauna genera or individual species.*
38. *There are no exact dates for the first human arrival in Australia.*
39. *There is no archaeological or palaeontological evidence for even basic hunting of marsupial megafauna in Australia, particularly in our earliest archaeological sites.*
40. *Glacial conditions resulted in limited numbers of sparsely distributed megafauna in a precarious position so that any pressure that did come from human arrival in Australia could have easily tipped the balance for those remnant species either sparsely distributed or confined to one or two small areas and already severely endangered.*

Appendix 1
Australian Tertiary Fauna

Tertiary and Quaternary Extinction Timing for 213 Species of Australian Marsupial fauna#

Species	Pre-Miocene	Miocene *	Pliocene*	Early	Mid-Late
		Tertiary		Quaternary	
Ankotarinja tirarensis	LO				
Apoktesis cuspis	LO				
Badjcinus turnbulli	LO				
Balbaroo gregoriensis	LO				
Balungamaya delicata	LO				
Burramys wakefieldi	LO				
Chunia illuminata	LO				
Chunia omega	LO				
Ektopodon stirtoni	LO				
Galanaria tessellata	LO				
Ganawamaya aediculis	LO				
Gummardee pascuali	LO				
Ilaria illumidens	LO				
Keeuna woodburnei	LO				
Kuterinjta ngama	LO				
Madakoala devisi	LO				
Madakoala wellsi	LO				
Marlu praecursor	LO				
Miralina doylei	LO				
Miralina minor	LO				
Muramura williamsi	LO				
Nambaroo couperi	LO				
Nambaroo novus	LO				
Nambaroo saltavus	LO				
Nambaroo tarrinyeri	LO				
Namilamadeta snideri	LO				
Ngapakaldia tedfordi	LO				
Ngapakaldia bonythoni	LO				
Palaeopotorous priscus	LO				
Perikoala palankarinnica	LO				
Perikoala robustsus	LO				
Pildra antiquus	LO				
Pildra secundus	LO				
Pilkipildra handae	LO				
Pitikantia dailyi	LO				
Priscileo pitikantensis	LO				
Purtia mosaicus	LO				
Raemeotherium yatkolai	LO				
Silvabestius johnnilandi	LO				
Wabularoo naughtoni	LO				
Wururoo daymayi	LO				
Balbaroo fangaroo		/			
Durudawiri anfractus		/			
Durudawiri inusitatus		/			
Ektopodon litolophus		/			
Ektopodon serratus		/			
Ganawamaya acris		/			
Ganawamaya ornata		/			
Ganguroo bilamina		/			
Litokoala kanunkaensis		/			
Marlu kutjamarpensis		/			
Nimbacinus dicksoni		/			
Ngamalacinus timmulvaneyi		/			
Nimiokoala greystanesi		/			
Nowidgee matrix		/			
Paljara tirarensis		/			
Pildra tertius		/			
Priscileo roskellyae		/			
Rhizophascolonus crowcrofti		/			
Thylacinus macknessi		/			
Wabulacinus ridei		/			
Wakiewakie lawsoni		/			
Wynyardia bassiana		/			

Taxon					
Yalkaparidon coheni	/				
Yarala burchfieldi	/				
Balbaroo camfieldensis		/			
Barinya wangala		/			
Bettongia moyesi		/			
Burramys brutyi		/			
Dasylurinja kokuminola		/			
Djilgaringa gillespiae		/			
Ekaltadeta ima		/			
Hypsiprymnodon bartholomaii		/			
Kolopsis yperus		/			
Litokoala kutjamarpensis		/			
Maximucinus muirheadae		/			
Muribacinus gadiyuli		/			
Mutpuracinus archibaldi		/			
Neohelos tirarensis		/			
Nimbadon lavarackorum		/			
Nimiokoala greystanesi		/			
Nimbadon lavarackorum		/			
Propalorchestes novaculacephalus		/			
Propalorchestes ponticulus		/			
Strigocuscus reidi		/			
Thylamcinus macknessi		/			
Trichosurus dicksoni		/			
Wakaleo vanderleueri		/			
Alkwertatherium webborum			/		
Dorcopsoides fossilis			/		
Ekaltadeta jamiemulvaneyi			/		
Ganbulanyi djadjinguli			/		
Hadronomus puckridgi			/		
Kolopsis torus			/		
Mayigriphus orbus			/		
Palorchestes annulus			/		
Palorchestes painei			/		
Plaisiodon centralis			/		
Pseudokoala curramulkensis			/		
Pyramios alcootense			/		
Thylacinus megiriani			/		
Thylacinus potens			/		
Tyarrpecinus rothi			/		
Wabulacinus ridei			/		
Wanburoo hilarus			/		
Zygomaturus gilli			/		
Archerium chinchillaensis				/	
Burramys triradiatus				/	
Darcius duggani				/	
Dasycercus cristicauda				/	
Dasyurus maculatus				/	
Dorcopsis winterecookorum				/	
Ischnodon australis				/	
Koobor notabilis				/	
Kurrabi mahoneyi				/	
Kurrabi merriwaensis				/	
Kurrabi pelchenorum				/	
Macropus pan				/	
Macropus pavana				/	
Meniscolophus mawsoni				/	
Milliyowi bunganditj				/	
Palorchestes selestiae				/	
Petauroides stirtoni				/	
Petauroides marshalli				/	
Phascolarctos yorkensis				/	
Prionotemnus palankarinnicus				/	
Protemnodon chinchillaensis				/	
Protemnodon devisi				/	
Pseudokoala erlita				/	
Simosthenurus antiquus				/	
Simosthenurus cegsai				/	
Sthenurus notabilis				/	
Trichosurus hamiltonensis				/	
Troposodon bluffensis				/	
Troposodon bowensis				/	
Troposodon gurar				/	
Wallabia indra				/	
Glaucodon ballaratensis					/
Kolopsoides cultridens					/
Nototherium watutense					/
Palorchestes parvus					/
Phascolarctos maris					/
Protemnodon bandharr					/

Species										
Protemnodon buloloensis PNG					/					
Protemnodon otibandus PNG					/					
Simosthenurus cegsai					/					
Watutia novaeguinae PNG					/					
Kolopsis rotundus						/				
Macropus mundjabus						/				
Simosthenurus brachyselenis						/				
Troposodon kenti						/				
Propleopus chillagoensis							/			
Jackmahoneya toxoniensis							/			
Pseudokoala catheysantamaria							/			
Baringa nelsonensis								/		
Euowenia grata								/		
Euryzygoma dunense								/		
Lasiorhinus augustidens								/		
Macropus thor								/		
Propleopus oscillans								/		
Sarcophilus moomaensis								/		
'Simosthenurus' baileyi								/		
Simosthenurus euryskaphus								/		
Sminthopsis floravillensis								/		
Antechinus puteus									/	
Baringa nelsonensis									/	
Bohra paulae									/	
Burungaboodie hatcheri									/	
Congruus congruus									/	
Congruus kitcheneri									/	
Dendrolagus noibano									/	
Euryzygoma dunense									/	
Propleopus wellingtonensis									/	
Ramsaya magna									/	
Tropsodon minor									/	
Warendja wakefieldi									/	
Metasthenurus newtonae									/	
Simosthenurus gilli									/	
Simosthenurus maddocki									/	
Diprotodon optatum										/
Eowenia grata										/
Genyornis newtoni										/
Lasiorhinus augustidens										/
Macropus ferragus										/
Macropus pearsoni										/
Macropus piltonensis										/
Megalania prisca										/
Palorchestes azeal										/
Phascolomys medius										/
Meiolania platyceps										/
Pallimnarchos pollens										/
Phascolomys gigas										/
'Procoptodon' browneorum										/
Procoptodon gilli										/
Procoptodon goliah										/
'Procoptodon' oreas										/
Procoptodon pusio										/
Procoptodon rapha										/
'Procoptodon' williamsi										/
Protemnodon anak										/
Protemnodon brehus										/
Protemnodon roechus										/
Simosthenurus mccoyi										/
Simosthenurus occidentalis										/
Simosthenurus orientalis										/
Simosthenurus newtoni										/
'Simosthenurus' baileyi										/
'Simosthenurus' pales										/
Sthenurus andersoni										/
Sthenurus atlas										/
Sthenurus stirlingi										/
Sthenurus tindalei										/
Thylacoleo carnifex										/
Wallabia kitcheneri										/
Wonambia naracoortensis										/
Zygomaturus trilobus										/
TOTAL	41	24	23	18	31	10	4	13	17	29

PNG - Only found in Papua New Guinea
 # - Quaternary listing includes large reptiles
* - forward slash (/) indicates early (left), mid (middle), late (right) epoch stage
MIS - Marine isotope stage
LO - Late Oligocene

Appendix 2
Ice Age Graphs

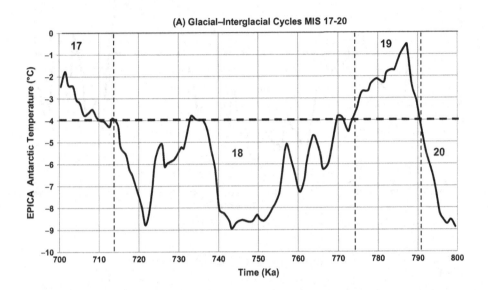

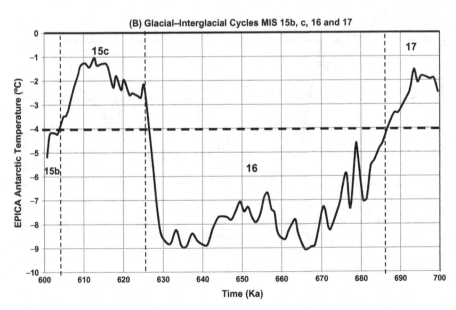

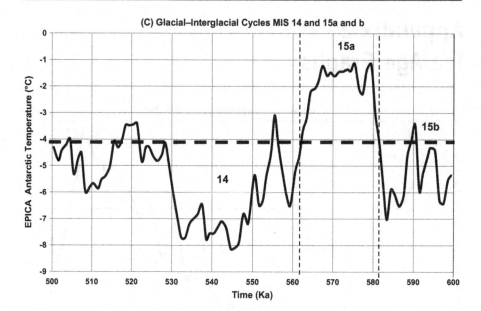

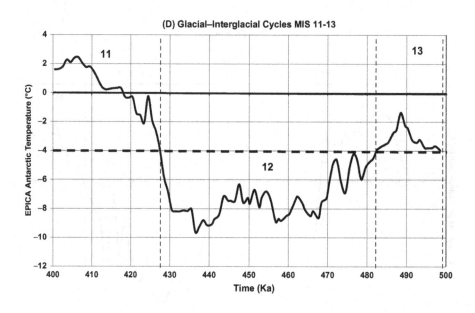

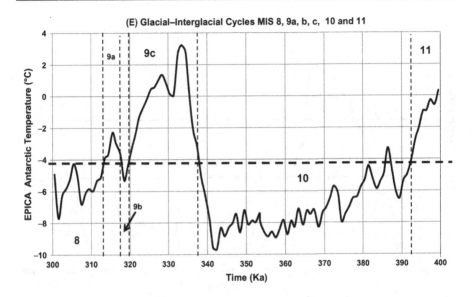

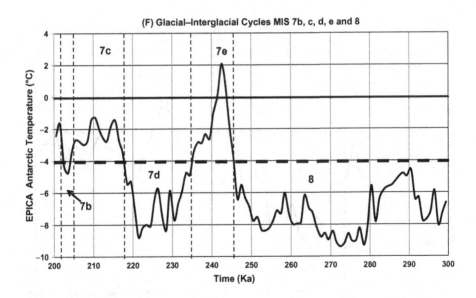

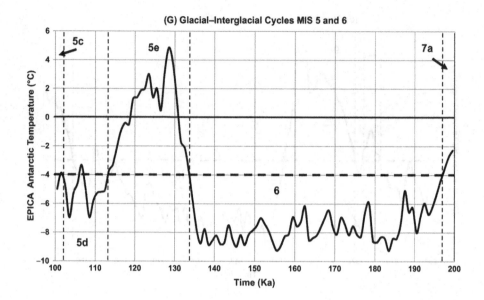

(G) Glacial–Interglacial Cycles MIS 5 and 6

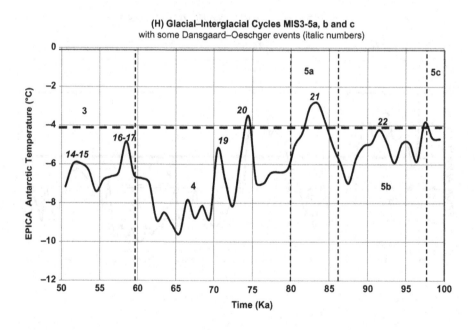

(H) Glacial–Interglacial Cycles MIS3-5a, b and c
with some Dansgaard–Oeschger events (italic numbers)

(I) Glacial Cycles MIS2 and 3

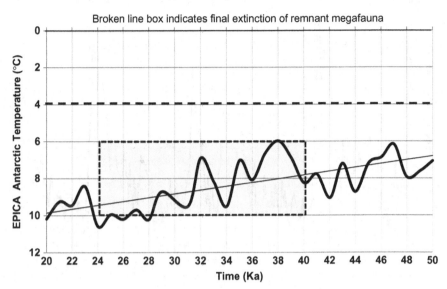

Broken line box indicates final extinction of remnant megafauna

(J) Glacial–Interglacial Cycles MIS 11-20

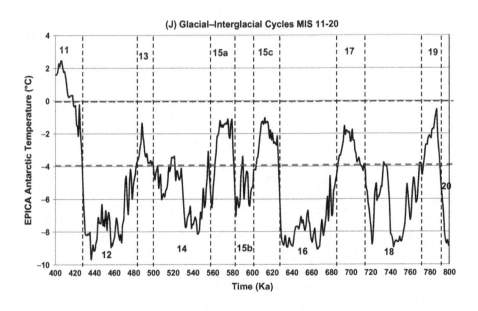

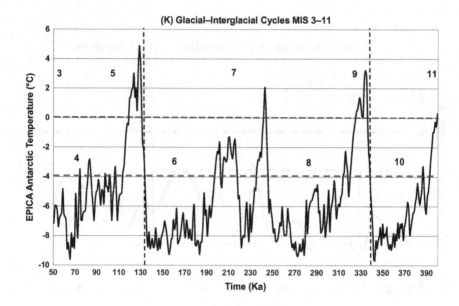

Appendix 3
Australian Mid-Late Quaternary Megafauna Sites

Queensland	
Sites	**Species**
Balonne R. Wycombe, Surat	*Diprotodon optatum*
Boubyjan, Goomeri	*Diprotodon optatum*
Brisbane River	*Simosthenurus pales*
Caiwarro, Paroo R.	*Genyornis newtoni*
Cape R. crossing, Lynd Hwy	*Megalania prisca*
Chillagoe Caves	*Phascolonus gigas*
Chinchilla (Charley Ck, Condamine R.)	*Procoptodon pusio, Diprotodon optatum, Thylacoleo carnifex, Megalania prisca, Palorchestes azeal, Sthenurus andersoni, Troposodon minor*
Clotherstone, Clermont	*Diprotodon optatum*
Collinvale, Longford Bowen	*Diprotodon optatum*
Coreena,	*Protemnodon brehus*
Diamantina R. Birdsville	*Genyornis newtoni, Diprotodon optatum*
Dulacca	*Megalania prisca*
Darling Downs:	*Thylacoleo carnifex, Zygomaturus trilobus,*
Warra Warra Stn., St Ruth's Stn. Bunya	*Macropus pearsoni, Macropus agilis siva,*
Ck, Clifton, Kings Creek Catchment,	*Macropus ferragus, Macropus piltonensis,*
Pilton, Freestone Ck, Gowrie, Gore	*Macropus rama, Diprotodon optatum,*
(Cement Mills), Ned's Gully, Spring Ck,	*Propleopus oscillans, Palorchestes azeal,*
(Warwick) Glengallan Ck	*Lasiorhinus augustidens, Simosthenurus pales, Sthenurus atlas, Simosthenurus orientalis, Metasthenurus newtonae, Sthenurus tindalei, Sthenurus andersoni, Sthenurus atlas, Protemnodon anak, Phascolonus gigas, Protemnodon brehus, Protemnodon roechus, Protemnodon anak, Propleopus oscillans, Procoptodon pusio, Procoptodon goliah, Procoptodon rapha, 'Procoptodon' oreas, 'Procoptodon' browneorum, Megalania prisca, Phascolarctos stirtoni, Troposodon minor, Troposodon kenti, Phascolomys medius, Ramsaya magna*
Eulo (Bundoona Stn)	*Diprotodon optatum, Protemnodon ?, megalania prisca, Pallimnarchos pollens*
Flinders R., Maxwelton	*Diprotodon optatum*

(Continued)

(Continued)

Sites	Species
Floraville Downs Stn (Leichardt R.)	*Pallimnarchus pollens*
Glen Garland (Cape York)	*Pallimnarchos pollens, Diprotodon optatum, Phascolonus gigas*
Glenhaughton, Taroom	*Diprotodon optatum*
Gogango Ck. nr Rockhampton	*Diprotodon optatum*
Gowrie	*'Procoptodon' browneorum, Procoptodon pusio, Sthenurus andersoni*
Jimboomba	*Protemnodon roechus, Sthenurus agilis siva*
Jimbour Ck (north Dalby)	*Procoptodon rapha*
Lansdowne Stn., Tambo	*Sthenurus agilis siva, Pallimnarchos pollens, Palorchestes azeal Protemnodon anak*
Mt Surprise	*Diprotodon optatum*
Macalister, Condamine R.	*Procoptodon pusio*
Marmor	*Thylacoleo carnifex, Protemnodon anak, Phascolarctos stirtoni, Diprotodon optatum*
Maryvale Ck, 150 km west	*Thylacoleo carnifex, Diprotodon optatum*
Monto	*Protemnodon roechus*
MucKadilla	*Diprotodon optatum*
Murgon	*Diprotodon optatum*
Orion Downs Stn, Springsure	*Megalania prisca*
Peak Downs Stn, Capella	*Thylacoleo carnifex*
Pelican Ck. Myross, Aramac	*Diprotodon optatum*
Planet Downs, Gulf	*Protemnodon brehus*
Riversleigh, Terrace Site L.F.	*Diprotodon optatum, Palorchestes azeal, Sthenurus sp., Pallimnarchos pollens*
Roma	*Diprotodon optatum, Sthenurus agilis siva*
Rosella Plains, Cairns	*Pallimnarchos pollens*
Rosewood	*Protemnodon roechus*
Sandhurst Ck, Fernlees	*Diprotodon optatum*
Sellheim R.	*Diprotodon optatum*
Springsure	*Megalania prisca*
Sutton Site, Kings Ck	*Diprotodon optatum, Megalania prisca, Protemnodon roechus, Protemnodon anak, Protemnodon brehus, Macropus siva, Macropus, pearsoni, Troposodon minor*
Texas Caves	*Simosthenurus maddocki, Procoptodon rapha*
Winton, Belmont Stn.	*Diprotodon optatum*
Winton, Eskdale Stn.	*Megalania prisca, Sthenurus sp.*
Wyandotte Stn/Ck	*Megalania prisca, Meiolania platyceps, Thylacoleo carnifex, Diprotodon optatum*

New South Wales

Abercrombie Caves	*Protemnodon anak*
Attunga Springs, Tamworth	*Protemnodon anak, Procoptodon rapha*
Bathurst (Cor Flat Quarries)	*Simosthenurus occidentalis*

(Continued)

(Continued)

Sites	Species
Bearbung, Gilgandra	*Diprotodon optatum*
Bingara Bone Camp Gully	*Sthenurus* sp., *Zygomaturus trilobus*, *Thylacoleo carnifex*, *Diprotodon optatum*, *Phascolonus* sp., *Simosthenurus pales*, *Sthenurus tindalei*, *Sthenurus andersoni*, *Protemnodon brehus*, *Procoptodon pusio*, *Procoptodon goliah*, *Procoptodon rapha*, *Simosthenurus maddocki*
Binnaway, Castlereagh R.	*Procoptodon goliah*
Bombala, Jincumbilly	*Simosthenurus pales*
Breeza nr Tamworth	*Protemnodon anak*
Bunyan siding, Cooma	*Sthenurus atlas, Procoptodon rapha*
Cox's Ck	*Diprotodon optatum*
Cuddie Springs	*Megalania prisca, Meiolania* sp., *Pallimnarchos pollens, Palorchestes azeal, Genyornis newtoniDiprotodon optatum*, *Simosthenurus pales, Phascolonus gigas*, *Protemnodon* sp., *Sthenurus* sp.
Cunningham Ck, Harden	*Protemnodon brehus*
Geurie	*Simosthenurus pales*
Gilgandra, Macquarie R.	*Simosthenurus pales*
Goodravale Caves (Monaro)	*Thylacoleo carnifex*
Huntsgrave , Lake Keepit	*Protemnodon anak*
Lake George	*Procoptodon goliah*
Lake Popilta	*Genyornis newtoni*
Lakes Menindee/Tandou	*Thylacoleo carnifex, Diprotodon optatum*, *Procoptodon goliah, Protemnodon anak*, *'Procoptodon' oreas, Sthenurus atlas*, *Sthenurus tindalei, Propleopus oscillans*, *Protemnodon brehus, Sthenurus andersoni*, *Sarcophilus* sp., *GE*
Lake Victoria	*Thylacoleo carnifex, Diprotodon optatum*, *Zygomaturus trilobus, Macropus ferragus*, *Protemnodon anak, Protemnodon brehus*, *Procoptodon goliah, 'Procoptodon' oreas*, *Sthenurus andersoni, Sthenurus murrayi*, *Sthenurus tindalei, Sthenurus atlas*, *Phascolonus gigas, Protemnodon brehus*
Lime Springs	*Diprotodon optatum, Procoptodon* sp., *Protemnodon* sp., *Sthenurus* sp.
Mendooran	*Metasthenurus newtonae*
Molong	*'Procoptodon' oreas*
Monaro, Teapot Ck	*Simosthenurus occidentalis, Procoptodon pusio*
Mooki R. and Cox's Ck	*Diprotodon optatum, Palorchestes azeal*, *Phascolonus gigas, Protemnodon* sp., *Sthenurus andersoni, Thylacoleo carnifex*

(Continued)

(Continued)

Sites	Species
Moree	*Thylacoleo carnifex*
Mt Fairy Caves	*Simosthenurus occidentalis*, *'Procoptodon' gilli*
Reddestone Ck., Glen Innes	*Thylacoleo carnifex*, *Protemnodon anak*, *'Simosthenurus' pales*, *Diprotodon optatum*, *Propleopus oscillans*, *Phascolonus* sp., *Procoptodon pusio*, *Procoptodon rapha*, *Procoptodon goliah*
Russenden Cave	*Protemnodon brehus*, *Zygomaturus trilobus*, *Sthenurus tindalei*, *Procoptodon pusio*, *Simosthenurus occidentalis*, *Sthenurus atlas*, *Sthenurus agilis siva*
Rylstone, Kandos Quarry	*Metasthenurus newtonae*
Talbragar	*Troposodon minor*
Tambar Springs	*Diprotodon* sp., *Macropus* sp., *Sthenurus andersoni*
Tocumwal	*Procoptodon goliah*
Warrah Ck, Quirindi	*Protemnodon anak*, *Megalania prisca*
Wellington Caves: Bone Cave, Mitchell Cave, Cathedral Cave	*Thylacoleo carnifex*, *Zygomaturus trilobus*, *Sthenurus agilis siva Simosthenurus orientalis,Simosthenurus occidentalis*, *Simosthenurus brachyselenis*, *Sthenurus andersoni*, *Sthenurus atlas*, *Simosthenurus pales*, *Metasthenurus newtonae*, *Protemnodon anak*, *Troposodon minor*, *Diprotodon optatum*, *Palorchestes azeal*, *Sarcophilus* sp., *Megalania prisca*, *,Procoptodon goliah*, *Propleopus oscillans*, *Protemnodon brehus*, *Procoptodon pusio*, *Procoptodon rapha*, *'Procoptodon' gilli*, *'Procoptodon' williamsi*, *Ramsaya magna*, *Genyornis newtoni*
Willandra Lakes	*Procoptodon goliah*, *Protemnodon* sp., *Zygomaturus trilobus*, GE
Willow Point	*Procoptodon goliah*
Wombeyan Caves	*Thylacoleo carnifex*, *Simosthenurus maddocki*

Victoria

Bacchus Marsh (Werribee R.)	*Diprotodon optatum*, *Simosthenurus occidentalis*
Buchan Caves (Foul Air, Pot Luck, Royal, Trogdip, Didgeridoo)	*Thylacoleo carnifex*, *Phascolarctos* sp., *Simosthenurus occidentalis*, *'Procoptodon' gilli*
Cloggs Cave	*Thylacoleo carnifex*, *Simosthenurus orientalis*
Colac	*Diprotodon optatum*, *Procoptodon rapha*
Duck Ponds	*Thylacoleo carnifex*
Footscray	*Protemnodon brehus*, *Protemnodon anak*, *Diprotodon optatum*, *Zygomaturus trilobus*

(Continued)

(*Continued*)

Sites	Species
Hamilton	*Troposodon minor*
Keilor	*Protemnodon anak, Protemnodon brehus, Diprotodon optatum Thylacoleo carnifex, Zygomaturus trilobus*
Kilcunda	*Simosthenurus occidentalis*
Lake Colongulac	*Diprotodon optatum, Zygomaturus trilobus,Thylacoleo carnifex, Procoptodon* sp., *Procoptodon rapha, Protemnodon anak, Protemnodon brehus, Simosthenurus* sp.
Lake Corangamite	*Thylacoleo carnifex, Diprotodon* sp., *Zygomaturus* sp., *Phascolonus gigas ,*
Lake Omeo	*Diprotodon optatum*
Lake Weeranganuk	*Phascolonus gigas, Protemnodon brehus*
Lancefield	*Thylacoleo carnifex, Protemnodon anak, Protemnodon brehus, Procoptodon rapha, 'Procoptodon' gilli, Propleopus oscillans, Simosthenurus occidentalis, Simosthenurus maddocki, Diprotodon optatum,Genyornis newtoni*
Limeburner's Point	*Simosthenurus maddocki*
McEachern's Cave	*'Procoptodon' gilli, Zygomaturus trilobus, Simosthenurus occidentalis, Metasthenurus newtonae, Procoptodon' browneorum*
Mt Fairy, Dolomite Quarry	*Simosthenurus maddocki*
Mt Hamilton Lava Cave	*Thylacoleo carnifex, Protemnodon roechus*
Minhamite, Spring Ck	*Simosthenurus occidentalis*
Fisherman's Cliff (Murray River)	*Sthenurus tindalei*
Murrindal (Honeycomb Cave)	*Simosthenurus occidentalis*
Nepean	*Zygomaturus trilobus, Palorchestes azeal, Protemnodon* sp.
Pejark Marsh	*Diprotodon optatum, Thylacoleo carnifex, Palorchestes azeal*
Puralka Caves	*Zygomaturus trilobus, Palorchestes azeal, Simosthenurus occidentalis*
Spring Ck.	*Protemnodon brehus, Protemnodon anak, Diprotodon optatum, Palorchestes azeal, Zygomaturus trilobus, 'Procoptodon' gilli, Simosthenurus occidentalis*
Strathdownie	*'Procoptodon' gilli, Simosthenurus occidentalis*
Talbot, Back Ck	*Diprotodon optatum*
Tocumwal	*Procoptodon goliah*
Warrnambool	*Thylacoleo carnifex, Genyornis newtoni* (footprint), *Procoptodon' browneorum, 'Procoptodon' gilli*
Watch Hill, Lake Corangamite	*Thylacoleo carnifex, Diprotodon optatum*

(*Continued*)

(*Continued*)

Sites	Species
South Australia	
Adelaide area	*Phascolonus* sp., *Diprotodon optatum*, *Palorchestes azeal*
Arcoona Stn, Rocky Ck, Lake Richardson, Red Lake	*Genyornis newtoni*, *Megalania prisca*, *Diprotodon* sp.
Arnold's Bank	*Diprotodon optatum*
Billa Kallina	*Genyornis newtoni*
Billeroo Creek, Frome Downs Stn	*Diprotodon optatum*, *Procoptodon goliah*, '*Procoptodon*' *oreas Propleopus oscillans*, *Thylacoleo carnifex*, *Macropus ferragus Sthenurus tindalei*, *Sthenurus stirlingi*, GE
Blackford Drain	*Diprotodon optatum*
Brothers Is. (Coffin Bay)	*Genyornis newtoni*, *Simosthenurus maddocki*, *Simosthenurus occidentalis*
Buckalowie Caves, Hawker	*Thylacoleo carnifex*
Burra District, Baldina Ck, Burra Ck, Bundey, Yam Ck, Ulooloo,	*Thylacoleo carnifex*,*Genyornis newtoni*, *Diprotodon optatum*, *Sthenurus atlas Metasthenurus newtonae*, *Sthenurus tindalei*, *Procoptodon goliah*
Calca (Streaky Bay)	*Procoptodon goliah*, *Sthenurus* sp.
Chowilla	*Phascolonus* sp
Chucka Bend	*Phascolonus* sp.
Linton, Yorke Peninsula	*Sthenurus tindalei*
Collinsville Stn, Newikie Ck.	*Diprotodon optatum*, *Procoptodon goliah*
Cootanoorina	*Diprotodon optatum*
Cooper Ck: Unkamilka, Tilla Tilla Cannatalkaninna, Piaranni , Illurunna, Malkuni, Margoona, Katipiri, Pataramordu, Unduwumpa, Cooper Sandbars, Kutjitara West Bluff, Cooper's Last	*Procoptodon goliah*, *Protemnodon brehus*, *Protemnodon anak*, *Sthenurus tindalei*, *Sthenurus stirlingi*, '*Procoptodon*' *browneorumi*,*Simosthenurus occidentalis*, '*Procoptodon*' *oreas*, *Procoptodon rapha*, *Diprotodon optatum*, *Thylacoleo carnifex*, *Megalania prisca*, *Genyornis newtoni* , *Phascolonus gigas*, *Phascolonus medius*, *Pallimnarchos pollens*, '*Procoptodon*' *williamsi*
Curramulka Quarry, York Pen. Town Well Cave	*Sthenurus atlas*, *Sthenurus stirlingi*, *Sthenurus andersoni*, *Metasthenurus newtonae*, *Simosthenurus maddocki*, *Simosthenurus occidentalis*, *Simosthenurus cf antiques*, *Procoptodon goliah*, *Procoptodon rapha*
Dempsey's Lagoon (Port Augusta)	*Genyornis newtoni* , *Diprotodon optatum*, *Protemnodon brehus*, *Sthenurus tindalei*, *Sthenurus andersoni*
Hookina Ck. (Parachillna)	*Diprotodon optatum*,*Genyornis newtoni*, *Propleopus oscillans*

(*Continued*)

(*Continued*)

Sites	Species
InnamincKa	*Megalania prisca*
Kallakoopah Ck: Murduwadinna ,	*Diprotodon optatum, Procoptodon*
Sleeping Dog Cliffs, Sleeping Dog	*goliah, Megalania prisca, Sthenurus*
Cliffs west, 'Ethiopia', Sam's Corner,	*tindalei, Sthenurus anderson , Sthenurus*
Jaws 1 & 2, First Bend, Dingo Bone	*atlas,Genyornis newtoniPhascolonus medius,*
Beds, Maroon Dog Bend, Dune Bluff	*Pallimnarchos pollens*
Bend	
Kangaroo Is. Seton rock shelter . Kelly	*Diprotodon optatum, „Simosthenurus*
Hill caves, Emu, Fossil and Mt Taylor	*occidentalis, 'Procoptodon' browneorum,*
caves, Rocky R.	*Zygomaturus trilobus, 'Procoptodon' gilli,*
	Protemnodon sp.
Kapunda	*Phascolonus* sp.
Kiana Cliff, Mt Misery, Eyre Pen	*Metasthenurus newtonae, Simosthenurus*
	occidentalis, Procoptodon' browneorum
Kingoonya	GE
Kingston, Blackford Ck	*Simosthenurus occidentalis*
Kyancutta (Eyre Penn.)	*Diprotodon optatum*
Lake Callabonna	*Diprotodon optatum, 'Procoptodon' oreas,*
	Sthenurus tindalei, Sthenurus andersoni,
	Sthenurus stirlingi, Protemnodon brehus,
	Phascolonus gigas, GE
Lake Eyre desert sites: Lakes Clayton,	*Diprotodon optatum, Megalania*
Hydra, Fly, Kalamurra, Kutjitara,	*prisca,Genyornis newtoniProcoptodon goliah,*
Clark, Horseshoe, Lost Shovel, 'Frank',	*Sthenurus tindalei*
Billicoorinna Ck, Giffbow and 'Bob'	
Sites, Madigan Gulf	
Lake Fowler (York Pen.)	*Thylacoleo carnifex*
Lake Kuturu, Frome Downs Stn	*Genyornis newtoni*
Lake Millyera, Frome Downs Stn	*Genyornis newtoni*
Marree (Hergot Springs)	*Diprotodon optatum*
Millicent	*Zygomaturus trilobus, Simosthenurus*
	maddocki
Morgan	*Phascolonus* sp.
Mt Eyre	*Diprotodon optatum*
Mt Gambier area: Drowned Cave,	*Genyornis newtoni, Propleopus oscillans,*
Kilsby's Hole, Wandillo Forest ave,	*Simosthenurus maddocki, Metasthenurus*
Moorak, Derrington and Grey Sts,	*newtonae, Sthenurus agilis siva, Diprotodon*
	optatum, 'Procoptodon' gilli, Simosthenurus
	occidentalis,Simosthenurus pales
Mt Shank, Goulden's Hole, Tankstand	*Metasthenurus newtonae,Simosthenurus*
Cave	*maddocki, 'Procoptodon' gilli*
Murnpeowie	*Diprotodon optatum*
Murrapaterinna (Mulka and L. Kan)	*Diprotodon optatum, Zygomaturus trilobus,*
	Macropus sp., *Protemnodon* sp.

(*Continued*)

(Continued)

Sites	Species
Naracoorte Caves: Victoria, Bat, Haystall, Alexandra Blanche, Fox, Specimen, Cathedral, SOS, Sand Funnel, Heschke's , Crawford's, Cornucopia, Wandillo, Robertson's, Specimen, Wet (Tomato Stick) , Possum, Rabbit, Wombat, Brown Snake, Jame's Quarry, Buckridge , Cheese and Putty Caves	*Congruus congruous,Genyornis newtoni Macropus greyi, Macropus rufogresius, Palorchestes azeal, Procoptodon goliah, Procoptodon gilli, Procoptodon rapha, Propleopus oscillans, Protemnodon anak, Protemnodon roechus, 'Simosthenurus' baileyi, 'Procoptodon' browneorum, 'Procoptodon' gilli,Simosthenurus maddocki, Simosthenurus newtonae, Simosthenurus occidentalis, 'Simosthenurus' pales, Sthenurus agilis siva, Sthenurus andersoni, Sthenurus atlas, Metasthenurus newtonae, Thylacoleo carnifex, Wallabia bicolor, Wonambi naracoortensis, Zygomaturus trilobus*
Nectar Brook	*Diprotodon optatum, Macropus sp.*
Normanville	*Diprotodon optatum*
Old Calca Stn, Eyre Pen.	*Procoptodon goliah, 'Procoptodon' browneorum*
Orroroo, Boolcunda Ck, Black Rock Gravel Pit, Pekina Ck	*Diprotodon optatum,Zygomaturus trilobus, Procoptodon goliah, Metasthenurus newtonae, Protemnodon sp.*
Pandie Pandie	*Diprotodon optatum*
Penola/Green Waterhole, Comaum Forest Cave, Monbulla Cave	*Genyornis newtoni , Sthenurus sp., Metasthenurus newtonae, Simosthenurus pales, Thylacoleo carnifex, Simosthenurus occidentalis, Procoptodon' browneorum, 'Procoptodon' gilli*
Pernatty Lagoon	*Diprotodon optatum*
Pitcairn Stn, NacKara	*Procoptodon goliah*
Port Pirie Gravel Pit, Bairstow's pit	*Sthenurus stirlingi, Procoptodon goliah*
Salt Ck, Normanville	*Diprotodon optatum, Phascolonus gigas, Thylacoleo carnifex Genyornis newtoni, Sthenurus andersoni*
Tantanoola Caves: Submerged, Green, Waterhole, Glencoe, Mt Burr, Millicent, Morgan's Cave	*Metasthenurus newtonae, Simosthenurus occidentalis,, 'Procoptodon' browneorum, 'Procoptodon' gilli*
Teetulpa, Brady's Gully	*Procoptodon rapha, Procoptodon goliah*
Terowie, Waupunya Ck,	*Procoptodon goliah*
Two Wells, Gawler	*Diprotodon optatum*
Warburton/Diamantina River: Cowarie, New Kalamurina, Cassidy (Piacoonannie), Toolapinna West, Green Bluff/Marcus, Lookout Site, Old Kalamurina, Toolapinna, Mulyanna,	*Procoptodon goliah,Diprotodon optatum, Zygomaturus trilobus, Megalania prisca, Procoptodon rapha, Protemnodon brehus, Protemnodon anak, 'Procoptodon' oreas, Simosthenurus orientalis, Simosthenurus*

(Continued)

(*Continued*)

Sites	Species
Punkrakadarinna, Keekalanna, Lake Kujitara, Wild Dog Waterhole, Lower Quana	*occidentalis, Simosthenurus pales, Sthenurus atlas, Sthenurus tindalei, Sthenurus stirlingi, Sthenurus andersoni, Phascolonus gigas and medius, Genyornis newtoni Meiolania sp., Pallimnarchos pollens, Wonambi naracoortensis*
Welcome Springs	*Diprotodon optatum*
Whydown Stn	*Procoptodon goliah*
Wood Point	*Genyornis newtoni*
Woomera	*Diprotodon optatum*
Yalpara, Hillpara Ck. nr Orroroo	*Diprotodon optatum*
Yunta, Whydown Stn	*Procoptodon goliah*

Western Australia

Billabalong, Murchison R.	*Zygomaturus trilobus*
Devil's Lair, Turner Brook and Skull Caves	*Thylacoleo carnifex, Macropus fuliginosus, Protemnodon brehus, Zygomaturus trilobus, Sthenurus brownie, Procoptodon' browneorum, Simosthenurus occidentalis, Phascolarctos cinerus, Wonambi naracoortensis*
Dingo Rock, Karonie (Goldfields)	*Diprotodon optatum*
Du Boulay Ck Dampier (Pilbara)	*Diprotodon optatum*
Foundation Cave	*Simosthenurus occidentalis, Procoptodon' browneorum*
Gingin, McIntyre's Cave	*Simosthenurus pales*
Greenough R.	*Zygomaturus trilobus*
Karatha	*Diprotodon optatum*
Koala Cave, Yanchep	*'Procoptodon' browneorum*
Kudjal Yolga Cave	*Macropus fuliginosus, Protemnodon sp., Simosthenurus occidentalis, Procoptodon' browneorum*
Labyrinth Cave	*Thylacoleo carnifex, 'Procoptodon' browneorum*
Leinster or 68 km west of L. Darlot	*Diprotodon optatum*
Macintyre Gully (Perth area)	*Zygomaturus trilobus*
Mammoth Cave	*Thylacoleo carnifex, Zygomaturus trilobus, Protemnodon brehus, 'Procoptodon' browneorum, Simosthenurus occidentalis, Palorchestes azeal,Genyornis newtoniWonambi naracoortensis, Macropus fuliginosus*
Melaleuca Cave, Yanchep	*Simosthenurus occidentalis*
Moondyne Cave	*Zygomaturus trilobus, Macropus fuliginosus, 'Procoptodon' browneorum*

(*Continued*)

(Continued)

Sites	Species
Nullarbor sites: Balladonia soak, Madura, Weebubie, Thylacoleo Caves (Leaena's Breath, Flightstar and Last Tree), Lindsay Hall Cave	*Thylacoleo carnifex, Sthenurus tindalei, Metasthenurus newtonae, Simosthenurus maddocki, Protemnodon brehus, Phascolonus gigas, Baringa* sp., *Bohra* sp., *Congruus kitcheneri, Macropus ferragus, Procoptodon goliah, Procoptodon browneorum, 'Procoptodon' williamsi, Protemnodon roechus, Sthenurus andersoni, 'Procoptodon' gilli*
Oakover River	*Diprotodon optatum*
Strong's Cave*	*Zygomaturus trilobus, Sthenurus* sp., *Simosthenurus occidentalis, Procoptodon' browneorum*
Terrible Cave	*Simosthenurus occidentalis*
Tight Entrance Cave	*Macropus fuliginosus, Simosthenurus occidentalis, Metasthenurus newtonae, Simosthenurus pales, Procoptodon' browneorum*
Wanneroo (Perth)	*'Procoptodon' browneorum*
Windjana Gorge and Tunnel Ck.	*Diprotodon optatum*
Wonberna Cave	*Diprotodon optatum, Thylacoleo carnifex, Sthenurus atlas, Phascolonus gigas*

Tasmania

Beginner's Luck Cave, Florentine Valley Titan's Shelter	*Simosthenurus occidentalis, Zygomaturus trilobus, Sthenurus* sp.
Egg Lagoon, King Island	*Thylacoleo carnifex, Diprotodon optatum*
Montagu and Scotch Town Caves, Main cave	*Simosthenurus occidentalis, Sthenurus andersoni, Simosthenurus orientalis, Protemnodon anak, Zygomaturus trilobus, Palorchestes azeal, Thylacoleo carnifex, Metasthenurus newtonae*
Mowbray, Pulbeena Swamp, Smithton	*Phascolonus* sp., *Zygomaturus trilobus, Palorchestes azeal*
Pleisto Scene Cave	*Simosthenurus occidentalis*
South East Lagoon, King Island	*Diprotodon optatum, Protemnodon anak, Simosthenurus occidentalis*
Surprise Bay, King Island	*Diprotodon optatum, Simosthenurus occidentalis*

Northern Territory

Katherine	*Diprotodon optatum*
Lake Lewis	*Genyornis newtoni*

Key* – Marsupial remains have been recovered from at least five caves in the southwest of WA and 30 caves in the Eucla region (Merrilees, 1967).

Bibliography

Abi-Rached, L., Jobin, M. J., McWhinnie, L. A., Green, R. E., Norman, P. J., & Parham, P. (2011). The shaping of modern human immune systems by multiregional admixture with archaic humans. *Science, 334*, 89–94.

Akerman, K. (1973). Two aboriginal charms incorporating fossil giant marsupial teeth. *The Western Australian Naturalist, 12*, 139–141.

Akerman, K. (1998). A rock painting, possibly of the now extinct marsupial *Thylacoleo* (marsupial lion). From the north Kimberley, Western Australia. *The Beagle, Records of the Museums and Art Galleries of the Northern Territory, 14*, 117–121.

Akerman, K. (2009). Interaction between humans and megafauna depicted in Australian rock art? *Antiquity, 083*, 319 (December).

Akerman, K., & Willing, T. (2009). An ancient rock painting of a marsupial lion, *Thylacoleo carnifex*, from the Kimberley, Western Australia. *Antiquity, 083*, 319 (March).

Ambrose, S. H. (1998). Late Pleistocene human population bottlenecks, volcanic winter, and the differentiation of modern humans. *Journal of Human Evolution, 34*, 623–651.

Ambrose, S. H. (2003). Did the supereruption of Toba cause a human population bottleneck? Reply to Gathorne-Hardy and Harcourt-Smith. *Journal of Human Evolution, 45*, 231–237.

Anderson, D. E., Goudie, A. S., & Parker, A. G. (2008). *Global environments through the Quaternary.* UK: Oxford University Press Oxford, UK.

Archer, M. (1984). The Australian marsupial radiation. In M. Archer & G. Clayton (Eds.), *Verteberate zoogeography and evolution in Australasia* (pp. 633–808). Australia: Hesperian Press, Dalkieth, Western Australia.

Archer, M., Hand, S. J., & Godthelp, H. (1991). *Riversleigh.* Sydney: Reed Books.

Auffenberg, W. (1972). Komodo dragons. *Natural History, 81*, 52–59.

Ayliffe, L. K., Marianelli, P. C., Moriarty, K. C., Wells, R. T., McCulloch, M. T., & Mortimer, G. E., et al. (1998). 500 ka precipitation record from southeastern Australia: Evidence for interglacial relative aridity. *Geology, 26*, 147–150.

Barnosky, A. D., Koch, P. L., Frenec, R. S., Wing, S. L., & Shabel, A. B. (2004). Assessing the causes of late Pleistocene extinctions on the continents. *Science, 306*, 70–75.

Bertrand, R., Lenoir, J., Peidallu, C., Riofrio-Dillon, G., de Ruffray, P., & Vidal, C., et al. (2011). Changes in plant community composition lag behind climate warming in lowland forests. *Nature, 479*, 517–520.

Bowler, J. M., Wyrwoll, K. -H., & Lu, Y. (2001). Variations of the northwest Australian summer monsoon over the last 300,000 years: The paleohydrological record of the Gregory (Mulan). Lakes system. *Quaternary International, 83–85*, 63–80.

Bowler, J. M., Duller, D. A. T., Perret, N., Prescott, J. R., & Wyroll, K. (1998). Hydrologic changes in monsoonal climates of the last glacial cycle: Stratigraphy and luminescence dating of Lake Woods, NT, Australia. *Palaeoclimates, 3*, 179–207.

Brook, B. W., & Bowman, M. J. S. (2002). Explaining the Pleistocene megafaunal extinctions: Models, chronologies, and assumptions. *Proceedings National Academic of Sciences, 99*, 14624–14627.

Brown, J. H. (1971). Mammals on mountaintops: Nonequilibrium insular biogeography. *American Naturalist, 105*, 467–478.

Brown, J. H. (1978). The theory of insular biogeography and the distribution of boreal birds and mammals: A symposium. *Great Basin Naturalist Memoirs, 2*, 209–227.

Callen, R. A., & Nanson, G. C. (1992). Formation and age of dunes in the Lake Eyre depocentres. *Geologische Rundschau, 81*, 589–593.

Cann, J. (1998). *Australian freshwater turtles* (p. 292). Singapore: Beaumont Publishing.

Casteel, R. W. (1976). *Fish remains in archaeology and palaeo-environmental studies* (p. 254). New York, NY: Academic Press.

Chaimanee, Y. (2007). Vertebrate records: Late Pleistocene of Southeast Asia. In S. A. Elias (Ed.), *Encyclopedia of Quaternary science* (pp. 3189–3197). USA: Elsevier.

Chaloupka, G. (1993). *Journey in Time*. Sydney: Reed Books. pp. 100–101.

Chappell, J. (1991). Late Quaternary environmental changes in eastern and central Australia, and their climatic interpretation. *Quaternary Science Reviews, 10*, 377–390.

Chen, X. Y., Bowler, J. M., & Magee, J. W. (1993). Late Cenozoic stratigraphy and hydrologic history of Lake Amadeus, a central Australian playa. *Australian Journal of Earth Sciences, 37*, 93–102.

Choquenot, D., & Bowman, D. M. (1998). Marsupial megafauna, Aborigines and the overkill hypothesis: Application of predator-prey models to the question of Pleistocene extinction in Australia. *Global Ecology and Biogeography Letters, 7*, 167–180.

Cohen, T. J., Nanson, G. C., Jansen, J. D., Jones, B. G., Jacobs, Z., & May, J. -H., et al. (2012). Late Quaternary mega-lakes fed by the northern and southern river systems of central Australia: Varying moisture sources and increased continental aridity. *Palaeogeography, Palaeoclimatology, Palaeoecology, 256–357*, 89–108.

Costelloe, J. F., Grayson, R. B., & McMahon, T. A. (2006). Modelling streamflow in a large anastomosing river of the arid zone, Diamantina River, Australia. *Journal of Hydrology, 323*, 138–153.

Croke, J. C., Magee, J. M., & Wallensky, E. P. (1999). The role of the Australian Monsoon in the western catchment of Lake Eyre, central Australia, during the last interglacial. *Quaternary International, 57*, 71–80.

Dansgaard, W., Johnsen, S. J., Clausen, H. B., Dahl-Jensen, D., Gundestrup, N. S., & Hammer, C. U., et al. (1993). Evidence for general instability of past climate from the 250-kyr ice-core record. *Nature, 364*, 218–220.

Dekker De, P., Norman, M., Goodwin, I. D., Wain, A., & Gingele, F. X. (2010). Lead isotopic evidence for an Australian source of Aeolian dust to Antarctica at times over the last 170,000 years. 170,000 years. *Palaeogeography, Palaeoclimatology, Palaeoecology, 285*, 205–223.

Derevianko, A. P., & Shunkov, M. V. (2009). Development of Early Human Culture in Northern Asia. *Paleontological Journal, 43(8)*, 881–889.

DeVogel, S. B., Magee, J. W., Manley, W. F., & Miller, G. H. (2004). A GIS-based reconstruction of late Quaternary paleohydrology: Lake Eyre, arid central Australia. *Palaeogeography, Palaeoclimatology, Palaeoecology, 204*, 1–13.

English, P., Spooner, N. A., Chappell, J., Questiaux, D. G., & Hill, N. G. (2001). Lake Lewis basin, central Australia: Environmental evolution and OSL chronology. *Quaternary International, 83–85*, 81–101.

EPICA Community Members, Augustin, L., Barbante, C., Barnes, P. R. F., Barnola, J. M., & Bigler, M., et al. (2004). Eight glacial cycles from an Antarctic ice core. *Nature, 429*, 623–628.

EPICA Community Members, Barbante, C., Barnola, J. -M., Becagli, S., Beer, J., & Bigler, M., et al. (2006). One-to-one coupling of glacial climate variability in Greenland and Antarctica. *Nature, 444*, 195–198.

Field, J., & Fullager, R. (2001). Archaeology and Australian megafauna. *Science, 294*, 7a.

Fike, D. A., Grotzinger, J. P., Pratt, L. M., & Summons, R. E. (2006). Oxidation of the Ediacaran Ocean. *Nature, 444*, 744–747.

Fillios, M., Field, J., & Bethan, C. (2010). Investigating human and megafauna co-occurrence in Australian prehistory: Mode and causality in fossil accumulations at Cuddie Springs. *Quaternary International, 211*, 123–143.

Finch, M. E. (1982). The discovery and interpretation of *Thylacoleo carnifex* (Thylacoleonidae, Marsupialia. In M. Archer (Ed.), *Carnivorous marsupials* (pp. 537–551). Sydney: Royal Zoological Society of New South Wales.

Fitzsimmons, K. E., Rhodes, E. J., Magee, J. W., & Barrows, T. T. (2007). The timing of linear dune activity in the Strzelecki and Tirari Deserts, Australia. *Quaternary Science Reviews, 26*, 2598–2616.

Flannery, T. F. (1990). Pleistocene faunal loss: Implications of the aftershock for Australia's past and future. *Archaeology in Oceania, 25*, 45–67.

Flannery, T. F. (1994). *The future eaters*. Sydney: Reed. pp. 117–129.

Flannery, T. F., & Schouten, P. (2001). *A gap in nature*. Melbourne: Text Publishing. pp. 11–25.

Fleagle, J. G. (2010). Out of Africa 1: The first hominin colonization of Eurasia. In J. G. Fleagle, J. J. Shea, A. J. Baden & R. E. Leakey (Eds.), *Vertebrate Paleobiology and Paleoanthropology Series*. London: Springer.

Gaffney, E. S., & McNamara, G. (1990). A meiolaniid turtle from the Pleistocene of northern Queensland. *Memoires of the Queensland Museum, 28*, 107–114.

Gaffney, E. S. (1981). A review of the fossil turtles of Australia. *American Museum Novitates, 2720*, 1–38.

Gaffney, E. S. (1985). *Meiolania platyceps*. The Lord Howe Island horned turtle. In P. V. Rich, G. F. van Tets & F. Knight (Eds.), *Kadimakara: Extinct vertebrates of Australia* (pp. 225–229). Canberra: Pioneer Design Studio.

Gaffney, E. S. (1996). The postcranial morphology of *Meiolania platyceps* and a review of the Meiolaniidae. *Bulletin of the American Museum of Natural History, 229*, 1–165.

Gillespie, R., & Brook, B. W. (2006). Is there a Pleistocene archaeological site at Cuddie Springs? *Archaeology in Oceania, 41*, 1–11.

Gilpin, M. E. (1987). Spatial structure and population vulnerability. In M. E. Soulé (Ed.), *Viable populations for conservation* (pp. 125–140). Cambridge: Cambridge University Press, Cambridge, UK.

Glover, C. J. M. (1989). Aquatic fauna. In C. W. Bonython & A. S. Fraser (Eds.), *The great filling of Lake Eyre in 1974* (pp. 94–96). Adelaide: Royal Geographic Society of Australia (South Australian Branch) Inc.

Gregory, J. W. (1906). *The dead heart of Australia*. (p. 233). London: John Murray.

Gröcke, D. R. (1997). Distribution of C_3 and C_4 plants in the late Pleistocene of South Australia recorded by isotope biogeochemistry of collagen in megafauna. *Australian Journal of Botany, 45*, 607–617.

Hanebuth, T., Stattegger, K., & Grootes, P. M. (2000). Rapid flooding of the sunda shelf a late-glacial sea-level record. *Science, 288*, 1033–1035.

Hansen, J. (2009). *Storms of my grandchildren*. London: Bloomsbury Press.

Harvey, K. R., & Hill, G. J. E. (2003). Mapping the nesting habits of saltwater crocodiles (*Crocodylus porosus*). In melalucca swamp and the Adelaide river wetlands, Northern Territory: An approach using remote sensing and GIS. *Wildlife Research, 30*, 365–375.

Hearty, P. J., Hollin, J. T., Neumann, A. C., O'Leary, M. J., & McCulloch, M. (2007). Global sea-level fluctuations during the last interglaciation (MIS 5e). *Quaternary Science Reviews, 26*, 2090–2112.

Heinrich, H. (1988). Origin and consequences of cyclic ice-rafting in the Northeast Atlantic Ocean during the past 130,000 years. *Quaternary Research, 29,* 142–152.

Hesse, P. P. (1994). The record of continental dust from Australia and Tasman sea sediments. *Quaternary Science Reviews, 13,* 257–272.

Hesse, P. P., & McTainsh, G. H. (2003). Australian dust deposits: Modern processes and the Quaternary record. *Quaternary Science Reviews, 22*(18e19), 2007c2035 (Note: Date changed in text).

Hobbs, T. (1996). Leviathan. In R.Tuck (Ed.), *Revised Student Edition (p.1651).* Cambridge, UK: Cambridge University Press.

Hocknull, S. A. (2005). Ecological succession during the late cainozoic of central eastern Queensland: Extinction of a diverse rainforest community. *Memoirs of the Queensland Museum, 51,* 39–122.

Hocknull, S. (2009). Dragon's paradise lost: Palaeobiogeography, evolution and extinction of the largest-ever terrestrial lizards (Varanidae). *PLoS One.*

Hocknull, S. A., Zhao, J. -x., Feng, Y. -x., & Webb, G. E. (2007). Responses of Quaternary rainforest vertebrates to climate change in Australia. *Earth and Planetary Sciences Letters, 264,* 317–331.

Holliday, R. (2005). Aging and the extinction of large animals. *Biogerontology, 6,* 151–156.

Horton, D. (1984). Red Kangaroos: Last of the Australian megafauna. In P. S. Martin & R. G. Klein (Eds.), *Quaternary extinctions: A prehistoric revolution (pp. 639–680).* Tucson, AZ: University of Arizona Press.

Horton, D. R., & Wright, R. V. S. (1981). Cuts on Lancefield bones: Carnivorous *Thylacoleo,* not humans, the cause. *Archaeology in Oceania, 16,* 73–80.

Huang, W. (1993). The skull, mandible and dentition of giant pandas (*Ailuropoda*). Morphological characters and their evolutionary implications. *Vertebrata PalAsiatica, 31,* 191–207.

Jablonski, N. G., & Whitfort, M. J. (1999). Environmental change during the Quaternary in East Asia and its consequences for mammals. *Records of the Western Australian Museum*(Suppl. *57*), 307–315.

Janis, C. M. (1993). Tertiary mammal evolution in the context of changing climates, vegetation, and tectonic events. *Annual Review of Ecological Systems, 24,* 467–500.

Janis, C. M. (2000). The radiation of the North American endemic ruminants. In E. S. Vrba & G. B. Schaller (Eds.), *Antelopes, Deer, and Relatives: Fossil Record, Behavioral Ecology, Systematics, and Conservation* (pp. 26–37). Yale University Press.

Johnson, B. J., Miller, G. H., Fogel, M. L., Magee, J. W., Gagan, M. K., & Chivas, A. R. (1999). 65,000 years of vegetation change in central Australia and the Australian summer monsoon. *Science, 284,* 1150–1152.

Johnson, C. N. (2002). Determinants of loss of mammal species during the late Quaternary 'megafauna' extinctions: Life, history and ecology, but not body size. *Proceedings of the Royal Society London B, 269,* 2221–2227.

Johnson, C. N. (2006). *Australia's mammal extinctions.* Melbourne: Cambridge University Press. pp. 36–53.

Johnson, C. N., & Prideaux, G. J. (2004). Extinctions of herbivorous mammals in the late Pleistocene of Australia in relation to their feeding ecology: No evidence for environmental change as cause of extinction. *Australian Ecology, 29,* 553–557.

Johnson, D. (2004). *The Geology of Australia.* UK: Cambridge University Press.

Jouzel, J., Lorius, C., Petit, J. R., Genthon, C., Barkov, N. I., & Kotlyakov, V. M., et al. (1987). Vostok ice core: A continuous isotope temperature record over the last climatic cycle (160,000 years). *Nature, 329,* 403–408.

Jouzel, J., Waelbroeck, C., Malaize, B., Bender, M., Petit, J. R., & Stievenard, M., et al. (1996). Climatic interpretation of the recently extended Vostok ice records. *Climate Dynamics, 12*, 513–521.

Jouzel, J., Barkov, N. I., Barnola, J. M., Bender, M., Chappellaz, J., & Genthon, C., et al. (1993). Extending the Vostok ice-core record of palaeoclimate to the penultimate glacial period. *Nature, 364*, 407–412.

Kemp, A. (1991). Australian mesozoic and cainozoic lungfish. In P. Vickers-Rich, J. M. Monaghan, R. F. Baird & T. H. Rich (Eds.), *Vertebrate palaeontology of Australia* (pp. 465–489). Melbourne: Monash University Publications.

Kershaw, A. P., & Whitlock, C. (2000). Palaeoecological records of the last glacia-interglacial cycle: Patterns and causes of change. *Palaeogeography, Palaeoclimatology, Palaeoecology, 155*, 1–5.

Kershaw, A. P., van der Kaars, S., & Moss, P. T. (2003b). Late Quaternary Milankovitch-scale climatic change and variability and its impact on monsoonal Australasia. *Marine Geology, 201*, 81–95.

Kershaw, P., Moss, P., & van der Kaars, S. (2003a). Causes and consequences of long-term climatic variability on the Australian continent. *Freshwater Biology, 48*, 1274–1283.

Lambert, F., Delmont, B., Petit, J. R., Bigler, M., Kaufmann, P. R., & Hutterli, M. A., et al. (2008). Dust-climate couplings over the past 800,000 years from the EPICA Dome C ice core. *Nature, 452*, 616–619.

Legler, J. M., & Georges, A. (1993). Biogeography and phylogeny of the Chelonia. In C. G. Glasby, G. J. B. Ross & P. L. Beesley (Eds.), *Fauna of Australia volume 2A amphibia and reptilia, section 18* (pp. 1–9). Canberra: *Australian Government Publishing Service*.

Lévêque, C., & Mounolou, J. -C. (2003). *Biodiversity*. (p. 284). Paris: Wiley.

Lisiecki, L. E., & Raymo, M. E. (2005). A Pliocene–Pleistocene stack of 57 globally distributed benthic $\delta^{18}O$ records. *Palaeoceanography, 20*, 1–17.

Lomolino, M. V., & Channel, R. (1995). Splendid isolation: Patterns of range collapse in endangered mammals. *Journal of Mammalogy, 76*, 335–347.

Long, J. A. (1995). *The Rise of Fishes: 500 Million Years of Evolution*. Sydney: University of New South Wales Press. (p.230). Baltimore, US: Johns Hopkins University Press.

Long, J. (2006). *Swimming in Stone - the Amazing Gogo Fossils of the Kimberley* (p. 320). Fremantle, Western Australia: Fremantle Arts Centre Press.

Long, J., Archer, M., Flannery, T., & Hand, S. (2002). *Prehistoric mammals of Australia and New Guinea* (p. 177). Sydney: UNSW Press.

Long, J., Young, G. C., Holland, T., Senden, T. J., & Fitzgerald, E. M. G. (2006). An exceptional Devonian fish from Australia sheds light on tetrapod origins. *Nature, 444*, 199–202.

Loulergue, L., Schilt, A., Spahni, R., Masson-Delmotte, V., Blunier, T., Lemieux, B., et al. (2008). Orbital and millennial-scale features of atmospheric CH_4 over the past 800,000 years. *Nature, 453*, 383–386.

Louys, J., & Meijaard, E. (2010). Paleoecology of Southeast Asian megafauna-bearing sites from the Pleistocene and a review of environmental changes in the region. *Journal of Biogeography, 2010*, 1–17. Published Online 20th April.

Louys, J., Curnoe, D., & Tong, H. (2007). Characteristics of Pleistocene megafauna extinctions in Southeast Asia. *Palaeogeography, Palaeoclimatology, Palaeoecology, 243*, 152–173.

Luly, J. G. (2001). On the equivocal fate of late Pleistocene *Callitris* Vent. (Cupressaceae). woodlands in arid South Australia. *Quaternary International, 83–85*, 155–168.

MacArthur, R. H., & Wilson, E. O. (1967). *The theory of island biogeography* (p. 224). Princeton, NJ: Princeton University Press.

MacDonald, G. (2003). *Biogeography, space, time and life*. (p. 518). New York, NY: Wiley.

Mackay, A. W. (2009). An introduction to late glacial-Holocene environments. In S. T. Turvey (Ed.), *Holocene extinctions* (p. 352). Oxford University Press.

MacPhee, R. D. (Ed.). (1999). *Extinctions in near time*. New York, NY: Kluwer. pp. 130–171.

Magee, J. W. (1997). *Late Quaternary environments and palaeohydrology of Lake Eyre, arid central Australia*. Canberra: Australian National University. Ph.D. dissertation, p. 295.

Magee, J. W., & Miller, G. H. (1998). Lake Eyre palaeohydrology from 60 ka to the present: Beach ridges and glacial maximum aridity. *Palaeogeography, Palaeoclimatology, Palaeoecology, 144*, 307–329.

Magee, J. W., Bowler, J. M., Miller, G. H., & Williams, D. L. G. (1995). Stratigraphy, sedimentology, chronology and palaeohydrology of Quaternary lacustrine deposits at Madigan Gulf, Lake Eyre, South Australia. *Palaeogeography, Palaeoclimatology, Palaeoecology, 113*, 3–42.

Magee, J. W., Miller, G. H., Spooner, N. A., & Questiaux, D. (2004). Continuous 150 k.y. monsoon record from Lake Eyre, Australia: Insolation-forcing implications and unexpected Holocene failure. *Geology, 32*, 885–888.

Martin, P. (1984). Prehistoric overkill: The global model. In P. S. Martin & R. G. Klein (Eds.), *Quaternary extinctions: A prehistoric revolution* (pp. 354–403). Tucson, AZ: University of Arizona Press.

Martin, P., & Guilday, J. E. (Eds.). (1967). *Pleistocene extinctions: The search for a cause*. New Haven, CT: Yale University Press.

Martin, P. S., & Klein, R. G. (Eds.). (1984). *Quaternary Extinctions: A Prehistoric Revolution*. Tucson: University of Arizona Press.

McBride, J. I., & Keenan, T. D. (1982). Climatology of tropical cyclone genesis in the Australian region. *Journal of Climatology, 2*, 13–33.

McLaren, S., & Wallace, M. W. (2010). Plio–Pleistocene climate change and the onset of aridity in southeastern Australia. *Global and Planetary Change, 71*, 55–72.

McManus, J. F. (2004). Palaeoclimate: A great grand-daddy of ice cores. *Nature, 429*, 611–612.

Meloro, C., Raia, P., & Corotenuto, F. (2008). Diversity and turnover of Plio–Pleistocene large mammal fauna from the Italian Peninsula. *Palaeogeography, Palaeoclimatology, Palaeoecology, 268*, 58–64.

Meltzer, D. J. (2004). Peopling of North America. *Developments in Quaternary Science, 1*, 539–563.

Messel, H., & Vorlicek, G. C. (1989). The *Ecology of* crocodylus porosus *in Northern Australia*: Crocodiles: Their ecology, management, and conservation. Gland, Switzerland: International Union for the Conservation of Nature and Natural Resources. A special publication of the Crocodile specialist group of the species survival commission of the international union for the conservation of nature and natural resources. pp. 164–183.

Miller, G. H., Magee, J. W., Johnson, B. J., Fogel, M. L., Spooner, N. A., & McCulloch, M. T., et al. (1999). Pleistocene extinction of *Genyornis newtoni*: Human impact on Australian megafauna. *Science, 283*, 205–208.

Molnar, R. (1991). Fossil reptiles in Australia. In P. Vickers-Rich, J. M. Monaghan, R. F. Baird & T. H. Rich (Eds.), *Vertebrate palaeontology of Australia* (pp. 606–688). Melbourne: Monash University Publications.

Molnar, R. (2004). *Dragons in the dust* (p. 210). Bloomington, IN: Indiana University Press.

Monaghan, J. M., Baird, R. F., & Rich, T. H. (Eds.), (1991). Melbourne: Monash University Publications.

Morton, S. R., Stafford Smith, D. M., Friedel, M. H., Griffin, G. F., & Pickup, G. (1995). The stewardship of arid Australia: Ecology and landscape management. *Journal of Environmental Management, 43*, 195–218.

Moss, P., & Kershaw, A. P. (2000). The last glacial cycle from the humid tropics of north-eastern Australia: Comparison of a terrestrial and marine record. *Palaeogeography, Palaeoclimatology, Palaeoecology, 155*, 155–176.

Mulvaney, D. J., & Kamminga, J. (1999). *Prehistory of Australia*. UK: Allen and Unwin.

Murray, P. (1984). Extinctions downunder: A bestiary of extinct Australian monotremes and marsupials. In P. S. Martin & R. G. Klein (Eds.), *Quaternary extinctions: A prehistoric revolution* (pp. 600–628). Tucson, AZ: University of Arizona Press.

Murray, P. (1996). The Pleistocene megafauna of Australia. In P. Vickers-Rich., J. M. Monaghan, R. F. Baird & T. H. Rich (Eds.). *Vertebrate Palaeontology of Australasia.* (pp. 1071–1164). Melbourne, Australia: Monash University Publications.

Murray, P., & Chaloupka, G. (1984). The Dreamtime animals: Extinct megafauna in Arnhem Land rock art. *Archaeology in Oceania, 19*, 105–116.

Murray, P., & Vickers-Rich, P. (2004). *Magnificent Mihirungs: The colossal flightless birds of the dreamtime.* Bloomington, IN: Indiana University Press. pp. 300–309.

Nanson, G. C., Callen, R. A., & Price, D. M. (1998). Hydroclimatic interpretation of Quaternary shoreleines on South Australian Playas. *Palaeogeography, Palaeoclimatology, Palaeoecology, 144*, 281–305.

Nanson, G. C., Callen, R. A., & Price, D. M. (1998). Hydroclimatic interpretation of Quaternary shoreleines on South Australian Playas. *Palaeogeography, Palaeoclimatology, Palaeoecology, 144*, 281–305.

Nanson, G. C., Chen, X. Y., & Price, D. M. (1995). Aeolian and fluvial evidence of changing climate and wind patterns during the past 100 ka in the western Simpson Desert, Australia. *Palaeogeography, Palaeoclimatology, Palaeoecology, 113*, 87–102.

Nanson, G. C., Jon East, T., & Roberts, R. G. (1993). Quaternary stratigraphy, geochronology and evolution of the Magela Creek catchment in the monsoon tropics of northern Australia. *Sedimentary Geology, 83*, 277–302.

Nanson, G. C., Price, D. M., & Short, S. A. (1992). Wetting and drying of Australia over the past 300 ka. *Geology, 20*, 791–794.

Nanson, G. C., Price, D. M., Jones, B. G., Maroulis, J. C., Coleman, M., & Bowman, H., et al. (2008). Alluvial evidence for major climate and flow regime changes during the middle and late Quaternary in eastern central Australia. *Geomorphology, 101*, 109–129.

Nanson, G. C., Price, D. M., Jones, B. G., Maroulis, J. C., Coleman, M., & Bowman, H., et al. (2008). Alluvial evidence for major climate and flow regime changes during the middle and late Quaternary in eastern central Australia. *Geomorphology, 101*, 109–129.

Nanson, G. C., Young, R. W., Price, D. M., & Rust, B. R. (1988). Stratigraphy, sedimentology and late Quaternary chronology of the channel country of southwest Queensland. In R. F. Warner (Ed.), *Fluvial geomorphology of Australia* (pp. 151–175). Sydney: Academic Press.

North Greenland Ice Core Project (NGRIP), Andersen, K. K., Azuma, N., Barnola, J. -M., Bigler, M., & Biscaye, P. (2004). High-resolution record of Northern Hemisphere climate extending into the last interglacial period. *Nature, 431*, 147–151.

O'Connell, J. F., & Allen, J. (2004). Dating the colonization of Sahul (Pleistocene Australia–New Guinea): A review of recent research. *Journal of Archaeological Science, 31*, 835–853.

ODP (2008). Ocean Drilling Program. *Final Technical Report 1983-2007* Published by Consortium for ocean Leadership; Lamont-Doherty Earth Observatory, Columbia University, New YorK and the National Science Foundation, USA.

Pack, S. M., Miller, G. H., Fogel, M. L., & Spooner, N. A. (2003). Carbon isotope evidence for increased aridity in northwestern Australia through the Quaternary. *Quaternary Science Reviews, 22,* 629–643.

Palombo, J. G., Alberdi., M. T., & Azanza, B. (2009). How did environmental disturbances affect carnivoran diversity? A case study of the Plio–Pleistocene Carnivora of the North-Western Mediterranean. *Evolutionary Ecology, 23,* 569–589.

Petit, J. R., Jouzel, J., Raynoud, D., Barkov, N. I., Barnola, J. -M., Basile, I., et al. (1999). Climate and atmosphere history of the past 420,000 years from the Vostok ice core, Antarctica. *Nature, 399,* 429–436.

Preston, F. W. (1962). The canonical distribution of commonness and rarity: Part 1. *Ecology, 43,* 185–215.

Price, G. J. (2006). *Pleistocene palaeoecology of the eastern Darling Downs.* University of Queensland. pp. 116–118. Ph.D. dissertation.

Price, G. J. (2008a). Taxonomy and palaeobiology of the largest-ever marsupial, *Diprotodon* Owen, 1838 (Diprotodontidae, Marsupialia). *Zoological Journal of the Linnean Society, 153,* 389–417.

Price, G. J. (2008b). Is the modern koala (*Phascolarctos cinereus*) a derived dwarf of a Pleistocene giant? Implications for testing megafauna extinction hypothesis. *Quaternary Science Reviews, 27,* 2516–2521.

Price, G. J., & Sobbe, I. H. (2005). Pleistocene palaeoecology and environmental change on the Darling Downs, southeastern Queensland, Australia. *Memoirs of the Queensland Museum, 51,* 171–201.

Price, G. J., & Webb, G. E. (2006). Late Pleistocene sedimentology, taphonomy and megafauna extinction on the Darling Downs, southeastern Australia. *Australian Journal of Earth Sciences, 53,* 947–970.

Price, G. J., Webb, G. E., Zhao, J. -x, Feng, Y. -x, Murray, A. S., Cooke, B. N., et al. (2011). Dating megafaunal extinction on the Pleistocene Darling Downs, eastern Australia: The promise and pitfalls of dating as a test of extinction hypotheses. *Quaternary Science Reviews, 30,* 899–914.

Prideaux, G. J. (2000). Tight entrance cave, Southwestern Australia: A late Pleistocene vertebrate deposit spanning more than 180 ka. *Journal of Vertebrate Palaeontology, 203,* 62A–63A.

Prideaux, G. J. (2004). Systematics and evolution of the sthenurine kangaroos. *University of California Publications, Geological Sciences, 146,* 326–355.

Prideaux, G. J., Gully, G. A., Ayliffe, L. K., Bird, M. I., & Roberts, R. G. (2000). Tight entrance cave, southwestern Australia: A late Pleistocene vertebrate deposit spanning more than 180 ka. *Journal of Vertebrate Paleontology, 20*(Suppl. 3), 62A–63A.

Prideaux, G. J., Roberts, R. G., Megirian, D., Westaway, K. E., Hellstrom, J. C., & Olley, J. M. (2007a). Mammalian responses to Pleistocene climate change in southeastern Australia. *Geology, 35,* 33–36.

Prideaux, G. J., Long, J. A., Ayliffe, L. K., Hellstron, J. C., Pillans, B., Boles, W. E., et al. (2007b). An arid-adapted middle Pleistocene vertebrate fauna from south-central Australia. *Nature, 445,* 422–425.

Pushkina, D., & Raia, P. (2008). Human influence on distribution and extinctions of the late Pleistocene Eurasian megafauna. *Journal of Human Evolution, 54,* 769–782.

Quammen, D. (1997). *The song of the dodo* p. 206. London: Pimlico Press.

Reichow, M. K., Pringle, M. S., Al'Mukhamedov, A. I., Allen, M. B., Andreichev, V. L., & Buslov, M. M., et al. (2009). The timing and extent of the eruption of the Siberian Traps large igneous province: Implications for the end-Permian environmental crisis. *Earth and Planetary Science Letters, 277,* 9–20.

Rich, P. (1985). Priest-geologists and knighted explorers. In P. V. Rich, G. F. van Tets & F. Knight (Eds.), *Kadimakara: Extinct vertebratesof Australia* (pp. 17–38). Melbourne, Victoria: Pioneer Design Studio.

Rich, P. V., van Tets, G. F., & Knight, F. (1985). *Extinct Vertebrates of Australia*. Victoria: Pioneer Design Studio.

Roberts, R. G., Jones, R., & Smith, M. A. (1990). Thermoluminescence dating of a 50,000-year-old human occupation site in northern Australia. *Nature, 345*, 153–156.

Roberts, R. G., Jones, R., Spooner, N. A., Head, M. J., Murray, A. S., & Smith, M. A. (1994). The human colonisation of Australia: Optical dates of 53,000 and 60,000 years bracket human arrival at Deaf Adder Gorge, Northern Territory. *Quaternary Science Reviews, 13*, 575–583.

Roberts, R. G., Yoshida, H., Galbraith, R., Laslett, G., Jones, R., & Smith, M. (1998). Single-aliquote and single-grain optical dating confirm thermoluminescence age estimates at Malakunanja II rock shelter in northern Australia. *Ancient TL, 16*, 19–24.

Roberts, R. G., Flannery, T. F., Ayliffe, L. K., Yoshida, H., Olley, J. M., & Prideau, G. J., et al. (2001). New ages for the last Australian megafauna: Continent-wide extinction about 46,000 years ago. *Science, 292*, 1888–1892.

Rohde, R. H., & Muller, R. A. (2005). Cycles in fossil Diversity. *Nature, 434*, 208–210.

Ross, C. A., & Magnusson, W. E. (1989). Living crocodilians. In C. A. Ross (Ed.), *Crocodiles and alligators* (pp. 58–75). Sydney: Golden Press.

Saunders, A., & Reichow, M. (2009). The Siberian traps and the end-Permian mass extinction: A critical review. *Chinese Science Bulletin, 54*, 20–37.

Shaffer, M. L. (1981). Minimum population sizes for species conservation. *BioScience, 31*, 131–134.

Shaffer, M. L. (1987). Minimum viable populations: Coping with uncertainty. In M. E. Soulé (Ed.), *Viable populations for conservation* (pp. 69–82). Cambridge, UK: Cambridge University Press.

Shaffer, M. L. (1990). Population viability analysis. *Conservation Biology, 4*, 39–40.

Shaffer, M. L., & Samson, F. B. (1985). Population size and extinction: A note on determining acritical population sizes. *American Naturalist, 125*, 144–152.

Simanjuntak, T., Semah, F., & Gaillard, C. (2010). The paleolithic of Indonesia: Nature and chronology. *Quaternary International, 223–224*, 418–421.

Slimak, L., Svendsen, J. I., Mangurud, J., Plisson, H., Heggen, H. P., Brugere, A, & Pavlov, P. V. (2011). Late Mousterian Persistence near the Arctic Circle. *Science, 332*, 841–845.

Stirling, E. C. (1896). The newly discovered extinct, gigantic bird of South Australia. *Ibis, 7*, 593.

Stirling, E. C. (1900). Physical features of Lake Callabonna. *Memoirs, Royal Society of South Australia, 1*, I–XV.

Stirling, E. C. (1913). Fossil remains of Lake Callabonna. Part IV. 1. Description of some further remains of *Genyornis newtoni*, Stirling and Zeitz. 2. On the identity of *Phascolomys (Phascolonus). gigas*, Owen, and *Sceparnodon ramsayi*, Owen, with a description of some remains. *Memoirs, Royal Society of South Australia, 1*, 111–178.

Stirling, E. C., & Zeitz, A. H. C. (1896). Preliminary notes on *Genyornis newtoni*; a new genus and species of fossil struthious bird found at Lake Callabonna, South Australia. *Transactions, Royal Society of South Australia, 20*, 171–190.

Stirling, E. C., & Zeitz, A. H. C. (1899). Fossil remains of Lake Callabonna. Part 1. Description of the manus and pes of *Diprotodon australis*, Owen. *Memoirs, Royal Society of South Australia, 1*, 1–40.

Stirling, E. C., & Zeitz, A. H. C. (1900). Fossil remains of Lake Callabonna. 1. *Genyornis newtoni*. A new genus and species of fossil struthious bird. *Memoirs, Royal Society of South Australia, 1*, 41–80.

Stirton, R. A., Tedford, R. H., & Woodburne, M. O. (1967). A new Tertiary formation and fauna from the Tirari Desert, South Australia. *Records of the South Australian Museum, 15*, 427–462.

Stirton, R. A. (1967). The Diprotodontidae from the Ngapakaldi fauna, South Australia. *Bureau Mineral Research, Bulletin, 8*, 1–44.

Stirton, R. A., Tedford, R. H., & Miller, A. H. (1961). Cenozoic stratigraphy and vertebrate palaeontology of the Tirari Desert, South Australia. *Records of the South Australian Museum, 14*, 19–61.

Strahan, R. (2000). *The mammals of Australia* (pp. 53–408). Sydney: Reed New Holland, Australian Museum.

Stuart, A. J., & Lister, A. M. (2012). Extinction chronology of the woolly rhinoceros *Coelodonta antiquitatis* in the context of late Quaternary megafaunal extinctions in northern Eurasia. *Quaternary Science Reviews, 51*, 1–17.

Suppiah, R. (1992). The Australian summer monsoon: A review. *Progress in Physical Geography, 16*, 283–318.

Tedford, R. H., & Wells, R. T. (1990). Pleistocene deposits and fossil vertebrates from the "dead heart of Australia". *Memoirs of the Queensland Museum, 28*, 263–284.

Tedford, R. H., Wells, R. T., & Barghoorn, F. (1992). Tirari formation and contained faunas, Pliocene of the Lake Eyre basin, South Australia. *The Beagle. Records of the Northern Territory Museum of Arts and Sciences, 9*, 173–194.

Tougard, C. (2001). Biogeography and migration routes of large mammal faunas in South-East Asia during the Late Middle Pleistocene focus on the fossil and extant faunas from Thailand. *Palaeogeography, Palaeoclimatology, Palaeoecology, 168*, 337–358.

Tougard, C., & Montuire, S. (2006). Pleistocene paleoenvironmental reconstructions and mammalian evolution in South-East Asia: Focus on fossil faunas from Thailand. *Quaternary Science Reviews, 25126*–25141.

Tresize, P. J. (1971). *Rock art of south-east Cape York.* (No. 24) (pp. 18–32). Canberra: Australian Institute of Aboriginal Studies.

Trueman, C. N. G., Field, J. H., Dortch, J., Charles, B., & Wroe, S. (2005). Prolonged co-existence of humans and megafauna in Pleistocene Australia. *Proceedings National Academy of Science, 102*, 8381–8385.

Turney, C. S. M., Bird, M. I., Fifield, L. K., Roberts, R. G., Smith, M., Dortch, C. E., et al. (2001). Early Human Occupation at Devil's Lair, Southwestern Australia 50,000 Years Ago. *Quaternary Research, 55*(1), 3–13. Available from http://www.sciencedirect.com/science/article/pii/S0033589400921951

Van der Kaars, S., & De Dekker, P. (2002). A late Quaternary pollen record from deep-sea core Fr10/95, GC17 offshore Cape Range Peninsula, northwestern Western Australia. *Review of Palaeobotany and Palynology, 120*, 17–39.

Van der Kaars, S., & De Dekker, P. (2003). Pollen distribution in marine surface sediments offshore Western Australia. *Review of Palaeobotany and Palynology, 124*, 113–129.

Van der Kaars, S., De Dekker, P., & Gingele, F. X. (2006). A 100,000-year record of annual and seasonal rainfall and temperature for northwestern Australia based on a pollen record obtained offshore. *Journal of Quaternary Science, 21*, 879–889.

Vanderwall, R., & Fullagar, R. (1989). Engraved *Diprotodon* tooth from the Spring Creek locality, Victoria. *Archaeology in Oceania, 24*, 13–16.

Vizcaino, S. F., Farrina, R. A., Zarateee, M. A., Bargo, M. S., & Schultz, P. (2004). Palaeoecological implications of the mid-Pliocene faunal turnover in the Pampean region (Argentina). *Palaeogeography, Palaeoclimatology, Palaeoecology, 213*, 101–113.

Vos de, J. (2007). Vertebrate records: Mid-Pleistocene of Southern Asia. *Encyclopedia of Quaternary Science.* pp. 3232–3249. USA: Elsevier.

Walker, G. (2004). Palaeoclimate: Frozen time. *Nature, 429*, 596–597.

Wang, X., Kaars, S., van der Kershaw, P., Bird, M., & Jansen, F. (1999). A record of fire, vegetation and climate through the last three glacial cycles from Lombok Ridge core G6-4, eastern Indian Ocean, Indonesia. *Palaeogeography, Palaeoeclimatology, Palaeoecology, 147*, 241–256.

Wasson, R. J. (1989). Desert dune building, dust raising and palaeoclimate in the Southern Hemisphere during the last 280,000 years. In T. H. Donnelly & R. J. Wasson (Eds.), *CLIMANZ 3: Proceedings of the 3rd Symposium. Late Quaternary climatic history of Australasia* (pp. 123–137). Melbourne: CSIRO.

Watanabe, O., Jouzel, J., Johnsen, S., Parrenin, F., Shoji, H., & Yoshida, N. (2003). Homogeneous climate variability across East Antarctica over the past three glacial cycles. *Nature, 422*, 509–512.

Webb, G., & Manolis, C. (1989). *Crocodiles of Australia* (p. 160). Sydney: Reed Books.

Webb, R. E. (1998). Megamarsupial extinction: The carrying capacity argument. *Antiquity, 72*, 46–55.

Webb, S. D. (1991). Ecogeography and the great American interchange. *Paleobiology, 17*, 266–280.

Webb, S. G. (2006). *The first boat people* (p. 318). Melbourne: Cambridge University Press.

Webb, S. G. (2008). Megafauna demography and late Quaternary climatic change in Australia: A predisposition to extinction. *Boreas, 37*, 329–345.

Webb, S. G. (2009). Late Quaternary distribution and biogeography of the southern Lake Eyre basin (SLEB). Megafauna, South Australia. *Boreas, 38*, 25–38.

Webb, S. G. (2010). Paleotrophic reconstruction and climate forcing of mega-Lake Eyre in late Quaternary Central Australia: A review. *Boreas, 39*, 312–324.

Wells, R. T. (1985). *Thylacoleo carnifex* a marsupial lion. In P. V. Rich, G. F. van Tets & F. Knight (Eds.), *Kadimakara: Extinct vertebrates of Australia.* (pp. 225–229). Canberra: Pioneer Design Studio. Wells, R. T., Moriarty, K., & Williams, L. G. (1984). The fossil vertebrate deposits of Victoria Fossil Cave Naracoorte: An introduction to the geology and fauna. *The Australian Zoologist, 21*, 305–333.

Wells, R. T., & Callan, R. A. (1986). The Lake Eyre Basin – Cainozoic sediments, fossil vertebrates and plants, landforms, silcretes and climatic implications: *Australasian sedimentologists group field guide series no. 4.* (p. 176). Sydney: Geological Society of Australia.

Wells, R. T., & Tedford, R. H. (1995). *Sthenurus* (Macropodidae: Marsupialia). from the Pleistocene of Lake Callabonna, South Australia. *Bulletin of the American Museum of Natural History, 225*, 1–111.

Wells, R. T., Horton, D. R., & Rogers, P. (1982). Thylacoleo carnifex *Owen (Thylacoleonidae): marsupial carnivore?* In M. Archer (Ed.), *Carnivorous marsupials* (pp. 573–586). Sydney: Publications of the Royal Zoological Society of New South Wales.

Wells, R. T., Grun, R., Sullivan, J., Forbes, M. S., Dalgairns, S., Bestland, E., et al. (2006). Late Pleistocene megafauna site at black creek, flinders chase National Park, Kangaroo Island, South Australia. *Alcheringa Special Issue, 1*, 367–387.

White, M. (1994). *The Greening of Gondwana* (2nd ed.). Sydney New South Wales: Reed Books.

Whiteside, J. H., Olsen, P. E., Eglinton, T., Brookfield, M. E., & Sambrotto, R. N. (2010). Compound-specific carbon isotopes from Earth's largest flood basalt eruptions directly linked to the end-Triassic mass extinction. *Proceedings of the National Academy of Sciences, 107*, 6721–6725.

Whitney-Smith, E. (1996). Second order overkill, New World late Pleistocene extinctions and increased continentality: A system dynamics perspective: *Abstracts of the XIV Biennial AMQUA Meeting*. USA: Flagstaff Arizona.

Williams, D. L. G. (1980). Catalogue of Pleistocene vertebrate fossils and sites in South Australia. *Transactions of the Royal Society of South Australia, 104*, 101–115.

Williams, M. A., Dunkerley, D., De Deckker, P., Kershaw, P., & Chappell, J. (1998). *Quaternary environments*. (p. 329). London: Arnold.

Willis, P. M. A., & Molnar, R. E. (1997). A review of the Plio–Pleistocene crocodilian genus. *Pallimnarchus. Proceedings of the Linnean Society of New South Wales, 117*, 224–242.

Winograd, I. J., Coplen, T. B., Landwehr, J. M., Riggs, A. C., Ludwig, K. R., Szabo, B. J., et al. (1992). Continuous 500,000 year climate record from vein calcite in Devils Hole, Nevada. *Science, 258*, 250–255.

Wright, R. V. S. (1986). New light on the extinction of the Australian megafauna. *Proceedings of the Linnean Society of New South Wales, 109*, 1–9.

Wroe, S. (2002). A review of terrestrial mammalian and reptilian carnivore ecology in Australian fossil faunas, and factors influencing their diversity: The myth of reptilian domination and its broader ramifications. *Australian Journal of Zoology, 50*, 1–24.

Wroe, S., & Field, J. (2006). A review of evidence for a human role in the extinction of Australian megafauna and an alternative interpretation. *Quaternary Science Reviews, 25*, 2692–2703.

Wroe, S., Field, J., Fullagar, R., & Jermin, L. S. (2004). Megafaunal extinction in the late Quaternary and the global overkill hypothesis. *Alcheringa, 28*, 291–331.

Wyneken, J., Godfrey, M. H., & Bels, V. (2008). *Biology of turtles*. (p. 387). New York, NY: CRC Press.

Wyrwoll, K. -H., & Dortch, C. E. (1978). Stone artifacts and an associated diprotodontid mandible from the Greenough River, Western Australia. *Search, 9*, 411–413.

Printed in the United States
By Bookmasters